Mathematical Modeling and Simulation of Complex Fluid Systems

This book explores modern fluid mechanics through mathematical modeling and advanced computational techniques, covering complex topics like compressible flows and fluid–structure interactions. It integrates computational techniques, multiscale modeling, and AI applications, with a focus on industry-specific challenges in energy, manufacturing, and environmental sciences. This book spans fundamental fluid mechanics to cutting-edge simulations of complex fluid systems, providing insights into a wide range of fluid behaviors in engineering, biological, environmental, and industrial contexts. It addresses key challenges such as non-Newtonian fluids, compressible flows, multiphase systems, and fluid–structure interactions. Contemporary research trends like AI-driven optimization and sustainable energy solutions are also discussed.

Main Features:

- Covers basic principles to specialized topics such as compressible flows, multiphase systems, and fluid–structure interactions.
- Focuses on numerical methods, optimization, and computational fluid dynamics (CFD) tools to solve complex fluid dynamics problems.
- Discusses applications relevant to industries such as petroleum, manufacturing, and energy, including heat exchangers, porous media, and lubrication systems.
- Covers modern advancements like machine learning, AI optimization, and sustainable energy solutions.
- Applies fluid dynamics across biological, environmental, and industrial contexts, making the book relevant to engineers, environmental scientists, and healthcare professionals.

This book is a valuable resource for researchers, engineers, and students in fluid mechanics, computational fluid dynamics, and applied mathematics, providing both theoretical depth and practical solutions.

Mathematical Modeling and Simulation of Complex Fluid Systems

Edited by
Satyvir Singh and Mukesh Kumar Awasthi

CRC Press
Taylor & Francis Group
Boca Raton London New York

CRC Press is an imprint of the
Taylor & Francis Group, an **informa** business
A CHAPMAN & HALL BOOK

Designed cover image: Shutterstock ID: 2022223799

First edition published 2027
by CRC Press
2385 NW Executive Center Drive, Suite 320, Boca Raton FL 33431

and by CRC Press
4 Park Square, Milton Park, Abingdon, Oxon, OX14 4RN

CRC Press is an imprint of Taylor & Francis Group, LLC

Library of Congress Cataloging-in-Publication Data
[Insert LoC Data here when available]

ISBN: 9781041121275 (hbk)
ISBN: 9781041123552 (pbk)
ISBN: 9781003664345 (ebk)

DOI: 10.1201/9781003664345

Typeset in Times
by Deanta Global Publishing Services, Chennai, India

Contents

Aim and Scope

The book *Mathematical Modeling and Simulation of Complex Fluid Systems* aims to present a coherent and in-depth treatment of modern fluid mechanics through rigorous mathematical modeling and advanced simulation techniques. It is designed to bridge fundamental fluid dynamic principles with cutting-edge computational and data-driven approaches, addressing the growing need for accurate modeling of complex fluid behavior in engineering, scientific, and biomedical applications.

The scope of the book spans classical Newtonian fluid theory to highly nonlinear non-Newtonian, magnetohydrodynamic, nanofluid, and viscoelastic systems. Emphasis is placed on deriving and interpreting governing equations, developing analytical and numerical solution strategies, and translating these results into practical engineering insights. The text highlights the role of modern tools such as MATLAB-based solvers, finite difference and finite element methods, high-fidelity simulation frameworks, and artificial intelligence in tackling multiscale and multiphysics fluid problems.

Chapter 1 lays the foundational framework by introducing the continuum hypothesis and key physical properties governing fluid behavior. It establishes the conservation laws of mass, momentum, and energy and presents the Navier–Stokes equations as the cornerstone for modeling real-world fluid systems. *Chapter 2* advances the discussion to high-speed compressible flows, focusing on nonlinearities and shock phenomena. It demonstrates the use of Lie symmetry methods to systematically reduce complex partial differential equations into similarity forms suitable for analytical and semi-analytical treatment.

Chapters 3 and *4* focus on numerical modeling of heat transfer and convection phenomena. These chapters explore three-dimensional hybrid nanofluid flows over permeable surfaces and buoyancy-driven convection in eccentric geometries, highlighting the influence of magnetic effects, suction, geometry, and entropy generation on flow stability and thermal performance. *Chapters 5* and *6* examine advanced nanofluid and non-Newtonian flow configurations. They analyze tri-hybrid nanofluids over stretching surfaces and the combined effects of thermophoresis, Brownian motion, and magnetohydrodynamics, offering insights relevant to thermal management, industrial processing, and biomedical transport systems.

Chapters 7–9 provide a deeper exploration of complex rheology and flow stability. These chapters discuss non-Newtonian fluid models, rotating and magnetized ferrofluids, and viscoelastic fluid instabilities, emphasizing how rotation, magnetic fields, and material properties influence flow behavior and delay or promote the onset of convection. *Chapters 10–12* extend the scope to porous media, dusty fluids, and instability-driven mixing processes. Through analytical, numerical, and high-resolution simulation approaches, these chapters investigate oscillatory convection, acceleration waves, and magnetically controlled Richtmyer–Meshkov instability, underscoring their relevance to geophysical, industrial, and high-energy flow systems.

Chapters 13–16 address emerging trends in fluid modeling and optimization. These chapters review single-phase non-Newtonian flow models, propose multi-objective optimization frameworks for energy-efficient fluid systems, analyze viscous dissipation effects in complex geometries, and introduce AI-enhanced simulation techniques using artificial neural networks integrated with classical numerical methods. Together, they highlight the future direction of intelligent, efficient, and predictive modeling of complex fluid systems.

Overall, this book is intended to serve as a valuable reference for graduate students, researchers, academicians, and practicing engineers seeking a unified perspective on the mathematical modeling and simulation of complex fluid systems, with strong relevance to aerodynamics, thermal engineering, biomedical flows, and advanced energy technologies.

Preface

The book *Mathematical Modeling and Simulation of Complex Fluid Systems* is designed to present a coherent and in-depth treatment of modern fluid mechanics, emphasizing the role of mathematical modeling and numerical simulation in understanding complex fluid behavior. It brings together classical theory, advanced analytical methods, and state-of-the-art computational techniques to address challenges encountered in contemporary engineering and scientific applications.

Fluid systems encountered in practice often exhibit nonlinear, multiscale, and multiphysics characteristics that go beyond the scope of conventional analytical solutions. This book responds to these challenges by systematically bridging fundamental governing equations with computational and data-driven approaches. Special emphasis is placed on Newtonian and non-Newtonian fluids, compressible and incompressible flows, and the integration of emerging fields such as magnetohydrodynamics, nanofluids, and artificial intelligence–assisted modeling.

The content is structured to serve graduate students, researchers, and practicing engineers in mechanical, aerospace, biomedical, and chemical engineering, as well as applied mathematics and computational science. Mathematical rigor is balanced with physical interpretation, ensuring that readers not only understand the equations governing fluid motion but also gain insight into their practical implications for real-world systems.

The early chapters establish a strong theoretical foundation and introduce modern mathematical tools for fluid analysis. Subsequent chapters focus on numerical simulation techniques and parametric studies of advanced fluid systems, highlighting the influence of geometry, material properties, external fields, and boundary conditions. The later chapters extend these ideas toward optimization and artificial intelligence–based frameworks, reflecting current research trends in fluid dynamics.

Chapter 1 introduces the continuum hypothesis, key thermophysical properties, and conservation laws, leading to the derivation of governing equations for engineering applications. *Chapter 2* addresses high-speed and nonlinear flows, demonstrating how Lie symmetry methods reduce complex partial differential equations into tractable forms. *Chapter 3* presents numerical simulations of three-dimensional hybrid nanofluid flow, highlighting the stabilizing effects of magnetization and suction. *Chapter 4* explores buoyancy-driven convection and entropy generation in eccentric geometries, revealing the role of Rayleigh number and geometric asymmetry.

Chapter 5 analyzes tri-hybrid nanofluid transport over nonlinear stretching surfaces and explains the crossover behavior associated with particle density. *Chapter 6* examines thermophoresis and Brownian motion in non-Newtonian MHD flows, offering insights relevant to biomedical and industrial control processes. *Chapter 7* provides a comprehensive discussion of non-Newtonian rheological models and their application to complex biological and engineering flows. *Chapter 8* investigates the stability of rotating ferrofluids, showing how rotation and boundary conditions govern thermal instability.

Chapter 9 focuses on magnetothermal instability in viscoelastic fluids and demonstrates the stabilizing influence of magnetic fields. *Chapter 10* studies convection in porous layers saturated with viscoelastic fluids, emphasizing the competing roles of temperature-dependent viscosity and throughflow. *Chapter 11* presents an analytical treatment of acceleration waves in dusty and rotating flows, highlighting the contrasting effects of dust loading and rotation. *Chapter 12* employs high-resolution numerical simulations to analyze magnetohydrodynamic control of Richtmyer–Meshkov instability.

Chapter 13 reviews single-phase models for Casson fluids in rotating systems, identifying the influence of permeability and magnetic effects. *Chapter 14* introduces multi-objective optimization strategies for fluid systems, integrating physical constraints with numerical optimization techniques. *Chapter 15* investigates viscous dissipation effects in hybrid nanofluid convection and discusses methods to mitigate thermal performance losses. *Chapter 16* concludes the book by presenting an AI-enhanced simulation framework that couples neural networks with finite element methods to predict complex flow and heat transfer behavior.

The editors hope that this volume will serve as both a reference and a source of inspiration for further research in mathematical fluid dynamics. By integrating theory, computation, optimization, and artificial intelligence, the book aims to contribute meaningfully to the ongoing advancement of modeling and simulation techniques for complex fluid systems.

Acknowledgement

The editors would like to express their sincere gratitude to all the contributors whose expertise, dedication, and scholarly efforts have made this volume, *Mathematical Modeling and Simulation of Complex Fluid Systems*, a comprehensive and authoritative work. The collective contributions of the authors have played a vital role in shaping a book that seamlessly integrates fundamental fluid mechanics with advanced mathematical modeling, numerical simulation, and modern computational methodologies.

We are deeply appreciative of the contributors for presenting rigorous theoretical developments alongside practical engineering applications, covering a broad spectrum of topics including Newtonian and non-Newtonian fluid behavior, compressible flows, magnetohydrodynamics, nanofluids, ferrofluids, viscoelastic systems, and emerging artificial intelligence–based modeling approaches. Their work successfully bridges classical governing equations with real-world applications in aerodynamics, biomedical engineering, thermal system optimization, and industrial fluid processes.

The editors also acknowledge the invaluable support of reviewers, whose insightful comments and constructive feedback significantly enhanced the clarity, depth, and technical quality of the chapters. Their contributions helped ensure the academic rigor and coherence of the volume.

Special thanks are extended to the editorial and production teams at *CRC Press, Taylor & Francis Group*, for their professional guidance, encouragement, and efficient handling of the publication process. Their continued support has been instrumental in bringing this work to fruition.

It is our sincere hope that this book will serve as a valuable reference for researchers, academicians, graduate students, and practicing engineers, and will inspire further advancements in the mathematical modeling and simulation of complex fluid systems.

About Editors

Satyvir Singh is a Research Associate Fellow in the Institute of Applied and Computational Mathematics at RWTH Aachen University, Germany. He earned his Ph.D. in Computational Fluid Mechanics with a thesis entitled "Development of 3D Discontinuous Galerkin Method for Solving Boltzmann-type Gas Kinetic Equations for Diatomic and Polyatomic Gases" in the School of Mechanical and Aerospace Engineering at Gyeongsang National University, South Korea. Subsequently, he worked as a Senior Research Fellow at the Research Center for Aircraft Parts Technology, Gyeongsang National University, South Korea. After that, Dr. Singh worked as a Research Fellow in the School of Physical and Mathematical Sciences at Nanyang Technological University Singapore. He completed his master's degree, M. Tech., in Industrial Mathematics & Scientific Computing at the Indian Institute of Technology Madras, India. He qualified for two highly competitive Indian examinations – Junior Research Fellowship and National Eligibility Test in Mathematical Sciences with All India Rank # 38, and Graduate Aptitude Test for Engineering in Mathematics with All India Rank # 244. For his teaching background, he worked in India as Assistant Professor at Galgotias College of Engineering & Technology Greater Noida, and IMS Engineering College Ghaziabad, where he taught various Mathematics courses to undergraduate and postgraduate students. Dr. Singh has a vast research area, including computational fluid dynamics, high-order numerical methods, hydrodynamic instability, gas kinetic theory, heat and mass transfer, and computational biology. His commitment to research is reflected in his more than **50** research articles (700+ citations) in reputable journals, including the *Physics of Fluids*, the *Journal of Computational Physics*, *Computers & Fluids*, the *International Journal of Heat and Mass Transfer*, *Physical Review Fluids*, *Applied Mathematics Computations*, etc. Also, he has published one book. Besides that, he has attended many international conferences and presented his research work in the USA, the UK, Italy, South Korea, Singapore, Germany, Greece, China, Japan, and India. As a Co-PI, Dr. Singh has received research funding for the project "Mathematical Modelling and High-Fidelity Simulations for Brain Tumor Dynamics" by the Deanship of Graduate Studies and Scientific Research Scheme, Jazan University, Saudi Arabia (2024). His Orcid is https://orcid.org/my-orcid?orcid=0000-0001-6669-5296, the Google Scholar web link is https://scholar.google.com/citations?user=sQT89LYAAAAJ&hl=en, and the ResearchGate web page is https://www.researchgate.net/profile/Satyvir-Singh.

Mukesh Kumar Awasthi did his Ph.D. on the topic "Viscous Correction for the Potential Flow Analysis of Capillary and Kelvin-Helmholtz Instability." He is an Assistant Professor in the Department of Mathematics at Babasaheb Bhimrao Ambedkar University, Lucknow. Dr. Awasthi is specialized in the mathematical modeling of flow problems. He has taught courses in fluid mechanics, discrete mathematics, partial differential equations, abstract algebra, mathematical methods, and measure theory to postgraduate students. He has acquired excellent knowledge in the mathematical modeling of flow problems, and he can solve these problems analytically as well as numerically. He has a good grasp of subjects like viscous potential flow, electro-hydrodynamics, magneto-hydrodynamics, heat, and mass transfer. He has excellent communication skills and leadership qualities. He is self-motivated and responds to suggestions in a more convincing manner. Dr. Awasthi has qualified the National Eligibility Test (NET) conducted at the all-India level in the year 2008 by the Council of Scientific and Industrial Research (CSIR) and got Junior Research Fellowship (JRF) and Senior Research Fellowship (SRF) for doing research. He has published over **150** research publications (journal articles/books/book chapters/conference articles) in international journals and conferences. Also, he has published **22** books. He is also a series editor of *Artificial Intelligence and Machine Learning for Intelligent Engineering Systems* published by CRC Press (Taylor & Francis USA). He has attended many symposia, workshops, and conferences in mathematics as well as fluid mechanics. He has received the "Research Awards" consecutively four times from 2013 to 2016 by the University of Petroleum and Energy Studies, Dehradun, India. He has also received the start-up research fund for his project "Nonlinear Study of the Interface in Multilayer Fluid System" from UGC, New Delhi. He is also listed among the **top 2% of influential researchers** in the world, as prepared by Stanford University based on Scopus data for the years 2022, 2023, and 2024. His Orcid is 0000-0002-6706-5226, Google Scholar web link is https://scholar.google.co.in/citations?user=Dj3ktGAAAAAJ, and ResearchGate web link is https://www.researchgate.net/profile/Mukesh-Awasthi-2.

List of Contributors

Mukesh Kumar Awasthi
Department of Mathematics
Babasaheb Bhimrao Ambedkar University
Lucknow, India

Sham Bansal
Department of Mathematics
Mata Gujri College, Fatehgarh Sahib
Punjab, India

Anirban Chattopadhyay
Education Directorate, Higher Education Department
Bikash Bhavan
Kolkata, West Bengal, India

Bikash Das
Department of Mathematics
Murkong Selek College
Assam, India

Satya Deo
Department of Mathematics
University of Allahabad
Uttar Pradesh, India

Reeta Devi
Department of Mathematics
Govt. College Nagrota Bagwan
Himachal Pradesh, India

Jayalatha G.
Department of Mathematics
RV College of Engineering
Bengaluru, India

Subrata Ghosh
Department of Mathematics
Rajiv Gandhi University
Itanagar, Arunachal Pradesh, India

Krishno D. Goswami
Department of Mathematics
SRM Institute of Science and Technology
Tamil Nadu, India

Hritik
Satyawati College
University of Delhi
Delhi, India

Deepak Kumar
Department of Mathematics and Scientific Computing
National Institute of Technology Hamirpur
Himachal Pradesh, India

Satish Kumar
Chhatrapati Sahu Ji Maharaj Government Polytechnic
Uttar Pradesh, India

Divya Mahajan
Satyawati College
University of Delhi
Delhi, India

Biswajit Mallick
Department of Mathematics
University of Allahabad
Uttar Pradesh, India

Rupchand Malo
Department of Mathematics
Lovely Professional University
Punjab, India

Suma N. Manjunath
Department of Mathematics
RV College of Engineering
Bengaluru, India

Deepak Kumar Maurya
Department of Mathematics, Prof. Rajendra Singh (Rajju Bhaiya) Institute of Physical Sciences for Study and Research
Veer Bahadur Singh Purvanchal University
Uttar Pradesh, India

Pankaj Kumar Maurya
Department of Mathematics
Prof. Rajendra Singh (Rajju Bhaiya) University
Uttar Pradesh, India

Nishi Deepa Palo
Department of Applied Sciences
Bharati Vidyapeeth College of Engineering
New Delhi,3 India

Pawan Kumar Patel
Department of Mathematics
University of Allahabad
Uttar Pradesh, India

Basit Rahman
Department of Mathematics
Rajiv Gandhi University
Arunachal Pradesh, India

Seetu Rana
Department of Mathematics
Government College Hisar
Hisar, India

Vanishree R. K.
Department of Mathematics
Maharani Cluster University
Bengaluru, India

Sahin Ahmed
Department of Mathematics
Rajiv Gandhi University
Arunachal Pradesh, India

Bidesh Sengupta
Department of Physics and Astronomy
University of Bologna
Bologna, Italy

Hema Sharma
Department of Mathematics
SGT University
Haryana, India

Sweta Sharma
Department of Mathematics and Scientific Computing
National Institute of Technology Hamirpur
Himachal Pradesh, India

Liwen Shen
Institute for Applied and Computational Mathematics
RWTH Aachen University
Aachen, Germany

Atul Kumar Shukla
Department of Mathematics
Ganga Singh College
Chapra, India

Satyvir Singh
Institute for Applied and Computational Mathematics
RWTH Aachen University
Aachen, Germany
and
Department of Mathematics
Graphic Era Deemed to be University
Dehradun, India

Sunil
Department of Mathematics and Scientific Computing
National Institute of Technology Hamirpur
Himachal Pradesh, India

Akanksha Thakur
Department of Mathematics and Scientific Computing
National Institute of Technology Hamirpur
Himachal Pradesh, India

Ashish Tiwari
Department of Mathematics
SGT University
Haryana, India

Manuel Torrilhon
Institute for Applied and Computational Mathematics
RWTH Aachen University
Aachen,, Germany

Dhananjay Yadav
Department of Mathematical & Physical Sciences
University of Nizwa
Oman

1 Fundamentals of Fluid Dynamics

Atul Kumar Shukla, Mukesh Kumar Awasthi, and Dhananjay Yadav

1.1 INTRODUCTION

Fluid mechanics, as the term suggests, deals with the behavior of fluids both at rest and in motion. Historically, the subject has found wide application in engineering practice, including the design and analysis of canals, levees, and dam systems. It also plays a central role in the development of pumps, compressors, and piping or ducting networks used in water supply and air-conditioning systems for residential and commercial buildings, as well as in process piping for chemical plants. In addition, fluid mechanics underpins the study of automobile aerodynamics, subsonic and supersonic aircraft design, and the development of a wide range of flow-measurement instruments, such as gas pump meters [1].

A fluid is defined as a substance that undergoes continuous deformation when subjected to a shear (tangential) stress, regardless of how small the applied stress may be. Fluids include both liquid and gaseous (or vapor) phases of matter. Owing to its wide range of practical applications, fluid mechanics is regarded as one of the most fundamental and important disciplines in engineering and applied sciences. It plays a crucial role in areas such as applied mathematics, plasma physics, geophysics, biophysics, and physical chemistry. The experimental aspects of fluid mechanics are investigated across various branches of engineering. In civil engineering, the study of fluid flow in pipes and open channels is of particular importance. Engineers apply principles of fluid mechanics to analyze river transport, irrigation systems, and air and water pollution, as well as to design pipelines, flood control structures, and dams. Furthermore, fluid mechanics forms the basis for the study and design of fluid machinery, including pumps, fans, blowers, air compressors, heat exchangers, jet and rocket engines, gas turbines, power plants, and pollution control equipment [2, 3].

1.2 FUNDAMENTAL LAWS

The solution of fluid-mechanics problems is based on the application of certain fundamental physical laws that describe the behavior of fluids in motion and at rest.

DOI: 10.1201/9781003664345-1

These laws are general in nature and are valid for all types of fluids [4]. The principal laws used in fluid-mechanics analysis are:

i. The law of conservation of mass
ii. Newton's second law of motion
iii. The law of conservation of angular momentum
iv. The first law of thermodynamics
v. The second law of thermodynamics

1.2.1 Fundamental Concepts and Definitions

1.2.1.1 Continuum Hypothesis

Fluids consist of large assemblies of molecules that are widely spaced in gases and relatively closely packed in liquids. Although the intermolecular distance is large compared to the molecular diameter, the total number of molecules involved in a fluid is enormous, and the separation between them is ordinarily negligible at engineering length scales. Under such conditions, a fluid may be treated as a continuum, and its properties at any point are regarded as representing the averaged, bulk behavior of the fluid. For the continuum assumption to be valid, the smallest volume of practical interest must contain a sufficiently large number of molecules so that meaningful average properties can be defined. For example, at sea-level conditions, a volume of air of the order of 10^{-9}mm^3 contains approximately 3×10^7 molecules. From an engineering standpoint, this volume is extremely small, and hence the continuum hypothesis is well justified [5].

However, in certain situations, such as very high-altitude flight, the molecular spacing becomes sufficiently large that a small control volume may contain only a few molecules, causing the continuum model to break down. In all situations considered in these lectures, the continuum assumption is assumed to be valid.

1.3 CLASSIFICATION OF FLUID FLOW

Fluid flow problems encountered in engineering practice may be broadly classified into the following categories:

1.3.1 Viscous and Inviscid Flow

When frictional effects within a fluid are significant, the flow is described as *viscous*. In such flows, relative motion between adjacent fluid layers gives rise to shear stresses, which are quantified by the fluid property known as viscosity. Boundary-layer flows are typical examples of viscous flow. If viscous effects are negligible and the viscous terms in the governing equations may be ignored, the flow can be approximated as *inviscid*.

1.3.2 Internal and External Flow

Flow over a surface immersed in an unbounded fluid is referred to as *external flow*, whereas flow confined entirely by solid boundaries is termed *internal flow*. For example, flow over a flat plate is an external flow, while flow through a pipe or duct is an internal flow. In special cases where the conduit is only partially filled and a free surface exists, the flow is classified as *open-channel flow*. Internal flows are generally dominated by viscous effects, whereas in external flows viscosity is primarily important within the boundary layer adjacent to the solid surface.

1.3.3 Compressible and Incompressible Flow

A flow is considered *incompressible* if the fluid density remains essentially constant throughout the flow field. When density variations exceed approximately 5%, the flow is treated as *compressible*. Under standard atmospheric conditions, this typically corresponds to a Mach number greater than about 0.3.

1.3.4 Laminar, Transitional, and Turbulent Flow

Laminar flow is characterized by smooth, orderly motion of fluid layers and is commonly observed at low velocities or in highly viscous fluids. At higher velocities, the flow often becomes *turbulent*, exhibiting irregular velocity fluctuations and enhanced mixing. The intermediate regime, in which the flow alternates between laminar and turbulent behavior, is known as *transitional flow*. The Reynolds number is the key dimensionless parameter governing this classification.

1.3.5 Steady and Unsteady Flow

A flow is said to be *steady* if fluid properties at any given point do not vary with time. If these properties change with time, the flow is described as *unsteady* or *transient*. A special class of unsteady flow, known as *periodic flow*, involves oscillations about a steady mean state.

1.3.6 Natural and Forced Flow

In *forced flow*, fluid motion is produced by external devices such as pumps, fans, or blowers. In contrast, *natural (or free) flow* arises due to density differences within the fluid, with buoyancy acting as the driving mechanism. A common example is the upward motion of a warmer fluid in a gravitational field.

1.4 PHYSICAL PROPERTIES OF FLUIDS

1.4.1 Density and Specific Volume

Density ρ is defined as the mass per unit volume of a substance and is expressed mathematically as

$$\rho = \frac{mass}{volume} = \frac{\nabla \tilde{m}}{\nabla \tilde{V}}$$

where ($\tilde{m}$) denotes the mass and ($\tilde{V}$) represents the volume of the fluid.

Specific volume is defined as volume per unit mass. Its formula is $\frac{1}{\rho} = \frac{\nabla \tilde{V}}{\nabla \tilde{m}}$.

1.4.2 Viscosity and Newton's Law of Viscosity

Viscosity is a fundamental property of a fluid that represents its resistance to shear deformation, i.e., resistance to tangential force per unit area acting between adjacent fluid layers in relative motion. It governs the internal friction within the fluid and thus plays a crucial role in controlling fluid flow behavior. In general, the viscosity of liquids decreases with an increase in temperature, whereas the viscosity of gases increases with temperature [6, 7].

Fluids in which the rate of deformation (or shear rate) is directly proportional to the applied shear stress are known as Newtonian fluids. For a one-dimensional flow system, this proportionality leads to a linear relationship between shear stress and rate of deformation, as illustrated in Figure 1.1. Accordingly, the shear stress τ may be expressed as:

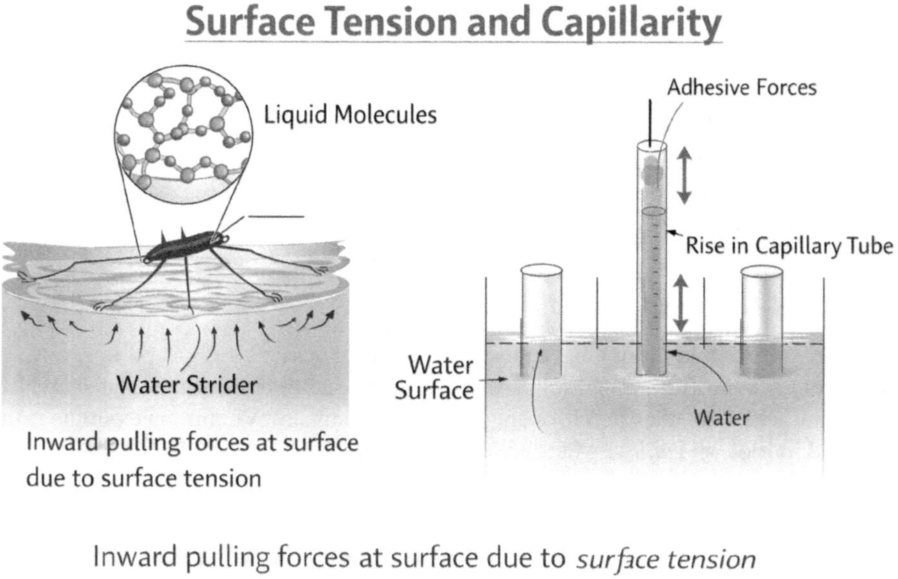

FIGURE 1.1 Viscosity and Newton's law of viscosity.

1.4.3 Compressibility and Bulk Modulus

It is a characteristic property of a fluid that indicates the manner in which its density changes with pressure while the temperature remains constant [8]:

$$E_{\alpha} = \rho \left[\frac{\partial \rho}{\partial T} \right]_T$$

1.4.4 Surface Tension and Capillarity

Surface tension is the property of a fluid that causes its free surface to behave like an elastic membrane, tending to minimize the surface area. This property enables certain insects, such as water striders, which are denser than water, to float and move on the water surface.

At a liquid–air interface, surface tension (Figure 1.2) arises because the cohesive forces between liquid molecules are stronger than the adhesive forces between the liquid molecules and the surrounding air. As a result, molecules at the surface experience a net inward force, causing the surface to contract. Consequently, the liquid surface behaves as if it were under tension due to these unbalanced molecular forces, giving rise to the term surface tension.

Water exhibits a relatively high surface tension because of strong intermolecular attraction arising from hydrogen bonding. At 20°C, the surface tension of water is

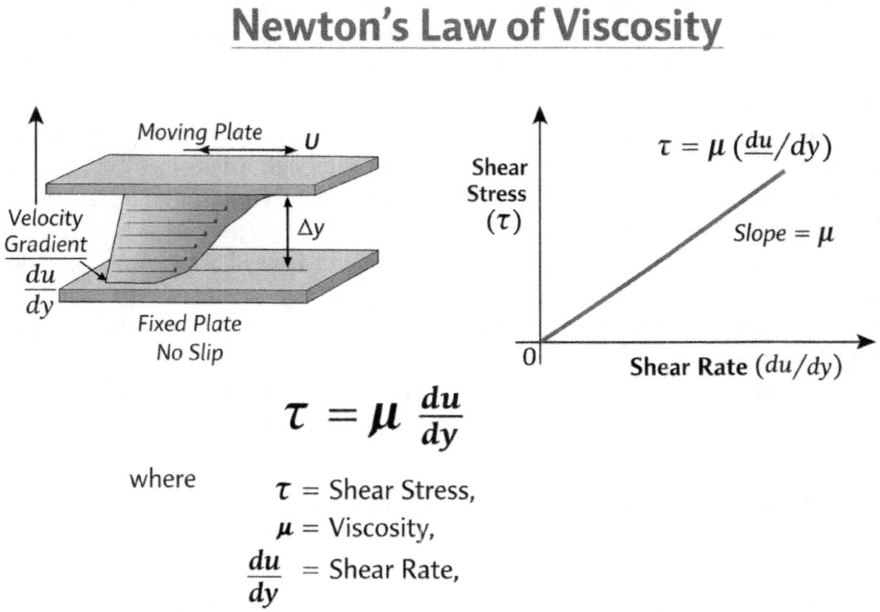

FIGURE 1.2 Surface tension and capillarity.

approximately 72.8 mN/m, which is higher than that of most other liquids. Surface tension plays a crucial role in various physical phenomena, particularly capillarity.

1.4.5 Kinematics of Fluid Motion

Kinematics is the branch of mechanics concerned with the description of motion without reference to the forces that produce it. In the case of rigid bodies, motion is characterized by translation and rotation. More generally, a vector field may be described in terms of its divergence, curl, and boundary values.

In fluid kinematics, the motion of a fluid is examined by considering microscopic or macroscopic fluid elements that undergo translation, rotation, and deformation with time. Fluids are treated as a continuum, meaning that a fluid element occupying a given position in space at one instant, together with its neighboring elements, moves to another position at a later time while largely retaining its neighborhood identity, except in cases such as bifurcation. This continuum description requires the motion to be deterministic rather than stochastic; therefore, random processes such as molecular diffusion and turbulent fluctuations are not explicitly considered. A fundamental concept in the kinematics of fluid motion is convection, which involves tracking or following the motion of an individual fluid "particle" as it moves through space and time.

1.4.6 Lagrangian and Eulerian Descriptions

In fluid mechanics, the fluid motion can be described in two fundamental ways: the Lagrangian and the Eulerian descriptions. In the Lagrangian description, individual fluid particles are followed as they move through space and time. Each particle is identified by its initial position, and its trajectory, velocity, and acceleration are obtained by tracking its motion continuously. This approach provides a clear physical picture of particle paths and is particularly useful in problems involving particle transport, interface motion, or multiphase flows, although it becomes cumbersome when dealing with a large number of particles. In contrast, the Eulerian description focuses on fixed points in space and examines how fluid properties such as velocity, pressure, and density vary with time at those locations. Instead of following particles, the flow is represented as a field, making this approach more suitable for continuum analysis and the formulation of governing equations such as the Navier–Stokes equations. The connection between the two descriptions is established through the material (or substantial) derivative, which expresses the rate of change of a fluid property following a moving particle while using Eulerian field variables [9, 10].

1.5 BASIC PRINCIPLES OF FLUID DYNAMICS

1.5.1 Conservation of Mass

From experimental observation, it can be postulated that mass is neither created nor destroyed; it is conserved. This statement is known as the **Principle of Conservation**

of Mass. The principle is applied to a fixed region in space of arbitrary shape containing a fluid, referred to as a **control volume**. Fluid is allowed to flow into and out of the control volume through its boundaries. So one can say:

Rate of increase of mass of material within the control volume = Net rate at which material enters the control volume.

Mathematically

$$A_1\vec{q}_1 = A_2\vec{q}_2$$

where A_1 and A_2 are the cross-sectional areas at two points in the pipe, and $\vec{q}_1$ and $\vec{q}_2$ are the fluid velocities at those two points.

1.5.2 Conservation of Momentum

The Navier–Stokes equations constitute a system of partial differential equations that govern the motion of viscous fluids. They form the foundation of fluid mechanics and are used to model a broad range of flow phenomena from simple laminar pipe flow to complex turbulent motions. These equations are derived by applying Newton's second law of motion to an infinitesimal fluid element, accounting for the forces acting on it, such as pressure forces, viscous stresses, and external body forces. The Navier–Stokes equations are commonly used in conjunction with the continuity equation, which represents the principle of conservation of mass.

Mathematically,

$$\rho\left(\frac{\partial \vec{q}}{\partial t} + (\vec{q}\cdot\nabla)\vec{q}\right) = \rho\boldsymbol{g} - \nabla p + \nabla\cdot\left\{\mu\left[\nabla\vec{q} + (\nabla\vec{q})^T - \frac{2}{3}(\nabla\cdot\vec{q})\boldsymbol{I}\right]\right\} + \nabla\left[\varsigma(\nabla\cdot\vec{q})\right]$$

where $\vec{U}$ is the velocity of fluid flow and ρ, p are the density and pressure respectively.

Although the Navier–Stokes equations are straightforward to formulate, obtaining analytical solutions is difficult except in a limited number of special cases. Consequently, most practical problems are addressed using numerical techniques implemented on powerful computers. This area of study is known as computational fluid dynamics (CFD).

Despite their mathematical complexity, the Navier–Stokes equations remain a cornerstone of fluid mechanics, offering a robust framework for analyzing, understanding, and predicting fluid behavior across a wide range of engineering applications.

1.5.3 Conservation of Energy

A physical system is defined as a clearly specified collection of bodies that interact with one another through applied forces. All bodies external to the prescribed system boundary are referred to as the surroundings. The state of a system is described

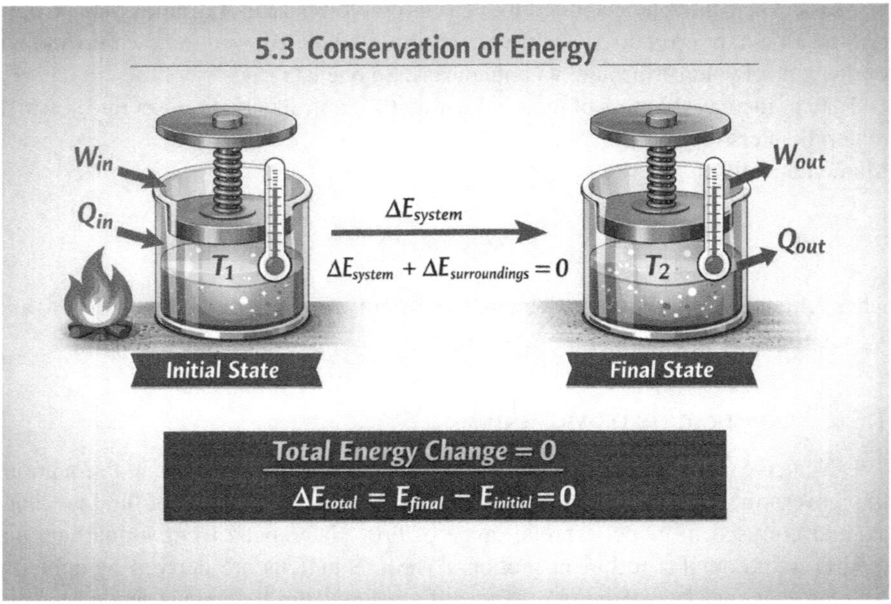

FIGURE 1.3 Conservation of energy.

by a set of measurable physical properties that uniquely and completely characterize its condition. The conceptual separation of the system, its boundary, and the surroundings are illustrated in Figure 1.3.

Thus far, the time-dependent dynamical behavior of the system has been examined through the application of Newton's laws of motion, which account for the effects of applied forces. We now introduce the principle of conservation of energy to analyze changes in the state of the system.

According to this principle, the net change in energy of the combined system and its surroundings between the initial and final states is zero.

$$\Delta \bar{E}_{\text{Total}} = \Delta \bar{E}_{\text{system}} + \Delta \bar{E}_{\text{Surroundings}} = 0$$

1.5.4 Forces Acting on a Fluid

According to Newton's second law of motion, the resultant force F_f acting on a fluid element in the x-direction is equal to the mass $\bar{m}$ of the fluid element multiplied by its acceleration a_f in the same direction. This can be written as

$$F_f = \bar{m} a_f$$

The motion of a fluid element is governed by several forces, which may include:

(i) gravitational force F_g,
(ii) pressure force F_p,
(iii) viscous force F_v,
(iv) turbulence force F_t,
(v) surface tensile (surface tension) force F_s.

1.6 GOVERNING EQUATIONS OF FLUID MOTION

1.6.1 Euler's Equations of Motion

Euler's equations of motion describe the dynamics of an ideal (inviscid) fluid by expressing the conservation of momentum in the absence of viscous effects. These equations are obtained by applying Newton's second law to an infinitesimal fluid element, considering only pressure forces and body forces such as gravity. They represent a simplified form of the general momentum equations and are applicable to flows where viscous stresses are negligible. In vector form,

Euler's equations may be written as:

$$\rho\left(\frac{\partial \vec{q}}{\partial t}+(\vec{q}\cdot\nabla)\vec{q}\right)=\rho\vec{F}-\nabla p$$

where p is the pressure, ρ is the fluid density, and $\vec{F}$ denotes the body force per unit mass. Euler's equations provide the theoretical foundation for inviscid and potential flow analysis and are widely used in aerodynamics and hydrodynamics, particularly in regions away from solid boundaries where viscous effects are small.

1.6.2 Navier–Stokes Equations

The equation that represents the conservation of momentum is given by

$$\rho\left(\frac{\partial \vec{q}}{\partial t}+\vec{q}\cdot\nabla\vec{q}\right)+\nabla p=\rho\vec{F}+\mu\nabla^2\vec{q}$$

This equation is the well-known Navier–Stokes equation, where (**f**) denotes the external force acting per unit mass.

1.7 INTRODUCTION TO SPECIAL FLOW MODELS

1.7.1 Inviscid Flow

An important characteristic of gases and liquids is that their molecules can move relatively freely. As these molecules move – even in a highly random manner – they carry mass, momentum, and energy from one point in the fluid to another. This molecular-level transport leads to observable effects such as mass diffusion, viscosity (internal friction), and heat conduction. All real fluid flows exhibit these transport processes to some extent, and such flows are referred to as viscous flows. On the other hand, a flow that is idealized by neglecting friction (i.e. viscosity $\mu = 0$), thermal conduction, and diffusion is known as an inviscid flow [11].

Although truly inviscid flows do not occur in nature, there are many practical situations, particularly in aerodynamics, where the effects of viscosity and other transport phenomena are so small that they can be safely ignored.

1.7.2 Viscous Flow

Viscous flow refers to fluid motion in which the effects of viscosity are significant and cannot be neglected. A fluid can be defined as a substance that undergoes continuous deformation when subjected to any amount of shear stress. This continuous deformation is known as flow, and the property that describes a fluid's resistance to such deformation is called viscosity.

Thus, viscosity measures how strongly a fluid resists shear while it is in motion. In general, the viscosity of fluids increases with increasing pressure at a constant temperature. Water, however, is an important exception to this behavior, as its viscosity decreases with increasing pressure at constant temperature.

1.7.3 Irrotational Flow

Irrotational flow describes a fluid motion in which individual fluid particles move without any rotation as they pass through the flow field. In such a flow, there are no vortices or swirling motions.

Mathematically,

$$\nabla \times \vec{q} = 0$$

Although the speed of the fluid may change from one point to another, the motion along a given streamline remains orderly, and the particles do not possess any angular or rotational motion. As a result, the flow is free from local spinning of fluid elements. Irrotational flow is commonly associated with laminar flow, where the fluid moves in orderly, parallel layers without mixing or disturbance. A key implication of irrotational flow is the absence of energy loss due to internal rotation of fluid particles. Since no rotational motion is present, there is minimal energy dissipation within the fluid. This feature makes irrotational flow especially important in many engineering applications, such as airfoil design in aerodynamics and the analysis of idealized flow in pipes and channels, where efficiency and simplicity of analysis are essential. Various authors have extensively investigated the instability of interfaces in irrotational flows, examining the mechanisms governing the onset, growth, and evolution of interfacial disturbances under different physical conditions [12, 13].

1.7.4 Potential Flow Concept

Viscous potential flow theory is an effective method for obtaining solutions of the Navier–Stokes equations when the flow is irrotational. The foundation of potential flow theory arises from kinematic principles. According to kinematic theory, for an irrotational flow the curl of the velocity field is zero, and hence a scalar velocity potential exists whose gradient represents the fluid velocity.

Mathematically,

$$\vec{q} = \nabla \varphi$$

For incompressible and irrotational flow, the continuity equation is given by

$$\nabla \cdot \vec{q} = 0$$

Under potential flow conditions, this equation reduces to the Laplace equation, shown as

$$\nabla^2 \varphi = 0$$

In potential flow theory, the viscous terms in the Navier–Stokes equations disappear, which is why it is commonly referred to as inviscid potential flow (IVP) theory [14–15]. However, since most practical flow problems are strongly influenced by viscosity, it is not realistic to set viscosity equal to zero. Therefore, while potential flow theory offers mathematical simplicity, viscosity must still be accounted for when applying it to real fluid flows.

1.8 CONCLUSION

The study of fluid dynamics was indeed made easy by the use of different methods. He has given us a great teaching tool with the treatment of fluids as continua and defining their behavior with the help of their basic properties such as density, viscosity, and compressibility. This has led to a systematic analysis of many different flow phenomena. The laws of conservation of mass, momentum, and energy are the foundation for the derivation of the governing equations of fluid motion, especially the Navier–Stokes equations, which are common to both theoretical and applied fluid mechanics. Flow classification and the application of idealized models make difficult problems easier and provide important physical insight. Together, these ideas form a strong basis not only for the study but also for the application of the concepts in the fields of hydraulics, aerodynamics, turbomachinery, and computational fluid dynamics.

REFERENCES

1. Chorin, A. J., Marsden, J. E., & Marsden, J. E. (1990). *A mathematical introduction to fluid mechanics* (Vol. 168). New York: Springer-Verlag.
2. Merci, B. (2016). Introduction to fluid mechanics. In *SFPE handbook of fire protection engineering* (pp. 1–24). New York, NY: Springer New York.
3. Hochstein, J. I., & Gerhart, A. L. (2021). *Young, Munson and Okiishi's a brief introduction to fluid mechanics*. John Wiley & Sons.
4. Elger, D. F., LeBret, B. A., Crowe, C. T., & Roberson, J. A. (2020). *Engineering fluid mechanics*. John Wiley & Sons.
5. Landau, L. D., & Lifshitz, E. M. (2013). *Fluid mechanics: Landau And Lifshitz: Course of theoretical physics* (Vol. 6). Elsevier.
6. Starkey, T. V. (1956). Viscosity and action. *British Journal of Applied Physics*, *7*(12), 448.
7. Abdulagatov, I. M., & Assael, M. (2009). Viscosity. In V. M. Valyashko (Ed.), *Hydrothermal properties of materials. Experimental data on aqueous phase equilibria and solution properties at elevated temperatures and pressures* (pp. 249–270). John Wiley & Sons.
8. Cheng, X., Zhang, Q., Yan, X., & Shi, C. (2015). Compressibility and equivalent bulk modulus of shelled corn. *Biosystems Engineering*, *140*, 91–97.
9. Ballio, F., Pokrajac, D., Radice, A., & HosseiniSadabadi, S. A. (2018). Lagrangian and Eulerian description of bed load transport. *Journal of Geophysical Research: Earth Surface*, *123*(2), 384–408.
10. Bontempi, F., & Faravelli, L. (1998). Lagrangian/Eulerian description of dynamic system. *Journal of Engineering Mechanics*, *124*(8), 901–911.
11. Landau, L., & Lifshitz, E. (1987). Chapter I—Ideal Fluids. *Fluid Mechanics*, *6*, 30–31.
12. Shukla, A. K., Awasthi, M. K., & Singh, S. (2023). Impact of heat and mass transport on Rayleigh–Taylor instability of Walter's B viscoelastic fluid layer. *Microgravity Science and Technology*, *35*(1), 3.
13. Shukla, A. K., Awasthi, M. K., & Agarwal, S. (2023). Stability analysis of spherical viscous fluid-Oldroyd B viscoelastic fluid interface: A viscous potential flow approach. *Chinese Journal of Physics*, *86*, 148–159.
14. Awasthi, M. K., & Shukla, A. K. (2024). Instability of power-law viscoelastic-viscous fluid interface with heat and mass transfer. *Chinese Journal of Physics*, *89*, 1441–1455.
15. Shukla, A. K., & Awasthi, M. K. (2022). Interfacial characteristics of power-law viscoelastic fluid with heat and mass transfer in planar configuration. *Journal of Fluids Engineering*, *144*(10), 101303.

2 Mathematical Tools for Fluid Modeling

Solutions of Nonlinear PDEs in High-Speed Fluid Flows

Hema Sharma and Ashish Tiwari

2.1 INTRODUCTION

Nonlinear partial differential equations established the foundation of mathematical modeling in a variety of scientific disciplines, including physics, biology, engineering, chemistry, finance, etc. The system of nonlinear partial differential equations (PDEs), such as *compressible Euler and Navier–Stokes equations,* determines high-speed fluid flows. In scientific and engineering domains, namely hypersonic vehicles, supersonic aircraft structures, aerospace applications, astrophysics, and high-speed fluid flows, it performs an integral contribution. High-speed flows occur when the flow velocities approach or exceed the speed of sound, leading to the rise in strong compressible effects, nonlinear wave interactions, formation of shocks, and rapid changes in thermodynamic properties [17].

The conservative form of high-speed flow [46]

$$\frac{\partial \boldsymbol{U}}{\partial t} + \nabla.\mathbf{F}\left(\mathbf{U}\right) = 0, \tag{2.1}$$

Here, **U** is a conserved quantity (density, momentum, energy), **F** represents the fluxes, crucial for accurately capturing phenomena like shocks in compressible, high-speed flows (supersonic/hypersonic) where density changes significantly, ensuring numerical stability and physical realism [56].

Analytic solutions to these equations are limited, especially in the presence of shock waves and complex boundary conditions. Consequently, advanced mathematical methods are required to seek qualitative and quantitative insights into the behavior of high-speed fluid flow. Exact solutions of these equations are rare and restricted to simplified geometries. As a result, there has been continuous interest in analytic methods capable of revealing invariant structure and similarity solutions.

Symmetry is a very elegant yet powerful concept in the understanding of laws of natural phenomena, which occur in scientific disciplines. Symmetry analysis

DOI: 10.1201/9781003664345-2

methods provide a comprehensive procedure for reducing higher nonlinear PDEs to simpler ordinary differential equations (ODEs). Symmetry methods enable the integration of differential equations and provide invariant solutions. Conservation laws can be derived, and equivalence relations can also be identified between different mathematical models. In a nonlinear system of ordinary and partial differential equations, symmetry methods generate a map of transformation that maps known solutions to new families of distinct solutions. Beyond theory, symmetry acts as a link between practical applications, impacting areas that include turbulent flow, energy system optimization, rotational invariance in pipelines, and wind turbines. Symmetry reduces computational costs and ensures scalable industrial designs. Thereby, symmetry-based solutions play an essential role in modern scientific understanding and ease rapid information decision-making in various industrial and engineering realms.

Lie symmetry methods provide a high-powered and structured framework for studying high-speed fluid flows by using the underlying invariance properties of the nonlinear PDEs such as *compressible Euler and Navier–Stokes equations* [7, 23, 54]. With the help of symmetry analysis, these complex PDE systems can be reduced to simpler ordinary or lower-dimensional equations, guiding fundamentally to self-similar and scaling solutions that represent key high-speed flow phenomena such as shock waves, boundary layers, blast waves, expansion fans, and stagnation-point flows. In compressible conservation laws, symmetries not only furnish a deep apprehension but also disclose fundamental scaling laws and conserved arrangements intrinsic in high-speed flows. Symmetry analysis enables a clear understanding of how compressibility and other physical phenomena influence shock formation, wave propagation, and energy transfer. It also distinguishes among subsonic, supersonic, and hypersonic attributes of flow, which are Mach-number-dependent behaviors. In the CFD solver, solutions obtained via symmetry analysis play an essential role in validating them. The combination of symmetry transformation and numerical scheme design enhances stability and robustness, improves conservation properties, and reduces numerical degeneracy. Symmetry methods link computation with analytic theory in high-speed flows. Adequately, that's how symmetry-based models empower accuracy and reliability in the prediction capability of the CFD solvers in supersonic and hypersonic flows. Lie symmetry methods, despite practical challenges such as turbulence and complex boundaries, remain a cornerstone for analyzing, interpreting, and modeling high-speed aerodynamics and plasma flow phenomena.

This work prioritizes physical understanding, methodological developments, and practical significance rather than detailed mathematical derivations. This chapter also highlights current challenges and potential directions for future research.

2.2 HISTORICAL BACKGROUND AND REVIEW OF LITERATURE

Lagrange, in the late 18th century, started studying groups. But the beginning of group theory was marked in 1846 by the publication of Augustin Louis Cauchy and Galois. Evariste Galois (1811–1832) was a French mathematician who determined a necessary and sufficient condition for a polynomial to be solved by a radical. His work laid the foundation of Galois theory and group theory, two major branches of abstract algebra. He was honored as the first mathematician to link group theory with field theory. The theory of symmetry analysis of differential equations came into existence in the 19th century with the pioneering work of a mathematician of Norwegian origin, **Marius Sophus Lie** (1842–1899) [28]. He started his study by investigating continuous groups (usually referred to as Lie Groups). Lie was the first mathematician to use the theory of continuous group transformations for finding solutions and describing symmetry properties of differential equations. He initiated a systematic study of *continuous transformation groups* that preserve the invariance of differential equations. Lie's work was developed between 1872 and 1899, which led to the foundation of what we call now "symmetry *analysis of differential* equations." Lie's theory is not only used in solving differential equations but is also used to identify symmetry properties systematically of the differential equations [26–29].

Emmy Noether was the first person who showed the implication of symmetry in physics and mathematics by its connection with *conservation laws*. The rigorous relationship was established in the year 1918, when Emmy Noether [33] proved a very famous theorem of her own, which established a relationship between *continuous symmetries and conservation laws*. In the year 1928 **Hermann Weyl** first introduced the term "**Lie group**." During the 1940s, notable development was seen in symmetry methods by the work of **Birkhoff** [8] and **I. Sedov** [40]. Their work demonstrated the practical applications of symmetry-based methods. From the 1960s onward, a decisive expansion of theories occurred when the Russian school, which was led by **L. V. Ovsiannikov** [36] systematically employed the symmetry methods to construct exact solutions in mathematics and physics. The *non–classical method* was first introduced by the very famous and renowned scientists *Bluman and Cole* [10] which was an important generalization of the classical approach. "Invariance of differential equations under transformations" of both dependent and independent variables was the central principle of Lie's symmetry theory. These transformations formed a "local group of point transformations" that induced a mapping of known solutions to new ones, diffeomorphism on the solution space. This demonstrated that the transformation of the variables naturally generates a corresponding transformation of their derivatives.

The interest in the work of Sophus Lie has been a center of attraction in recent decades and, more broadly, in "symmetries of differential equations" has increased significantly. Numerous textbooks on this subject have appeared within a relatively short period. Notably, the influential monographs by **Ibragimov** [22] and **Olver** [34, 35], following the earlier foundational work of **Ovsiannikov** [36], have become standard references. In the mid-1970s, a very famous book on "similarity methods" was published, inspired by the work of **Bluman and Cole**. **Anderson and Ibragimov** addressed specialized topics on generalized symmetries, Vladimirov devoted texts to ordinary differential equations (ODEs), and Ames contributed by writing a two-volume work on differential field theory. Brian Cantwell's [14] work was focused on finding symmetries in differential equations with the help of Lie groups. He worked on simplifying complex physics problems like fluid dynamics and magnetohydrodynamics (MHD). He reduced their complexity and obtained exact solutions through symmetry methods. In the main, Cantwell's work in the context of Lie symmetry provided a systematic and computational concept to unravel the structure and exact solutions of differential equations in science and engineering.

Lie symmetry analysis is a very useful tool with broad applicability that can be studied in [14, 16, 19, 34, 36]. In fluid dynamics, Lie symmetry groups are generally used to acquire invariant solutions and conservation laws for differential equations. Symmetry groups are also used to reduce complex higher-order PDEs to simpler ODEs. Some of the very well-known examples are where self-similar solutions are derived with the help of similarity variables through symmetry analysis associated with explicitly boundary layer systems.

Recent research has extended Lie symmetry methods beyond classical and nonlinear wave models. For instance, Sharma *et al.* [43] applied the Lie symmetry reduction method to a two-phase compressible flow with the Chaplygin Pressure Law and derived invariant and exact solutions, conservation laws, and weak discontinuity evolution characteristics. This clearly shows the importance of symmetry methods in relevance to high-speed compressible flow and wave dynamics. Topno *et al.* [54] investigated "hyperbolic quasilinear partial differential equations (PDEs) system" that arises in the mathematical description of "one-dimensional shallow water magnetohydrodynamics (SMHD)" by using the Lie symmetry method and examined its evolutionary dynamics of nonlinear wave propagation, constructed the optimal system of one-dimensional subalgebras, and obtained various closed-form exact solutions. She established several nonlocal symmetries through conservation laws and symmetry-based methods. She analyzed the evolution of weak discontinuities and their interaction with characteristic shock behavior using the analytical solution of the governing equations. Mandal *et al.* carried out a complete symmetry classification of the Navier–Stokes equation describing the three-dimensional incompressible plane shear flows. Firstly, they obtained the infinitesimal generators of the generalized system by using the classical Lie symmetry approach. After that, for each sub-case, they determined the symmetries, and the corresponding solution behavior was analyzed over a range of parameter values [30].

The authors Patel and Garg in their paper used the Lie group of symmetry analysis to study planar shock wave propagation in an ideal gas influenced by viscous stress and magnetic field, and the systems of partial differential equations (PDEs) were reduced into a system of ordinary differential equations (ODEs). It showed symmetry-based techniques are applicable to high-speed fluid flow PDEs, such as shock waves, and shows high-dimensional PDEs are reduced systematically [37]. Singhal *et al.* focused on obtaining invariant solutions and conservation laws of the (3+1)-dimensional Zakharov–Kuznetsov (ZK) equation [23, 44, 57] by using Lie symmetry analysis. The Zakharov–Kuznetsov (ZK) equation [57] plays an important role in modeling nonlinear dispersive wave phenomena in magnetized media. These equations describe the phenomena of wave stability and soliton propagation. This paper showed the current applications of symmetry-based techniques on dispersive waves [45]. Samina *et al.* [39] in their study presented the complete analysis of (2+1)-dimensional nonlinear wave-type PDEs characterized by anisotropic power law nonlinearities and a general power law source term, which arises in diverse physical environments, such as nonlinear media, fluid dynamics, and elastic wave propagation. In their study, they found that the behavior of exact solutions and symmetry analysis had not been thoroughly investigated for arbitrary values of nonlinear parameters. To overcome the limitations, they performed a complete Lie symmetry analysis for the parameters m and n and obtained a full set of generators. By the help of these symmetries, they reduced the PDEs into ODEs and enabled the exact analytical solutions.

Tiwari *et al.* explored the behavior of blast waves in a self-gravitating, non-ideal gaseous medium under the influence of an azimuthal magnetic field [48–53]. He also established exact solutions of some nonlinear PDEs using Lie symmetry techniques. Many researchers used Lie symmetry reduction methods to obtain invariant solutions of higher-order nonlinear PDEs, which are essential in high-speed fluid flow. One can find such papers in References [8, 11, 13, 15, 20, 22, 30, 31, 41, 42, 43, 45]. Boutrous *et al.* [12] investigated nonlinear equations of motion characterizing laminar, incompressible flow in a rectangular domain with porous walls regarding successional contraction and expansion using the symmetry method. The differential equations were reduced to simpler equations with the help of symmetry properties of the given differential equation. Rehman *et al.* [38] presented a study through their paper to determine symmetries for mixed convective flow over an inclined porous stretching surface of viscous fluids. He used the Lie symmetry approach for the cases of flows to obtain the transformation. Symmetries were developed in the most generalized form for the nonlinear PDE system. Msomi *et al.* [32] applied the Lie symmetry analysis method to a radiating shear-free spherically symmetric metric in the higher-dimensional gravitating flows. He found new solutions to Einstein's equation with the help of symmetry analysis. He founded the five-point symmetry of the equation. He extended known solutions to the new solutions.

Msomi *et al.* [32] demonstrated the applicability of Lie symmetry analysis in generating rough wave and periodic solutions of a nonlinear diffusion equation. He also found the invariant structures and exact analytic solution of the associated problem, which shows the physical admissibility. This work broadens the significance of symmetry analysis for extreme events in dissipative systems. Karna and Satapathy [24] in their paper investigated a nonlinear hyperbolic system, the Cargo–Leroux model in a compressible fluid system. They applied Lie symmetry techniques to derive exact solutions and the invariant structure of the model. Their primary focus was on the isentropic perturbation pressure equation of state. This study reinforces the essence of Lie symmetry in exploring conservation laws and their physical applications. It also provides the propagation and formation of nonlinear waves, scaling behavior in isentropic perturbation, and the effects of pressure-density in wave dynamics. Alshammari *et al.* [4] examined the generalized $A^{\mp}$equation, which is a form of the evolution equation, expressing bidirectional wave propagation with dispersive effects. He operated Lie symmetry to reduce the equation into a lower dimension for constructing an invariant structure and a conservation law. This paper demonstrated that the coexistence of solitons and chaotic dynamics in the generalized $A^{\mp}$equation can be guided by Lie symmetry and the conservation law. The authors Aliyu *et al.* [2]

explored a detailed study of a sixth-order nonlinear Ramani equation, which expresses classical integrable models by gathering higher-order spatial derivatives. In this paper, they used Lie group theory to determine continuous symmetries, derive explicit exact solutions, and construct conservation laws. This study showed the rich structural and dynamical properties of the sixth-order nonlinear Ramani equation by the Lie symmetry approach.

Huang *et al.* [21] presented a study on the generalized nonlinear beam equation, extending Euler-Bernoulli and Timoshenko models, with Lie symmetry analysis and revealed invariant structure and exact solutions in some cases. Almusawa and Jhangeer [3] conducted an investigation into dust size distribution shocks in quantum dusty plasma using Lie symmetry analysis and self-adjoint theory. They discovered hidden conservative laws and quantum-verified invariants that were not identified by classical models. Asghar *et al.* [6] derived conserved vectors with Lie symmetry infinitesimals through nonlinear self-adjointness. This controlled the emergence and formation of solitons and their stability. They showed the balance between nonlinearity and quantum dispersion produced by solitons. Vats *et al.* discovered similarity solutions for shock propagating in cylindrical, self-gravitating, rotating gas under monochromatic and magnetic fields through Lie symmetry analysis. This study explained the combined effect of gravity, rotation, radiation, and magnetism on shock wave propagation. Ali *et al.* [1] showed the effective part of Lie symmetry in studying wave propagation in a nonlinear system of variable coefficients. He determined an invariant structure through symmetry methods that control similarity reduction and conserve quantities. He showed how this discovery supports stable wave propagation in designing media.

Biswas *et al.* [9] explored the influence of surface tension on wave propagation dominated by the Boussinesq equation. The authors derived exact or semi-analytical solitary wave solutions and conservation laws. Tanwar and Wazwaz [47] applied the Lie symmetry technique to the dissipative KdV equation, significant to dusty plasma systems. They obtained exact analytical solutions, including traveling wave solutions.

2.3 FUNDAMENTALS OF LIE SYMMETRY METHOD

In the field of mathematics, an expression acquires symmetry if every variable related to the given expression can be transformed according to some transformations, and the following expression shows explicitly the same as in the original form of the expression. A Lie symmetry group consists of a set of continuous symmetries that harmonize to smooth transformations. When harnessed, these transformations map a system to itself. That being so, the Lie symmetry group is a smooth manifold. In the classical context, a Lie group consisted of geometric transformations on the space of independent and dependent variables of the system. They act on the solution by transforming their graphs. Some of the typical examples are groups of scaling symmetries, translation, and rotation. But these did not tire out the possibilities. In the case of continuous groups, complex nonlinear conditions of invariance of the system under the group transformations can be replaced by equivalent linear conditions in the form of "infinitesimal" invariance of the system under the generators of the groups. These infinitesimal symmetry conditions, which are known as the defining equations of the symmetry group, can be solved positively in closed form. Therefore, a complete continuous symmetry group can be obtained directly. This task can be performed using either mechanical or symbolic computations. Many operations can be possible after the determination of the symmetry group of the differential equation. One can start with the defining property of the symmetry group and determine new solutions from the existing ones. Two solutions are considered equivalent if one solution can be transformed into the other by a symmetry transformation. Thus, the symmetry groups help in classifying the solutions. The families of differential groups that depend on some function or parameter can be classified with symmetry groups. Another important application is to determine whether a differential equation allows a symmetry group. This problem was solved by infinitesimal methods and the theory of differential invariants. ODEs can be reduced to order one with the invariance under a one-parameter symmetry group. The solution of the original equation can be obtained through the reduced one by a single integration. But in the case of the reduction of multi-parameter groups, it is complex. For PDE systems, symmetry rarely yields general solutions except for some special cases where they reveal simpler invariant transformations. Though symmetry methods are used to determine group-invariant solutions. These "group-invariant" solutions are obtained by reducing the independent variables. Classical similarity solutions from groups of scaling symmetries, traveling wave solutions, and many other explicit solutions are some of the important examples of general group-invariant solutions.

The infinitesimal method is based on the important concept "prolongation," that is, extending a group action to the derivatives of dependent variables appearing in the system.

Lie symmetry method relies on the principle that invariant properties of differential equations under continuous transformation groups [35].

Let there be a system of partial differential equations (PDEs) with independent variable x^i and dependent variable u^α. A one-parameter Lie group of transformations is induced by the vector field [10, 11, 35]

$$X = \xi^i \left(u, x\right)\frac{\partial}{\partial x^i} + \eta^i \left(u, x\right)\frac{\partial}{\partial u^\alpha} \tag{2.2}$$

where ξ^i and η^i are the infinitesimal generators of the group [11, 28, 35].

The invariance condition requires that the prolonged vector field leave the governing equations unchanged on the solution manifold. This leads to a system of determining equations for the infinitesimals, whose solutions define the admitted symmetries of the system. Once identified, these symmetries may be used to reduce the number of independent variables, construct similarity variables, or generate invariant solutions.

In the context of fluid dynamics, Lie symmetries often correspond to physical invariances such as scaling transformations, Galilean invariance, or translational symmetries. Their systematic exploitation provides a powerful alternative to ad hoc similarity assumptions.

In the absence of viscosity, high-speed fluid motion is described by the one-dimensional compressible Euler equations, which represent the conservation equations of mass, energy, and momentum.

One-dimensional Euler equations are very well known in the field of compressible fluids and can be expressed as [5, 25, 55]

$$\rho_t + (\rho u)_x = 0, \tag{2.3}$$

$$(\rho u)_t + (\rho u^2 + p)_x = 0, \tag{2.4}$$

$$E_t + u(E + p)_x = 0, \tag{2.5}$$

where density is represented by ρ, velocity by u, pressure by p, and

$$E = \frac{p}{\gamma - 1} + \frac{1}{2}\rho u^2 \tag{2.6}$$

is the total energy per unit volume, and the specific heats are represented by $\gamma > 1$.

One-parameter Lie group of infinitesimal transformations can be demonstrated as [35]

$$\tilde{x} = x + \varepsilon\xi(x, t, \rho, u, p), \tag{2.7}$$

$$\tilde{t} = t + \varepsilon\tau(x, t, \rho, u, p), \tag{2.8}$$

$$\tilde{\rho} = \rho + \varepsilon\phi\rho(x, t, \rho, u, p), \tag{2.9}$$

$$\tilde{u} = u + \varepsilon\phi u(x, t, \rho, u, p), \tag{2.10}$$

$$\tilde{p} = p + \varepsilon\phi p(x, t, \rho, u, p), \tag{2.11}$$

where ε represents an infinitesimal group parameter. The corresponding infinitesimal generator for the given system of equations can be written as follows:

$$X = \xi\frac{\partial}{\partial x} + \tau\frac{\partial}{\partial t} + \phi^{\rho}\frac{\partial}{\partial \rho} + \phi^{u}\frac{\partial}{\partial u} + \phi^{p}\frac{\partial}{\partial p}. \tag{2.12}$$

Under the Lie group Equations (2.3)–(2.6), the Euler equations are invariant if

$$pr^{(1)}X(\Delta_i) = 0, \text{ whenever } \Delta_i = 0, \tag{2.13}$$

where Δ_i symbolizes the governing equations and $pr^{(1)}X$ represents the first prolongation of the generator.

The solution of the associated determining equations can give the complete set of admissible systems of Lie symmetries. Several physically important Lie symmetries are admitted by the one-dimensional Euler equations, including:

- **Time translation**

$$X_1 = \frac{\partial}{\partial t}.$$

- **Space translation**

$$X_2 = \frac{\partial}{\partial x}.$$

- **Galilean invariance**

$$X_3 = t\frac{\partial}{\partial x} + \frac{\partial}{\partial u}.$$

- **Scaling (self-similar) symmetry**

$$X_4 = x\frac{\partial}{\partial x} + t\frac{\partial}{\partial t} + \rho\frac{\partial}{\partial \rho}.$$

The scaling symmetry plays an important role in the analysis of high-speed fluid flows such as shock-dominated flows.

The similarity variables by using the scaling generator $\mathbf{X}_4$ are demonstrated as

$$\eta = \frac{x}{t}, \quad \rho(x,t) = t^{-1}R(\eta), \quad u(x,t) = U(\eta), \quad p(x,t) = t^{-1}P(\eta), \tag{2.14}$$

After substituting the above equations into the equation of mass and energy, the non-linear PDE system is converted into a corresponding ODE system:

$$(U-\eta)R' + RU' + R = 0, \quad (U-\eta)U' + \frac{1}{R}P' = 0, \quad (U-\eta)\left(\frac{P}{R^{\gamma}}\right)' = 0. \tag{2.15}$$

In the above equations primes correspond to derivatives with respect to η. Self-similar shock and blast-wave behavior emerges from the reduced system, yielding invariant solutions that form the core of many classical high-speed flow theories, including Sedov–Taylor blast waves and supersonic shock dynamics. The introduction of the similarity variable $\eta = x/t$ reflects wave motion with constant characteristic speed and reveals the dominant structure of compressible high-speed flows.

2.4 LIE SYMMETRY AND CONSERVATION LAWS

Lie symmetry methods are used to reduce nonlinear PDEs into simpler ODEs and to find the exact solutions. Conservation laws, a mathematical formulation of the acquainted physical laws of conservation of energy, momentum, mass, and so on play an integral part in the evaluation of basic properties of the solutions [34, 35]. Conservation laws provide a deep connection to the physical laws of nature. The symmetries are directly linked to the conservation laws. This was proved by Noether's theorem, which states that every continuous symmetry of a physical system corresponds to a specific conserved quantity. For example, invariance of a variational principle under a group of time translations implies the conservation of energy for the solution of the related Euler–Lagrange equations, and invariance under a group of spatial translations implies conservation of momentum [35]. The first fundamental result was constituted by this basic principle in the field of classical and quantum-mechanical arrangements with distinct groups of symmetries. For the complex system of PDEs, Noether's technique is the foundation of the systematic method for deriving conservation laws. It showed that symmetries are as fundamental as the conservation laws in nature itself. This highlights the need for symmetries in the analysis of physical phenomena, a major progression in modern physics [14].

2.5 SYMMETRY REDUCTION IN HIGH-SPEED FLOWS

Similarity reductions represent one of the most influential contributions of Lie symmetry analysis, providing a systematic mechanism for reducing complex nonlinear partial differential equations to lower-dimensional ordinary differential systems. In regimes often characterized by high-speed fluids, strategies of analytic reduction become more advantageous. One of the reasons is the invariant patterns that continuously arise in flow regions enforced by shock interaction, stagnation effects, and compressible boundary layers. The similarity solutions provide deep rigorous insight into the scaling behavior of velocity, temperature, and density fields and have provided the fundamental foundation for many classical models of supersonic and hypersonic flows. Nevertheless, the scope of similarity-driven techniques has inherent invariance limitations, restricting their use to describe fully unsteady or multidimensional flows. Similarity solutions will remain an integral part of high-speed compressible flow analysis, despite the fact that they cannot entirely replace large-scale numerical simulation. The fundamental flow phenomena can be understood with the help of symmetry solutions, without the full cost of numerical simulations. In the study of high-speed compressible flow phenomena, symmetry solutions will continue to complement modern computational techniques by linking theoretical analysis and numerical modeling [18].

2.6 SYMMETRY-BASED APPROACHES IN SHOCK-DOMINATED HIGH-SPEED FLOWS

Shock wave formation is a governing feature in supersonic and hypersonic flows. Shock waves are studied through conservational laws and standardized by

Rankine-Hugoniot jump conditions. Symmetry analysis aggravated these classical approaches. Symmetry methods empower the derivation of invariant and self-similar shock solutions. Shock wave appearances, including planar, cylindrical, and viscous shocks, are adequately employed by symmetry-based techniques. In high-speed compressible flows, symmetry methods tender analytical perceptiveness into elementary shock characteristics, entailing shock structure, propagation speed, upstream and downstream flow behavior, and internal shock profile [8, 17].

Thermal effects and real-gas phenomena considerably intensify the complexity of flow modeling in hypersonic domains. Symmetry-based studies unveiled scaling relationships that influence pressure distributions and aerodynamic heating. Symmetry approaches reduce the nonlinear complexities of the system without rectification in dominant physical processes. Regardless of the limitations arising from the assumption of invariance, symmetry-reduced models propose heave insight into flow behavior. Numerical and computational simulations, such as rapid preliminary design analysis, are validated effectively through symmetry methods.

2.7 SYMMETRY-BASED RECENT CONCEPTUAL INSIGHTS

The aspect of symmetry-based techniques has been enhanced by the new advances in the field of high-speed flow, particularly in compressible and shock-dominated environments. Symmetry techniques were conventionally used as solution-constructing techniques. However, the symmetry methods are accepted as an important technique to illustrate invariant structures of complex nonlinear physical problems. The classical Lie symmetry is extended to generalized symmetries. This widened the relevance of the symmetries in practical flow problems.

Analytical and numerical techniques in complex fluid flow routines have been intensified by the precocious research in nonclassical and approximate symmetry. Weak nonlinearities and small perturbations are treated specifically with approximate symmetry methods. They facilitate approximate invariants and conservation laws pertinent in fluid dynamics. The classical Lie method has been expanded with nonclassical symmetries through invariant surface conditions. It concedes solutions that are unattainable with standard methods. Contemporaneously, symmetry-based numerical methods improve stability and consistency in computational simulations. Essential invariants and conservation laws are patched through symmetry-preserving numerical schemes. These enhancements corroborate the increasing importance of symmetry-based approaches in consolidating theoretical analysis with numerical modeling of high-speed compressible flows.

The data-driven and machine learning-based new research is playing an essential part in analyzing symmetries in fluid dynamics. These broaden the approach of symmetry methods as data-driven techniques speculate invariant structure directly from experimental abstracts or numerical simulations. This is beneficial in complex flow regimes partially characterized by governing equations. Data-driven methods normally contain approximate symmetries that represent realistic shock wave phenomena. This implication is important in the comprehensiveness of high-speed fluid flows. For example, despite strong symmetry breaking in instantaneous flows, ensemble shock data reveal self-similar scaling. The governing equation grew more complicated in multiphysics systems such as chemical radiations, reactions, and phase changes and usually refused symmetry classification. Advanced machine learning techniques provide an invariant structure for complex high-dimensional systems. This reduces model complexity without affecting the physical significance.

Data-driven analysis completely shifted the equation-based invariance to data-based invariance. Computational modeling is beginning to be influenced by the data-driven symmetries. These techniques identify invariant structure from the data, which can be used in numerical solvers, turbulence, and shock-capturing schemes. These discoveries preserve observed symmetries.

2.8 LIE SYMMETRY IN COMPARISON TO OTHER ANALYTICAL METHODS

Traditional analytic methods such as homotopy or perturbation methods generally depend on small or large parameters and use the asymptotic expansion method to give approximate results. These methods are only reliable for weak nonlinear media and not suitable in the presence of strong nonlinearity or in the absence of appropriate parameters. While decomposition methods give solutions through convergent iterative expansions without linearization. They can work in nonlinear systems, but their convergence and physical explicability are not assured. Scaling methods or classic similarities are usually based on dimensional analysis. Assumed scaling laws are used to reduce equations in these techniques. In high-speed flows, Lie symmetry analysis offers deeper and more desirable outcomes compared to the mentioned classical analytical methods. Variational methods are dependent on Lagrangian formulation. Therefore, they lack conservative systems. One of the advantages of the symmetry method over the numerical methods is that it reduces the computational complexity of the problem. The Lie symmetry method can produce exact solutions and expose self-similar structures that cannot be obtained by other conventional methods. The Lie symmetry methods avoid assumptions and provide solutions in closed form with physical significance. Symmetry techniques can be used for both conservative and non-conservative systems. Symmetry methods usually lead to conservative laws of physical nature. Although this method requires nonclassical or symmetry extensions in complex or perturbed flows, this method's lesser effectiveness in turbulent flows and highly irregular systems can be a disadvantage. But recent research in hybrid analytic-numerical techniques and advances in computation can overcome this limitation and broaden the scope. In spite of this, the symmetry methods will remain a powerful and essential tool for shock waves and other high-speed fluid flows.

2.9 FUTURE SCOPE AND CONCLUSION

Symmetry principles dispense logical and persistent prediction of aerodynamic load under a wide range of transport by identifying the scaling laws and invariant structure. The flow dynamics are dominated by conditions such as strong shock waves and extreme compressible environments. Lie symmetry methods are effective and useful tools in studying and providing applications for predictive modeling of high-speed fluid flow that includes hypersonic and high-speed compressible flow. But symmetry methods are limited by the ability to handle turbulence multi-species interactions and three-dimensional complexities. Overcoming these limitations requires advanced integrated computational insights and data-driven techniques that are AI-powered. Future research is expected to focus on symmetry techniques with high-fidelity simulations. This improves predictive modeling for hypersonic vehicle re-entry and aerodynamics. This can also include coupled thermochemical systems, approximate and non-classical symmetries for turbulence modeling, and perturbed flows. Emerging high-powered AI and quantum computing are expected to extend symmetry methods to multiphysical, highly nonlinear flow phenomena.

The integration of Lie symmetry methods with advanced computational techniques such as artificial intelligence (AI) and quantum computing offers an optimistic approach to exploiting symmetry techniques in fluid dynamics. Although quantum computing could provide new methodological opportunities to reduce complicated algebraic structures inherited in symmetry analysis, more particularly linear algebra, symbolic computation and optimization, and quantum algorithms could possibly enhance faster symmetry detection and reduce-order modeling. Furthermore, new computational paradigms like quantum computing, adaptive meshing, and artificial intelligence have the potential to expand symmetry-based analysis to highly nonlinear, multi-scale, and multi-physics flow problems, connecting analytical understanding with contemporary simulation capabilities.

Symmetry techniques provide a powerful intrigue for the abridgment of nonlinear high-speed compressible equations. Symmetry generates invariant and self-similar solutions. These solutions apprehend essential features of flow. After being integrated with modern computation techniques, Lie symmetry methods enhance prediction in thermal effects during hypersonic flights and atmospheric re-entry, shock interactions, and vehicle aerodynamics. Lie symmetry analysis can bridge analytical theories and modern simulations, which will help in the study of extreme flow environments. In hypersonic and high-speed compressible flow modeling, symmetry methods can be a major enhancement. From this perspective, the integration of Lie symmetry techniques with AI and emerging quantum computing could possibly establish classical symmetry methods in a purely technical role in the next generation of computational fluid dynamics and complex compressible flows.

REFERENCES

1. Ali, M. R., Ma, W.-X., and Sadat, R. (2022). Lie symmetry analysis and wave propagation in variable-coefficient nonlinear physical phenomena. *East Asian Journal on Applied Mathematics*, 12(1):201–212.
2. Aliyu, A. I., Inc, M., Yusuf, A., and Baleanu, D. (2018). Symmetry analysis, explicit solutions, and conservation laws of a sixth-order nonlinear ramani equation. *Symmetry*, 10(8):341.

3. Almusawa, H. and Jhangeer, A. (2022). Nonlinear self-adjointness, conserved quantities and lie symmetry of dust size distribution on a shock wave in quantum dusty plasma. *Communications in Nonlinear Science and Numerical Simulation*, 114:106660.
4. Alshammari, F. S., et al. (2025). Exploring chaotic behavior, conservation laws, lie symmetry, and soliton dynamics in the generalized $A\mp$equation. *AIMS Mathematics*, 10(9):22150–22179.
5. Anderson, R. L. and Ibragimov, N. H. (1979). *Lie-Bäcklund transformations in applications*. SIAM.
6. Asghar, U., Asjad, M. I., Faridi, W. A., and Muhammad, T. (2024). The conserved vectors and solitonic propagating wave patterns formation with lie symmetry infinitesimal algebra. *Optical and Quantum Electronics*, 56(4):540.
7. Batchelor, G. K. (1967). *An introduction to fluid dynamics*. Cambridge University Press.
8. Birkhoff, G. (1960). *Hydrodynamics: A study in logic, fact, and similitude*. Princeton University Press.
9. Biswas, A., Vega-Guzman, J., Bansal, A., Kara, A. H., Aphane, M., Yıldırım, Y., and Alshehri, H. M. (2023). Solitary waves, shock waves and conservation laws with the surface tension effect in the boussinesq equation. *Proceedings of the Estonian Academy of Sciences*, 72(1):17–29.
10. Bluman, G. W. and Cole, J. D. (1969). The general similarity solution of the heat equation. *Journal of Mathematics and Mechanics*, 18(11):1025–1042.
11. Bluman, G. W. and Cole, J. D. (2012). *Similarity methods for differential equations*, volume 13. Springer Science & Business Media.
12. Boutros, Y. Z., Abd-el Malek, M. B., Badran, N. A., and Hassan, H. S. (2007). Lie-group method solution for two-dimensional viscous flow between slowly expanding or contracting walls with weak permeability. *Applied Mathematical Modelling*, 31(6):1092–1108.
13. Bozhkov, Y., Freire, I. L., and Ibragimov, N. H. (2014). Group analysis of the novikov equation. *Computational and Applied Mathematics*, 33(1):193–202.
14. Cantwell, B. J. (2002). *Introduction to symmetry analysis*. Cambridge University Press.
15. Chauhan, A., Tiwari, A., Sharma, K., and Arora, R. (2022). Steepening of waves in non-ideal radiative magnetogasdynamic flow with dust particles. *Pramana*, 96(3):139.
16. El Naschie, M. (2008). The exceptional lie symmetry groups hierarchy and the expected number of higgs bosons. *Chaos, Solitons & Fractals*, 35(2):268–273.
17. Evans, L. C. (2022). *Partial differential equations*, volume 19. American mathematical society.
18. Gardner, C. S., Greene, J. M., Kruskal, M. D., and Miura, R. M. (1967). Method for solving the korteweg-devries equation. *Physical Review Letters*, 19(19):1095.
19. Gazizov, R. K. and Ibragimov, N. H. (1998). Lie symmetry analysis of differential equations in finance. *Nonlinear Dynamics*, 17(4):387–407.
20. Hu, X., Li, Y., and Chen, Y. (2015). A direct algorithm of one-dimensional optimal system for the group invariant solutions. *Journal of Mathematical Physics*, 56(5).
21. Huang, D., Li, X., and Yu, S. (2017). Lie symmetry classification of the generalized nonlinear beam equation. *Symmetry*, 9(7):115.
22. Ibragimov, N. H. (1984). *Transformation groups applied to mathematical physics*, volume 3. Springer Science & Business Media.
23. Infeld, E. and Rowlands, G. (2000). *Nonlinear waves, solitons and chaos*. Cambridge University Press.
24. Karna, A. K. and Satapathy, P. (2023). Lie symmetry analysis for the cargo–leroux model with isentropic perturbation pressure equation of state. *Chaos, Solitons & Fractals*, 177:114241.

25. Landau, L. D. and Lifshitz, E. M. (1987). *Fluid mechanics*, 2nd edition. Pergamon Press.
26. Lie, S. (1880). *Über die Integration durch bestimmte Integrale von einer Classe linearer partieller Differentialgleichungen*. Cammermeyer.
27. Lie, S. (1881). On integration of a class of linear partial differential equations by means of definite integrals. *Archiv for Mathematik og Naturvidenskab*, 6(3):328–368.
28. Lie, S. (1891). *Vorlesungen über differentialgleichungen mit bekannten infinitesimalen transformationen*. BG Teubner.
29. Lie, S. and Engel, F. (1893). *Theorie der transformationsgruppen*, volume 3. Teubner.
30. Mandal, S., Sil, S., and Ghosh, S. (2025). On the lie symmetry analysis of three-dimensional perturbed shear flows. *Chaos, Solitons & Fractals*, 191:115875.
31. Meinhardt, J. (1981). Symmetries and differential equations. *Journal of Physics A: Mathematical and General*, 14(8):1893.
32. Msomi, A., Govinder, K., and Maharaj, S. (2012). Applications of lie symmetries to higher dimensional gravitating fluids. *International Journal of Theoretical Physics*, 51(4):1290–1299.
33. Noether, E. (1983). Invariante variationsprobleme. In *Gesammelte Abhandlungen—Collected papers*, pages 231–239. Springer.
34. Oliveri, F. (2010). Lie symmetries of differential equations: Classical results and recent contributions. *Symmetry*, 2(2):658–706.
35. Olver, P. J. (1993). *Applications of Lie groups to differential equations*, volume 107. Springer Science & Business Media.
36. Ovsiannikov, L. V. (2014). *Group analysis of differential equations*. Academic Press.
37. Patel, A. and Garg, Y. (2024). Lie group of similarity analysis of shock waves in viscous flow under magnetic field. *Chinese Journal of Physics*, 92:1531–1547.
38. Rehman, K. U., Malik, A. A., Malik, M., and Hayat, T. (2017). Generalized lie symmetry analysis for non-linear differential equations: A purely viscous fluid model. *Results in Physics*, 7:3537–3542.
39. Samina, S., Almusawa, H., Arif, F., and Jhangeer, A. (2025). Lie symmetries, solitary waves, and noether conservation laws for (2+1) dimensional anisotropic power-law nonlinear wave systems. *Symmetry*, 17(9):1445.
40. Sedov, L. I. (2018). *Similarity and dimensional methods in mechanics*. CRC Press.
41. Shagolshem, S., Bira, B., and Nagaraja, K. (2024). Analysis of shock wave propagation in two-layered blood flow model via lie symmetry. *International Journal of Non-Linear Mechanics*, 164:104761.
42. Shagolshem, S., Bira, B., and Nagaraja, K. (2025). Analysis of soliton wave structure for coupled higgs equation via lie symmetry, paulpainlevé approach and the unified method. *Nonlinear Dynamics*, 113(10):11999–12020.
43. Sharma, A. K., Shagolshem, S., and Arora, R. (2025). Shock wave dynamics via symmetrydriven analysis of a two-phase flow with the chaplygin pressure law. *Chaos, Solitons & Fractals*, 198:116512.
44. Shukla, P. K. and Mamun, A. A. (2002). *Introduction to Dusty Plasma physics*. Institute of Physics Publishing.
45. Singhal, A., Joshi, U., and Arora, R. (2025). Lie symmetry analysis, parametric reduction and conservation laws of (3+1) dimensional nonlinear dispersive soliton equation. *arXiv preprint arXiv:2502.07682*.
46. Smoller, J. (2012). *Shock waves and reaction—diffusion equations*, volume 258. Springer Science & Business Media.
47. Tanwar, D. V. and Wazwaz, A.-M. (2022). Lie symmetries and exact solutions of kdv–burgers equation with dissipation in dusty plasma. *Qualitative Theory of Dynamical Systems*, 21(4):164.

48. Tiwari, A. and Arora, R. (2020). An impact of forcing terms on the delta shock front. *The European Physical Journal Plus*, 135(7):1–20.
49. Tiwari, A. and Arora, R. (2022). Lie symmetry analysis, optimal system and exact solutions of a new (2+1)-dimensional kdv equation. *Modern Physics Letters B*, 36(12):2250056.
50. Tiwari, A. and Arora, R. (2024). A self-similar solution for unsteady isothermal and adiabatic flows behind the shock wave in a dusty non-ideal rotating medium under the action of magnetic field. *Physica D: Nonlinear Phenomena*, 458:134006.
51. Tiwari, A. and Arora, R. (2025). Analytical study of self-similar motion following shock waves in magnetized, non-ideal, self-gravitating gas with variable density. *Physics of Fluids*, 37(12):126119.
52. Tiwari, A., Sharma, K., and Arora, R. (2021). Lie symmetry analysis, optimal system, and new exact solutions of a (3+1) dimensional nonlinear evolution equation. *Nonlinear Engineering*, 10(1):132–145.
53. Tiwari, A. and Trivedi, L. M. (2023). A review on some applications of Lie symmetry groups in nonlinear PDE. *International Journal of Engineering Sciences & Emerging Technologies,* 11(2):382–386.
54. Topno, A., Sil, S., and Raja Sekhar, T. (2025). Nonlocal symmetry analysis, exact solutions, and evolution of nonlinear waves for one-dimensional shallow water magnetohydrodynamics equations. *Physics of Fluids*, 37(8).
55. Toro, E. F. (2009). *Riemann solvers and numerical methods for fluid dynamics*, 3rd edition. Springer.
56. Whitham, G. B. (2011). *Linear and nonlinear waves*. John Wiley & Sons.
57. Zakharov, V. E. and Kuznetsov, E. A. (1974). On three-dimensional solitons. Soviet Physics JETP, 39(2):285–286.

3 Computational Fluid Dynamics Techniques

Finite Difference Modeling of Magnetic Nanofluid Convection

Basit Rahman and Sahin Ahmed

3.1 INTRODUCTION

In present-day industries and technologies, the augmentation of heat transfer is essential, which involves energy production, indoor climate regulation, and energy transportation because of its applications in heating and cooling processes. In a wide range of applications, water, ethylene glycol, and oil serve as pure fluids, but their low thermal conductivity reduces the potential for heat transfer improvement. Initially, the concept of distributing micro-sized solid particles with superior thermal conductivity into the traditional pure fluids was employed [26], and this idea was further developed with an aim to achieve escalating thermal conductivity [17]. However, to intensify the effect of this kind of thermal performance, a novel heat transfer fluid known as "nanofluid" has been introduced by suspending metallic nanoparticles into the conventional base fluids [13]. In current times, many investigators have explored the behavior and properties of these nanofluids in various fields like thermal engineering, which involves automobiles, energy plants, and even biomedical applications, capturing nanofluids to intensify the thermal transport performance [24, 29, 31]. To represent nanofluids' rheological and thermal behavior, two standard mathematical models have been proposed [10, 42]. The Tiwari-Das single-phase approach accounts for nanoparticle volume fraction effects on heat transfer, with thermophysical properties dependent on the nanoparticle and base fluid. Unlike the single-phase approach, Buongiorno's two-phase model incorporates nanoparticle dynamics, including the impacts due to Brownian motion and thermophoresis.

Alongside the progress of technology, an innovative heat transfer fluid called "hybrid nanofluid" has been engineered by the suspension of multiple nanosized particles into the pure traditional fluid, commonly known as base fluids [39]. This novel type of fluid was developed to deliver enhanced thermal efficiency relative to single-nanosized dispersed fluids. It has been revealed that the appropriate choice of nanoparticles also

DOI: 10.1201/9781003664345-3

plays a key role in ensuring the flow steadiness [34]. Many researchers explored the ability of increased thermal performance hybrid nanofluids in terms of practical-world applicability [5, 25]. The recent works considerably focus on modeling hybrid nanofluids for attaining precise simulation algorithms [28, 44, 46].

The domain of porosity grew significant consideration due to its ability to regulate the thermal transportation process. Previous studies indicated that fluid motion behavior is highly influenced by the porous structures and hence considered as significant factors in the relevant domain [8, 28]. Moreover, it has been established that the association of porous media notably raises the efficacy of thermal transportation devices [22]. Earlier works on the Darcy–Forchheimer hybrid nanofluid model suggested that a resistive flow motion is observed in the porous materials when enhanced porosity is encountered [3].

Permeability of a medium is generally indicated by the pulling force that demonstrates the ability of the shallow to pass the flow. For acquiring steadiness in the BL flow, the application of suction on a permeable sheet is essential [27]. Within this framework, the behavior of Marangoni hybrid nanofluid across an infinite-permeable porous medium was analyzed, and results indicated that the rate of thermal transportation improved and BL bifurcation was delayed with a greater suction parameter within the contracted case [43]. A similar conclusion regarding the role of suction in improving warmth transfer while studying chemically reactive flow across a permeable surface has been made [32].

Due to its broad industrial and engineering applications, the phenomenon of Newtonian or viscous flow over an elongated surface has remained a subject of research interest. A foundational analysis of the steady 2D BL motion of Newtonian fluid across an elongated sheet was investigated initially [14]. After the pioneering research of Crane, the study on stretching surfaces sparked significant scientific interest among the researchers, incorporating various physical influences – including suction, blowing, magnetism, and convection phenomena. Regarding this framework, the impact of suction was analyzed; early foundation for convective flow research was reported [11, 12, 16, 30].

To investigate the configuration of momentum transportation near the surface, the concept of slip in the surface wall under a permeable structure was introduced [9]. Additionally, the impact of velocity slip in the BL flow was examined [6]. A research work on MHD-driven flow with pore sections was scrutinized, and an interrelation of magnetization with porosity effects was established, regarding their effect on thermal transportation systems [19]. In another relevant contribution, a computational analysis of an MHD-driven flow model of nanofluid with copper, silver, and iron oxide nanoparticles past a permeable elongated surface with porous configurations was reported [19]. Additional works have focused on hybrid nanofluid motion, considering the role of MHD across stretching/shrinking boundaries, subject to distinct physical aspects [20, 37]. A 3D hybrid nanofluid model across a permeable elongating/contracting surface incorporating slip velocity and convective heating conditions was explored along with suction [23]. In connection, a similar flow configuration was explored under the influence of quadratic or nonlinear velocity conditions [47]. In a similar context, the hybrid nanofluid flow possessing the significance of thermal generation–absorption and warmth emission past boundary layers was examined [45]. Additional studies have contributed to this field by analyzing hybrid nanofluid flow, incorporating various physical flow parameters under different feasible conditions such as Darcy–Forchheimer law, Lorentz forces, etc. [15, 21].

The current numerical study investigates the three-dimensional model comprising hybrid $MoS_2 - SiO_2 / H_2O$ nanofluid flow and thermal performance passing through a permeable and stretching surface. However, several features of their behavior under physical conditions remain insufficiently explored. Motivated by these gaps, the originality of this work focuses on examining the combined influence of magnetic field intensity, porous medium resistance, velocity slip, and suction or injection in modifying the thermofluidic response. Through appropriate similarity conversions, the coupled nonlinear governing partial differential equations are converted into a nondimensional form, which is further solved by employing a reliable finite difference algorithm. The adopted numerical method provides stable and convergent results, effectively demonstrating the impact of hybrid nanoparticles on temperature distribution and flow dynamics. The outcomes of this computational scrutiny deliver useful insights for improving thermal emission efficacy in industrial activities such as cooling, extrusion, and coating processes, where stretching and permeability effects are of significant importance.

3.2 MATHEMATICAL FORMULATIONS

In this work, a 3D incompressible time-independent boundary layer motion, involving hybrid ($MoS_2 - SiO_2 / H_2O$) nanofluid model, driven by a permeable elongated surface boundary over the xy-surface subject to velocity slip and convective boundary effects is considered. The study inculcates the following flow assumptions:

- The xy-plane represents a permeable stretching sheet subjected to convective surface conditions.
- Flow domain is restricted to the area $z \geq 0$, where the motion of hybrid nanofluid is considered.
- The stretching surface moves with velocities $u_w(x) = ax$ in the x-direction, $v_w(x) = ay$ in the y-direction, with $a > 0$ indicating the stretching rate.
- The bottom boundary surface is heated at a fixed temperature T_f, convectively by an external fluid, resulting in a finite thermal emission coefficient h_f at the boundary.
- The agglomeration or clustering effects of hybrid nanoparticles are neglected, and it is assumed that the nanoparticles have uniform dimensions in all regions within the fluid.
- The fluid motion is considered to be steady, incompressible, and laminar, neglecting pressure gradients and body forces other than those explicitly considered (e.g., magnetic or porous resistance).
- All the thermophysical axioms of the nanoparticles and base fluid are fixed, except for density variations considered in buoyancy or external forces, if applicable.

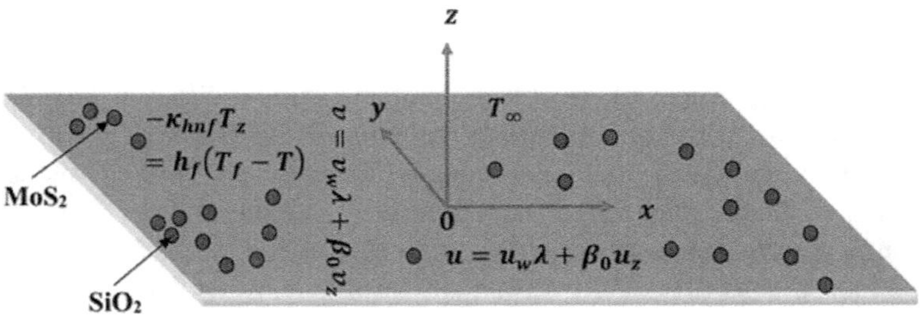

FIGURE 3.1 Model geometry.

With the aforementioned assumptions, the model is represented mathematically by the following governing equations [23, 44, 45]:

$$u_x + v_y + w_z = 0 \tag{3.1}$$

$$uu_x + vu_y + wu_z = \frac{\mu_{hnf}}{\rho_{hnf}}u_{zz} - \frac{\mu_{hnf}}{\rho_{hnf}}\frac{1}{K}u - \frac{\sigma_{hnf}}{\rho_{hnf}}B^2u \tag{3.2}$$

$$uv_x + vv_y + wv_z = \frac{\mu_{hnf}}{\rho_{hnf}}v_{zz} - \frac{\mu_{hnf}}{\rho_{hnf}}\frac{1}{K}v - \frac{\sigma_{hnf}}{\rho_{hnf}}B^2v \tag{3.3}$$

$$uT_x + vT_y + wT_z = \frac{\kappa_{hnf}}{(\rho C_p)_{hnf}}T_{zz} - \frac{Q}{(\rho C_p)_{hnf}}(T - T_\infty) \tag{3.4}$$

Boundary conditions

$$\left.\begin{array}{c} u = u_w(x)\lambda + \beta_0 u_z,\ v = v_w(y)\lambda + \beta_0 v_z,\ w = w_0 \\ -\kappa_{hnf}T_z = h_f(T_f - T) \end{array}\right\} at\ z = 0 \tag{3.5}$$

$$u \to 0, \quad v \to 0, T \to T_\infty, \quad \text{as } z \to \infty \tag{3.6}$$

Here u, v, and w represents the velocities of the flow along x, y, and z directions, with T denoting the temperature of the hybrid nanofluid. The parameter w_0 signifies the constant mass flux, with $w_0 < 0$ for suction and $w_0 > 0$ for fluid injection, and λ indicates the constant elongating/contracting variable with $\lambda > 0$ for elongated surface, $\lambda < 0$ for contracting surface, and $\lambda = 0$ for static surface.

Table 3.1 highlights the thermophysical characteristics of mono and hybrid nanofluids, where ρ, μ, κ, ρ, and σ represent density, dynamic viscosity, thermal conductivity, heat capacitance, and electric conductivity [4, 45]. Here, the indices "hnf," "nf," "*f*," "s_1," "s_2," respectively, specify the hybrid nanofluid, nanofluid, base fluid,

TABLE 3.1
Thermophysical Features [4, 45]

Properties	Nanofluid	Hybrid Nanofluid
Density	$\rho_{nf} = (1-\Phi)\rho_f + \Phi\rho_s$	$\rho_{hnf} = (1-\Phi_2)\left[(1-\Phi_1)\rho_f + \Phi_1\rho_{s1}\right] + \Phi_2\rho_{s2}$
Heat capacity	$(\rho C_p)_{nf} = (1-\Phi)(\rho C_p)_f + \Phi(\rho C_p)_s$	$(\rho C_p)_{hnf} = (1-\Phi_2)\left[(1-\Phi_1)(\rho C_p)_f + \Phi_1(\rho C_p)_{s1}\right] + +\Phi_2(\rho C_p)_{s2}$
Dynamic viscosity	$\frac{\mu_{nf}}{\mu_f} = \frac{1}{(1-\Phi)^{2.5}}$	$\frac{\mu_{hnf}}{\mu_f} = \frac{1}{(1-\Phi_1)^{2.5}(1-\Phi_2)^{2.5}}$
Thermal conductivity	$\frac{\kappa_{nf}}{\kappa_f} = \left[\frac{\kappa_s + 2\kappa_f - 2\Phi(\kappa_f - \kappa_s)}{\kappa_s + 2\kappa_f + \Phi(\kappa_f - \kappa_s)}\right]$	$\frac{\kappa_{hnf}}{\kappa_{nf}} = \left[\frac{\kappa_{s2} + 2\kappa_{nf} - 2\Phi_2(\kappa_{nf} - \kappa_{s2})}{\kappa_{s2} + 2\kappa_{nf} + \Phi_2(\kappa_{nf} - \kappa_{s2})}\right]$ where, $\frac{\kappa_{nf}}{\kappa_f} = \left[\frac{\kappa_{s1} + 2\kappa_f - 2\Phi_1(\kappa_f - \kappa_{s1})}{\kappa_{s1} + 2\kappa_f + \Phi_1(\kappa_f - \kappa_{s1})}\right]$
Electric conductivity	$\frac{\sigma_{nf}}{\sigma_f} = \left[\frac{\sigma_{s1} + 2\sigma_f - 2\Phi_1(\sigma_f - \sigma_{s1})}{\sigma_{s1} + 2\sigma_f + \Phi_1(\sigma_f - \sigma_{s1})}\right]$	$\frac{\sigma_{hnf}}{\sigma_{nf}} = \left[\frac{\sigma_{s2} + 2\sigma_{nf} + 2\Phi_2(\sigma_{s2} - \sigma_{nf})}{\sigma_{s2} + 2\sigma_{nf} - \Phi_2(\sigma_{s2} - \sigma_{nf})}\right]$ where, $\frac{\sigma_{nf}}{\sigma_f} = \left[\frac{\sigma_{s1} + 2\sigma_f + 2\Phi_1(\sigma_{s1} - \sigma_f)}{\sigma_{s1} + 2\sigma_f - \Phi_1(\sigma_{s1} - \sigma_f)}\right]$

MoS_2 solid nanoparticle, and SiO_2 solid nanoparticle, and and represent the MoS_2 and SiO_2 concentrations, respectively (Tables 3.2 and 3.3).

3.2.1 Similarity Transformations

The following similarity transformations have been employed [23].

$$\begin{gathered} u = axf'(\eta), v = axg'(\eta), w = -(a\upsilon_f)^{0.5}\left[f(\eta)+g(\eta)\right] \\ \theta(\eta) = \frac{T-T_\infty}{T_f-T_\infty}, \eta = z\left(a\upsilon_f^{-1}\right)^{0.5} \end{gathered} \tag{3.7}$$

So that $w_0 = -\left(a\upsilon_f^{-1}\right)^{0.5} S$. Here S represents the constant mass flux parameter.

TABLE 3.2
Thermophysical Quantities [7, 2, 33, 35, 38]

Physical Properties	MoS_2	SiO_2	Pure Water
$\rho\ (\mathrm{kg/m^3})$	5060	2220	997.1
$C_p\ (\mathrm{J/kgK})$	397.21	745	4179
$\kappa\ (\mathrm{W/mK})$	904.4	1.4	0.6130
$\sigma\ (\mathrm{S.m^{-1}})$	1990	10^{-17}	0.000005

TABLE 3.3
The Nondimensional Physical Flow Parameters

$Pr = \dfrac{\mu_f (C_p)_f}{\kappa_f}$	Prandtl number
$M = \dfrac{\sigma_f B^2}{a\rho_f}$	Magnetic parameter
$\Omega = \dfrac{Q}{a\left(\rho C_p\right)_{hnf}}$	Heat source/sink parameter
$Bi = \dfrac{h_f}{\kappa_f}\sqrt{\dfrac{\vartheta_f}{a}}$	Biot number
$Da = \dfrac{\vartheta_f}{K.a}$	Porosity parameter
$\beta = \sqrt{\dfrac{a}{\vartheta_f}}\beta_0$	Velocity slip parameter

3.2.2 Transformed Governing Equations

Through the employment of similarity variables (see Equation 3.7), Equation (3.1) holds inherently, whereas Equations (3.2)–(3.4) are reduced to the following ordinary differential equations (ODEs):

$$\frac{\alpha_1}{\alpha_2} f''' + (f+g) f'' - \left(\frac{\alpha_1}{\alpha_2} Da + \frac{\alpha_5}{\alpha_2} M \right) f' - f'^2 = 0 \tag{3.8}$$

$$\frac{\alpha_1}{\alpha_2} g''' + (f+g) g'' - \left(\frac{\alpha_1}{\alpha_2} Da + \frac{\alpha_5}{\alpha_2} M \right) g' - g'^2 = 0 \tag{3.9}$$

$$\frac{1}{Pr} \frac{\alpha_3}{\alpha_4} \theta'' + (f+g)\theta' - \Omega\theta = 0 \tag{3.10}$$

With

$$\begin{aligned} f(0) = S, f'(0) = \lambda + \beta f''(0), g(0) = 0, g'(0) = \lambda + \beta g''(0), \\ -\alpha_3\theta'(0) = Bi\left[1 - \theta(0)\right], \\ f'(\eta) \to 0,\ g'(\eta) \to 0,\ \theta(\eta) \to 0,\ as\ \eta \to \infty \end{aligned} \tag{3.11}$$

where

$$\alpha_1 = \frac{\mu_{hnf}}{\mu_f}, \alpha_2 = \frac{\rho_{hnf}}{\rho_f}, \alpha_3 = \frac{\kappa_{hnf}}{\kappa_f}, \alpha_4 = \frac{(\rho C_p)_{hnf}}{(\rho C_p)_f}, \alpha_5 = \frac{\sigma_{hnf}}{\sigma_f}. \tag{3.12}$$

3.2.3 Engineering Parameters

$$\left. \begin{aligned} \text{Skin friction coefficient along x axis: } C_{fx} = \frac{\tau_{wx}}{\rho_f {u_w}^2}, \\ \text{Skin friction coefficient along y axis: } C_{fy} = \frac{\tau_{wy}}{\rho_f {v_w}^2}, \\ \text{Local Nusselt number: } Nu_x = \frac{x q_w}{\kappa_f (T_f - T_\infty)}, \end{aligned} \right\} \tag{3.13}$$

where τ_{wx} and τ_{wy} denote skin frictions (shear stress) in the directions of x and y, respectively, and q_w denotes the thermal flux from the surface.

$$\tau_{wx} = \mu_{hnf} \left(\frac{\partial u}{\partial z} \right)_{z=0}, \tau_{wy} = \mu_{hnf} \left(\frac{\partial v}{\partial z} \right)_{z=0}, q_w = -\kappa_{hnf} \left(\frac{\partial T}{\partial z} \right)_{z=0} \tag{3.14}$$

Using Equations (3.7), (3.13), and (3.14), the following expression can be obtained:

$$Re_x^{1/2} C_{fx} = \frac{\mu_{hnf}}{\mu_f} f''(0),\ Re_y^{1/2} C_{fy} = \frac{\mu_{hnf}}{\mu_f} g''(0), \text{ and } Re_x^{-1/2} Nu_x = -\frac{\kappa_{hnf}}{\kappa_f} \theta'(0) \quad (3.15)$$

Here, $Re_x = \frac{u_w(x)x}{\vartheta_f}$ and $Re_y = \frac{v_w(y)y}{\vartheta_f}$ indicates local Reynolds number.

3.3 NUMERICAL APPROACH

The strong nonlinearity of the ODEs, together with the specified boundary conditions (Equations 3.8–3.11), makes obtaining an analytical solution impractical. To address this, the nonlinear system has been solved numerically by adopting MATLAB's built-in **bvp4c** in combination with the finite difference algorithm. The **bvp4c** boundary value problem solver, originally developed by Shampine [36], employs a three-stage Lobatto IIIA scheme, which provides an efficient and accurate framework for tackling highly nonlinear models.

By employing the variables introduced below, initially the nonlinear DEs were transformed into an arrangement of first-order linear differential equations:

$$\begin{pmatrix} y(1) = f, y(2) = f', y(3) = f'', y(4) = g, y(5) = g', \\ y(6) = g'', y(7) = \theta, y(8) = \theta' \end{pmatrix} \quad (3.16)$$

Thus, utilizing Equations (3.8)–(3.12) with Equation (3.16), we have

$$\begin{pmatrix} y_1' \\ y_2' \\ y_3' \\ y_4' \\ y_5' \\ y_6' \\ y_7' \\ y_8' \end{pmatrix} = \begin{pmatrix} y_2 \\ y_3 \\ \frac{\alpha_2}{\alpha_1}\left[\left(\frac{\alpha_1}{\alpha_2} Da + \frac{\alpha_5}{\alpha_2} M\right) y(2) - \left(y(1) + y(4)\right) y(3) + y(2)^2\right] \\ y_5 \\ y_6 \\ \frac{\alpha_2}{\alpha_1}\left[\left(\frac{\alpha_1}{\alpha_2} Da + \frac{\alpha_5}{\alpha_2} M\right) y(5) - \left(y(1) + y(4)\right) y(6) + y(5)^2\right] \\ y_8 \\ \frac{pr.\alpha_4}{\alpha_3}\left[\Omega y(7) - \left(y(1) + y(4)\right) y(8)\right] \end{pmatrix} \quad (3.17)$$

Newly transformed boundary equations can be expressed as

$$\begin{Bmatrix} ya(1) - S, ya(2) - \lambda - \beta ya(3), ya(4), ya(5) - \lambda - \beta ya(6), \\ -\alpha_3 ya(8) - Bi\left[1 - ya(7)\right], yb(2), yb(5), yb(7) \end{Bmatrix} \quad (3.18)$$

where, ya = initial and yb = boundary conditions.

3.4 VALIDITY AND ACCURACY

For assuring the reliability of the computed numerical results for the current, it is matched with findings of [23] and [18].

The values obtained in the current study, as shown in Table 3.4, are very much similar to results from previous studies [23] and [18]. The minor differences observed are negligible, which confirms both the validity and accuracy of the current numerical approach using the **bvp4c** solver. This comparison demonstrates that the present method reliably reproduces established results for the chosen limiting case.

TABLE 3.4
Comparative Analysis of Present and Previous Works [23] and [18] When $S = \beta = Bi = \varphi_1 = \varphi_2 = 0$ and $\lambda = 1$.

Gradients at $\eta = 0$	Present Study (bvp4c)	[23] (bvp4c)	[18] (HAM)
$-f''(0)$	1.179824	1.173721	1.173721
$-g''(0)$	1.179824	1.173721	1.173721

3.4.1 *Stability Analysis*

A stability and convergence test was conducted for the velocity profile $f'(\eta)$ by varying the step size h within the computational domain, as shown in Table 3.5. The

TABLE 3.5
Stability Test for Velocity Profile

Stability Test

η	$f'(\eta)$		
	h (step size)		
	$h = 0.1$	$h = 0.01$	$h = 0.001$
0.0	0.012616	0.012617	0.012617
0.1	0.0098496	0.0098497	0.0098497
0.2	0.0076891	0.0076893	0.0076893
0.3	0.0060023	0.0060024	0.0060024
0.4	0.0046853	0.0046854	0.0046854
0.5	0.0036571	0.0036572	0.0036572
0.6	0.0028544	0.0028545	0.0028545
0.7	0.0022278	0.0022279	0.0022279
0.8	0.0017386	0.0017387	0.0017387
0.9	0.0013568	0.0013569	0.0013569
1.0	0.0010587	0.0010587	0.0010587
1.1	0.00082592	0.00082599	0.00082599
1.2	0.00064421	0.00064427	0.00064427
1.3	0.00050234	0.00050239	0.00050239
1.4	0.00039156	0.0003916	0.0003916

results demonstrate that reducing the step size from 0.1 to 0.01 and further to 0.001 has a negligible effect on the computed velocity values across all η. This highlights that the numerical solution is both stable and grid-independent, confirming the reliability and convergence of the proposed model.

3.5 RESULTS AND DISCUSSION

Figures 3.2–3.7 have been plotted for the stretching (elongated) case ($\lambda > 0$) and they are investigated for the variation of nanoparticle volume fraction φ_1 *or* φ_{MoS_2} corresponding to the nanoparticle MoS_2, which is of higher thermal conductivity. As shown in the graph of Figures 3.2–3.5 the profile $f'(\eta)$ decreased as φ_1 *or* φ_{MoS_2} increased. The behavior is observed as increasing the volume fraction values results in a growth in the nanofluid's density. Thereby, the induced intermolecular forces between the nanofluid particles intensify, causing a velocity drop. The temperature profiles of both cases of Figures 3.6 and 3.7 boosts as the molybdenum disulfide (MoS_2) solid volume fraction enhances because of the energy dissipation (heat) from the nanoparticles.

According to Figure 3.2, an increase in slip velocity ($\beta = 0.6$) consistently reduces the motion of the molecules within the model, while the lower slip velocity ($\beta = 0.1$) accelerates the motion of the molecules. In Figure 3.3 it is seen that the higher porosity parameter ($\text{Da} = 1.5$) signifies a drag force which always retards the motion of the particles of the hybrid nanofluid, while the lower porosity parameter ($\text{Da} = 0.05$) means that the diameter of the porous material is higher than the case of $\text{Da} = 1.5$, and thus $\text{Da} = 0.05$ creates inertial forces which lead to higher velocity compared to the $\text{Da} = 1.5$. Figure 3.4 demonstrates that the inclusion of M generates the Lorentz

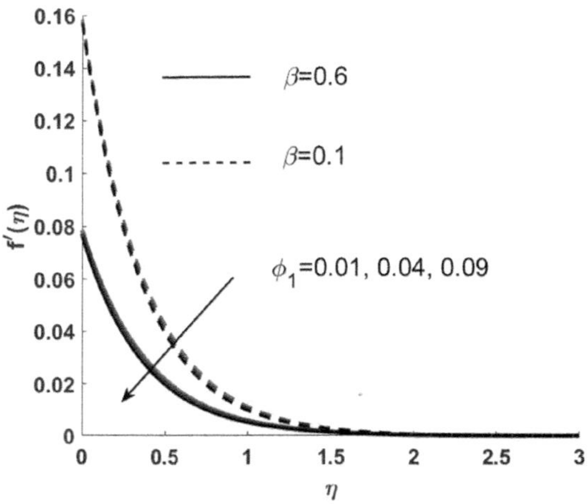

FIGURE 3.2 Variation of $f'(\eta)$ by β and ϕ_1.

forces which resist the motion of the nanoparticle fluid. The development of frictional forces between the fluid particles is expedited by these increasing Lorentz forces which eventually reduce the overall velocity. Higher magnetic parameter (M = 1.2) associated with the strong resisting force in comparison to the lower magnetic parameter (M = 0.03). Figure 3.5 reveals that the higher suction parameter (S = 2.5) retards the motion of the molecules as the greater suction absorbs the molecules near the pores of the stretching sheet, while lower suction (S = 1.3) offers smooth flow of motion of the molecules compared to S = 2.5, which indicates that the pulling intensity fails to exert adequate influence to move the molecules towards the boundary

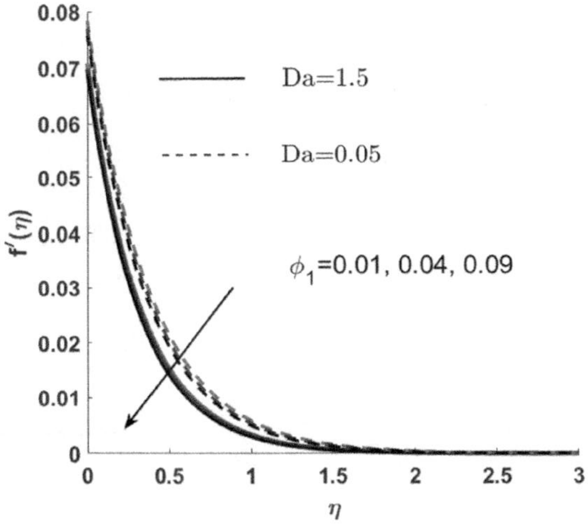

FIGURE 3.3 Variation of $f'(\eta)$ by Da and ϕ_1.

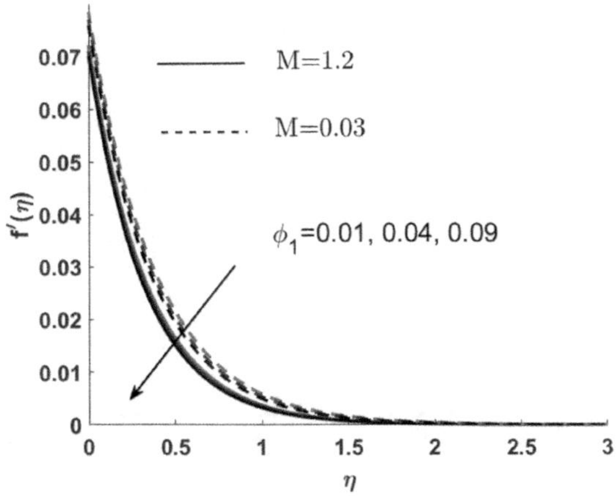

FIGURE 3.4 Variation of $f'(\eta)$ by M and ϕ_1.

surface. The Biot number represents the ratio of conductive heat transfer drag to the convective heat transfer resistance within the solid. As shown in Figure 3.6 both Bi < 1, which indicates convective heat transfer is the dominant factor over the conductive heat transfer. Moreover, for higher Bi, θ is maximum due to maximum convection. In Figure 3.7 the heat source sink parameter ($\Omega = 5.0$) indicates that the heat capacitance $(\rho c_p)_{hnf}$ is very small and this results in a reduction of the hybrid nanofluid temperature. However, for lower value of heat generation–absorption parameter ($\Omega = 0.1$) the heat capacitance $(\rho c_p)_{hnf}$ becomes higher which generates the higher temperature of hybrid nanofluid.

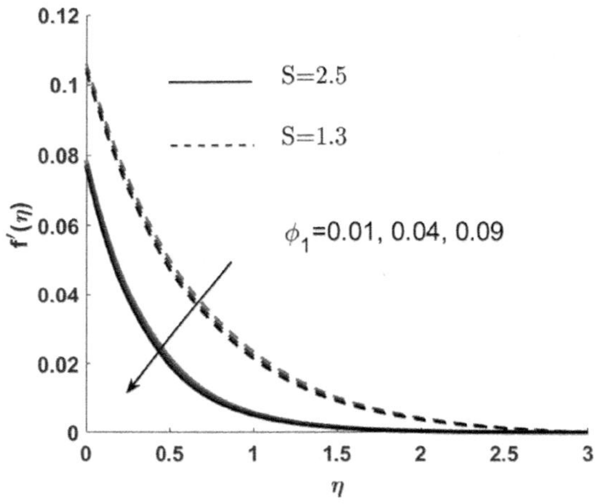

FIGURE 3.5 Variation of $f'(\eta)$ by S and ϕ_1.

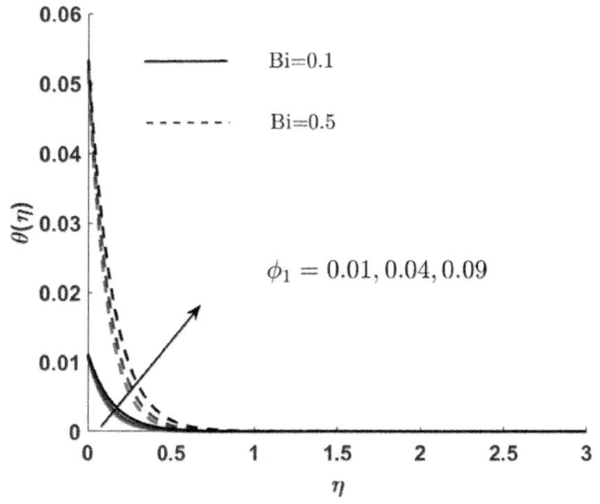

FIGURE 3.6 Variation of $\theta(\eta)$ by Bi and ϕ_1.

The examination of fluid motion and thermal emission for both MoS_2 and SiO_2 under the influence of numerous feasible variations by M, S, Bi, and heat generation–absorption parameter (Ω) is presented in Figures 3.8–3.11. Figures 3.8 and 3.9 explain the slower motion of MoS_2 nanoparticle compared to SiO_2 nanoparticle, as the higher density of MoS_2 nanoparticle increases the viscosity significantly more than SiO_2, which depicts viscous forces resulting in a resisting flow in the case of MoS_2 nanoparticles. On the contrary, Figures 3.10 and 3.11 explain the higher thermal diffusion rate for the MoS_2 nanoparticle compared to SiO_2 nanoparticle. This is

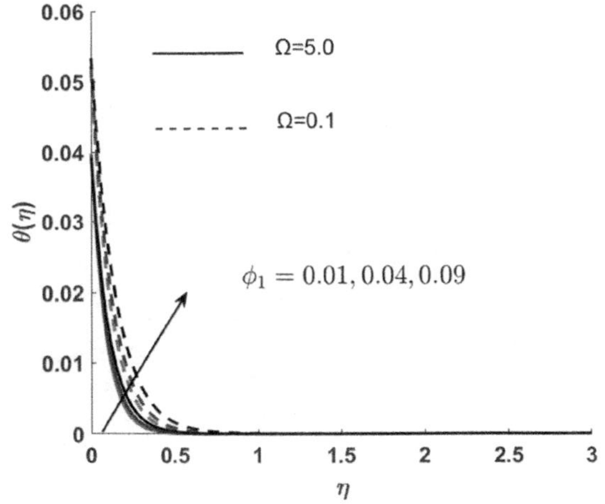

FIGURE 3.7 Variation of $\theta(\eta)$ by Ω and ϕ_1.

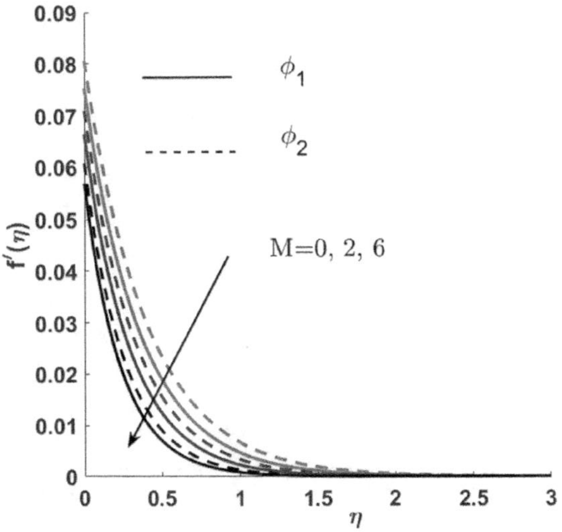

FIGURE 3.8 Variation of $f'(\eta)$ by M and ϕ_1, ϕ_2.

because the higher thermal conductivity of MoS_2 directly enhances the heat diffusion rate and despite the faster flow of SiO_2 nanoparticle it expresses minimal heat diffusion rate.

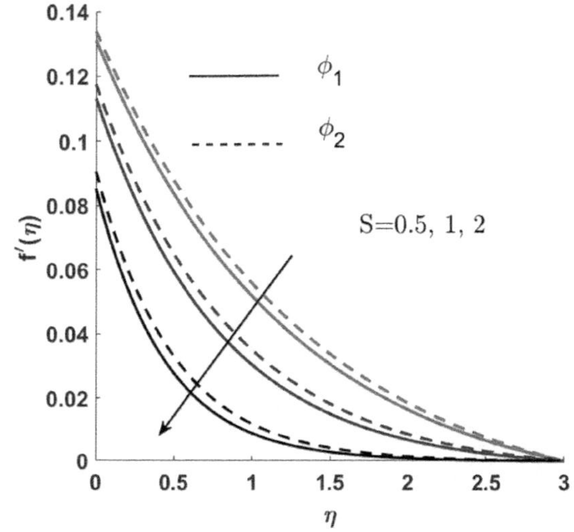

FIGURE 3.9 Variation of $f'(\eta)$ by S and ϕ_1, ϕ_2.

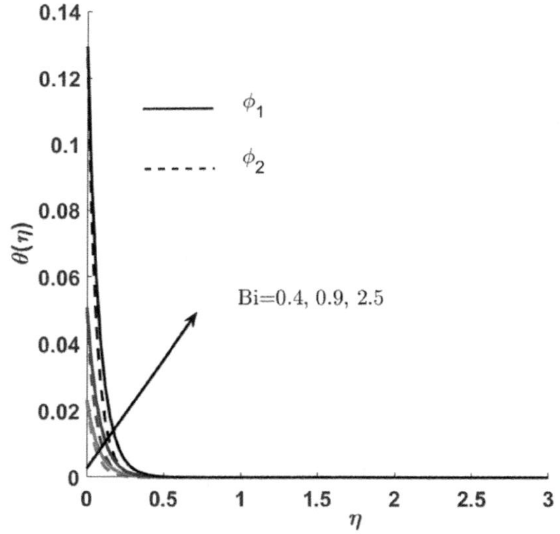

FIGURE 3.10 Variation of $\theta(\eta)$ by Bi and ϕ_1, ϕ_2.

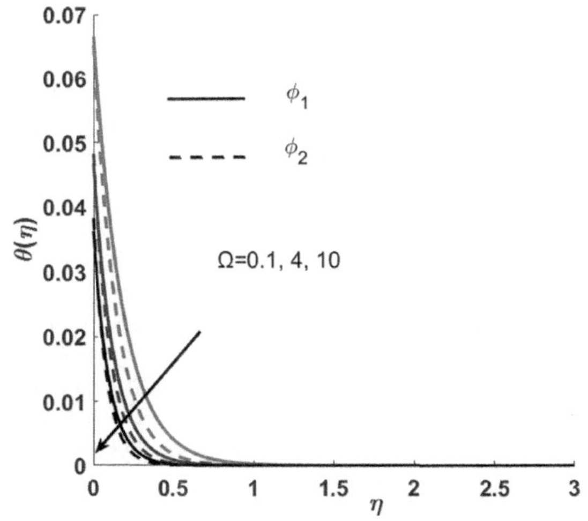

FIGURE 3.11 Variation of $\theta(\eta)$ by Ω and ϕ_1, ϕ_2.

3.6 CONCLUSION

Hybrid nanofluids have emerged as a key advancement in thermal engineering, improving the efficacy of heat transfer systems and aiding in substantial energy conservation. The study uniquely presents a parametric analysis, involving the joint influence of magnetization effects, suction-driven effect, velocity slip effect along with warmth radiation effect, across an infinitesimal elongated surface for a three-dimensional hybrid MoS_2–SiO_2/H_2O nanofluid model, under feasible boundary conditions, indicating the following mentioned observations:

- The porous drag force due to the Darcy resistance significantly reduces fluid momentum, decreasing velocity and modifying the boundary layer structure.
- The Lorentz force from the imposed M retards the motion of the fluid, enhancing control and stability of the flow.
- Surface suction, in combination with MHD effects, helps maintain boundary layer stability and limits flow separation.
- The inclusion of the porosity parameter highlights the model's relevance for Darcy-type creeping flows, with potential applications in filtration systems and thermal insulation.

3.7 FUTURE PROSPECTS

The present research may be implicated in different areas and they are as follows:

- The current study thoroughly analyzed the linear-magnetic effects; however, there is also scope for studying the impact of non-linear magnetization.

- The thermophysical axioms of the nanoparticles are presumed to be fixed during the study; however, further work could be built on investigating the thermal transportation properties using variable heat conductivity.
- The cylindrical/diamond shaped nanoparticles can also be introduced to study their impact in the above-presented model.

REFERENCES

1. Ahmed, S., & Hazarika, S. (2025). Insights into two-dimensional hydromagnetic flow in porous and non-porous media: Horizontal channel behaviour with COMSOL multiphysics. *Latin American Applied Research-An International Journal*, *55*(1), 113–122.
2. Al Kalbani, K. S., Alam, M. S., & Rahman, M. M. (2016). Finite element analysis of unsteady natural convective heat transfer and fluid flow of nanofluids inside a tilted square enclosure in the presence of oriented magnetic field. *American Journal of Heat and Mass Transfer*, *3*(3), 186–224.
3. Alshehri, A., & Shah, Z. (2022). Computational analysis of viscous dissipation and Darcy-Forchheimer porous medium on radioactive hybrid nanofluid. *Case Studies in Thermal Engineering*, *30*, 101728.
4. Aly, E. H., & Pop, I. (2020). MHD flow and heat transfer near stagnation point over a stretching/shrinking surface with partial slip and viscous dissipation: Hybrid nanofluid versus nanofluid. *Powder Technology*, *367*, 192–205.
5. Ambreen, T., Saleem, A., Ali, H. M., Shehzad, S. A., & Park, C. W. (2019). Performance analysis of hybrid nanofluid in a heat sink equipped with sharp and streamlined micro pin-fins. *Powder Technology*, *355*, 552–563.
6. Andersson, H. I. (2002). Slip flow past a stretching surface. *Acta Mechanica*, *158*(1), 121–125.
7. Arif, M., Kumam, P., Khan, D., & Watthayu, W. (2021). Thermal performance of GO-MoS2/engine oil as Maxwell hybrid nanofluid flow with heat transfer in oscillating vertical cylinder. *Case Studies in Thermal Engineering*, *27*, 101290.
8. Badruddin, I. A., Khan, T. Y., & Baig, M. A. A. (2020). Heat transfer in porous media: A mini review. *Materials Today: Proceedings*, *24*, 1318–1321.
9. Beavers, G. S., & Joseph, D. D. (1967). Boundary conditions at a naturally permeable wall. *Journal of Fluid Mechanics*, *30*(1), 197–207.
10. Buongiorno, J. (2006). Convective transport in nanofluids. *Journal of Heat Transfer, 128*(3), 240–250
11. Char, M. I. (1988). Heat transfer of a continuous, stretching surface with suction or blowing. *Journal of Mathematical Analysis and Applications*, *135*(2), 568–580.
12. Chen, C. H. (1998). Laminar mixed convection adjacent to vertical, continuously stretching sheets. *Heat and Mass Transfer*, *33*(5), 471–476.
13. Choi, S. U., & Eastman, J. A. (1995). *Enhancing thermal conductivity of fluids with nanoparticles* (No. ANL/MSD/CP-84938; CONF-951135–29). Argonne, IL: Argonne National Lab (ANL).
14. Crane, L. J. (1970). Flow past a stretching plate. *Zeitschrift für angewandte Mathematik und Physik ZAMP*, *21*(4), 645–647.
15. Das, B., & Ahmed, S. (2025). Numerical modelling in heat transfer analysis of SWCNT/EG nanofluid in a stretching sheet: A darcy–forchheimer model. In *Advances in topology, dynamical systems and interdisciplinary applications* (pp. 211–224). Singapore: Springer Nature Singapore.
16. Gupta, P. S., & Gupta, A. S. (1977). Heat and mass transfer on a stretching sheet with suction or blowing. *The Canadian Journal of Chemical Engineering*, *55*(6), 744–746.

17. Hamilton, R. L., & Crosser, O. K. (1962). Thermal conductivity of heterogeneous two-component systems. *Industrial & Engineering Chemistry Fundamentals*, *1*(3), 187–191.
18. Hayat, T., Ali Shehzad, S., Qasim, M., & Asghar, S. (2014). Three-dimensional stretched flow via convective boundary condition and heat generation/absorption. *International Journal of Numerical Methods for Heat & Fluid Flow*, *24*(2), 342–358.
19. Hazarika, S., Ahmed, S., & Chamkha, A. J. (2021). Investigation of nanoparticles Cu, Ag and Fe_3O_4 on thermophoresis and viscous dissipation of MHD nanofluid over a stretching sheet in a porous regime: A numerical modelling. *Mathematics and Computers in Simulation, 182,* 819–837.
20. Jamrus, F. N., Waini, I., Khan, U., & Ishak, A. (2024). Effects of magnetohydrodynamics and velocity slip on mixed convective flow of thermally stratified ternary hybrid nanofluid over a stretching/shrinking sheet. *Case Studies in Thermal Engineering, 55,* 104161.
21. Jan, R. U., Ullah, I., Khan, H., Nisar, K. S., Kouki, M., & Alam, M. M. (2024). Heat transfer analysis in hybrid nano-composite flow in a stretchable convergent/divergent channel in the presence of Darcy-Forchheimer law and Lorentz force. *Ain Shams Engineering Journal*, *15*(8), 102828.
22. Kasaeian, A., Daneshazarian, R., Mahian, O., Kolsi, L., Chamkha, A. J., Wongwises, S., & Pop, I. (2017). Nanofluid flow and heat transfer in porous media: A review of the latest developments. *International Journal of Heat and Mass Transfer*, *107*, 778–791.
23. Khashi'ie, N. S., Arifin, N. M., Pop, I., Nazar, R., Hafidzuddin, E. H., & Wahi, N. (2020). Three-dimensional hybrid nanofluid flow and heat transfer past a permeable stretching/shrinking sheet with velocity slip and convective condition. *Chinese Journal of Physics, 66*, 157–171.
24. Kumam, P., Tassaddiq, A., Watthayu, W., Shah, Z., & Anwar, T. (2022). Modeling and simulation based investigation of unsteady MHD radiative flow of rate type fluid; a comparative fractional analysis. *Mathematics and Computers in Simulation*, *201*, 486–507.
25. Kumar, V., & Sarkar, J. (2019). Numerical and experimental investigations on heat transfer and pressure drop characteristics of Al2O3-TiO2 hybrid nanofluid in minichannel heat sink with different mixture ratio. *Powder Technology*, *345*, 717–727.
26. Maxwell, J. C. (1873). *A treatise on electricity and magnetism* (Vol. 1). Clarendon press.
27. Miklavčič, M., & Wang, C. (2006). Viscous flow due to a shrinking sheet. *Quarterly of Applied Mathematics*, *64*(2), 283–290.
28. Qureshi, M. Z. A., Raza, Q., Eldin, S. M., Zafar, M., Ali, B., & Siddique, I. (2023). Thermal performance of hybrid magnetized nanofluids flow subject to joint impact of ferro oxides/CNT nanomaterials with radiative and porous factors. *Case Studies in Thermal Engineering*, *41*, 102648.
29. Raza Shah Naqvi, S. M., Farooq, U., Waqas, H., Muhammad, T., & Alshehri, A. (2022). Inspection of thermal jump conditions on nanofluids with nanoparticles and multiple slip effects. *Scientific Reports*, *12*(1), 5586.
30. Reddy Gorla, R. S., & Sidawi, I. (1994). Free convection on a vertical stretching surface with suction and blowing. *Applied Scientific Research*, *52*(3), 247–257.
31. Rooman, M., Jan, M. A., Shah, Z., Vrinceanu, N., Ferrándiz Bou, S., Iqbal, S., & Deebani, W. (2022). Entropy optimization on axisymmetric darcy–forchheimer powell–eyring nanofluid over a horizontally stretching cylinder with viscous dissipation effect. *Coatings*, *12*(6), 749.
32. Roy, N. C., & Pop, I. (2022). Heat and mass transfer of a hybrid nanofluid flow with binary chemical reaction over a permeable shrinking surface. *Chinese Journal of Physics*, *76*, 283–298.

33. Safie, N. E., Azam, M. A., Arith, F., Anand, T. J. S., Mohamad, N. R., & Takasaki, A. (2024). Electrical conductivity enhancement of rGO-MoS2 composite electrode for inverted PSC by parameter optimization. *Results in Engineering*, *22*, 102237.
34. Sajid, M. U., & Ali, H. M. (2018). Thermal conductivity of hybrid nanofluids: a critical review. *International Journal of Heat and Mass Transfer*, *126*, 211–234.
35. Sekrani, G., & Poncet, S. (2016). Further investigation on laminar forced convection of nanofluid flows in a uniformly heated pipe using direct numerical simulations. *Applied Sciences*, *6*(11), 332.
36. Shampine, L. F., Gladwell, I., & Thompson, S. (2003). *Solving ODEs with matlab.* Cambridge University Press.
37. Sharma, R. P., Prakash, O., & Shafi, S. (2024). MHD hybrid nanofluid and slip flow towards an exponentially stretching/shrinking sheet with nonlinear thermal radiation and nonuniform heat source/sink. *Modern Physics Letters B*, *38*(23), 2450195.
38. Srivastava, J. K., Prasad, M., & Wagner, J. B. (1985). Electrical conductivity of silicon dioxide thermally grown on silicon. *Journal of The Electrochemical Society*, *132*(4), 955.
39. Sundar, L.S., Singh, M. K., & Sousa, A. C. (2014). Enhanced heat transfer and friction factor of MWCNT-Fe_3O_4/water hybrid nanofluids. *International Communications in Heat and Mass Transfer*, *52*, 73–83.
40. Suresh, S., Venkitaraj, K. P., Selvakumar, P., & Chandrasekar, M. (2012). Effect of A12O3–Cu/water hybrid nanofluid in heat transfer. *Experimental Thermal and Fluid Science*, *38*, 54–60.
41. Swain, K., Mebarek-Oudina, F., & Abo-Dahab, S. M. (2022). Influence of MWCNT/Fe_3O_4 hybrid nanoparticles on an exponentially porous shrinking sheet with chemical reaction and slip boundary conditions. *Journal of Thermal Analysis & Calorimetry*, *147*(2), 1561–1570.
42. Tiwari, R. K., & Das, M. K. (2007). Heat transfer augmentation in a two-sided lid-driven differentially heated square cavity utilizing nanofluids. *International Journal of Heat and Mass Transfer*, *50*(9–10), 2002–2018.
43. Wahid, N. S., Arifin, N. M., Khashi'ie, N. S., & Pop, I. (2021). Marangoni hybrid nanofluid flow over a permeable infinite disk embedded in a porous medium. *International Communications in Heat and Mass Transfer*, *126*, 105421.
44. Wahid, N. S., Arifin, N. M., Khashi'ie, N. S., Pop, I., Bachok, N., & Hafidzuddin, E. H. (2022). MHD hybrid nanofluid flow with convective heat transfer over a permeable stretching/shrinking surface with radiation. *International Journal of Numerical Methods for Heat & Fluid Flow*, *32*(5), 1706–1727.
45. Waqas, H., Farooq, U., Liu, D., Abid, M., Imran, M., & Muhammad, T. (2022). Heat transfer analysis of hybrid nanofluid flow with thermal radiation through a stretching sheet: A comparative study. *International Communications in Heat and Mass Transfer*, *138*, 106303.
46. Yahaya, R. I., Arifin, N. M., Pop, I., Ali, F. M., & Isa, S. S. P. M. (2023). Dual solutions for the oblique stagnation-point flow of hybrid nanofluid towards a shrinking surface: Effects of suction. *Chinese Journal of Physics*, *81*, 193–205.
47. Zainal, N.A., Nazar, R., Naganthran, K., & Pop, I. (2020). Stability analysis of MHD hybrid nanofluid flow over a stretching/shrinking sheet with quadratic velocity. *Alexandria Engineering Journal*, *60*, 915–926.

4 Numerical Investigation of Heat Flow and Entropy Generation in Vertical Eccentric Annular Convection

Satyvir Singh, Seetu Rana, and Bidesh Sengupta

4.1 INTRODUCTION

Natural convection in annular geometries is a fundamental heat transfer mechanism encountered in a wide range of engineering and industrial applications, including nuclear fuel assemblies, double-pipe heat exchangers, thermal insulation of electrical cables, geothermal systems, and energy storage devices [1–3]. In such configurations, heat transfer is governed by buoyancy-driven flow arising from temperature-induced density variations, leading to complex interactions between momentum and thermal fields. Among annular systems, vertical configurations are of particular importance due to the strong coupling between gravity and axial temperature gradients, which significantly affects flow structure and heat transfer performance [4, 5].

Most classical investigations of natural convection in annuli have focused on concentric geometries, where geometric symmetry simplifies the analysis of flow and thermal transport [3–5]. However, in practical engineering systems, perfect concentricity is rarely achieved. Manufacturing tolerances, thermal expansion, mechanical deformation, and installation constraints often result in eccentric arrangements in which the inner and outer cylinders are offset [6, 7]. This geometric eccentricity breaks the symmetry of the flow and temperature fields, leading to non-uniform boundary layer development, localized heat transfer enhancement or degradation, and the emergence of complex circulation patterns [8, 9].

Several experimental and numerical studies have demonstrated that eccentricity strongly influences the onset of convection, the transition from conduction-dominated to convection-dominated regimes, and the overall heat transfer rate in annular systems [7, 9]. In vertical configurations, these effects are further amplified by buoyancy-induced axial motion, which alters plume formation, thermal stratification, and mixing characteristics [10, 11]. Recent investigations have further shown

DOI: 10.1201/9781003664345-4

that geometric eccentricity plays a critical role in governing thermal mixing and entropy generation patterns in annular enclosures, emphasizing the need for combined heat transfer and irreversibility analyses in such systems [20].

Beyond conventional heat transfer metrics, thermodynamic irreversibility has emerged as a key criterion for evaluating the efficiency of buoyancy-driven thermal systems [2, 12]. Entropy generation analysis provides a rigorous framework for quantifying losses associated with heat transfer across finite temperature differences and viscous dissipation due to fluid friction. In natural convection flows, these mechanisms coexist and vary spatially depending on flow intensity, thermal gradients, and geometric constraints [13, 14]. The Bejan number has therefore been widely employed to identify dominant irreversibility sources and to guide the optimal design of convection-dominated thermal systems [12, 13].

With increasing geometric and physical complexity, numerical simulation has become an indispensable tool for investigating natural convection phenomena where analytical solutions are intractable and experimental studies are often limited [15, 16]. Computational fluid dynamics enables detailed resolution of velocity, temperature, and entropy generation fields, facilitating systematic parametric studies. The open-source finite-volume framework OpenFOAM has gained widespread adoption for buoyancy-driven flow simulations due to its flexibility, numerical robustness, and suitability for customized thermodynamic post-processing [17, 18]. Recent OpenFOAM-based studies have successfully demonstrated its capability for simulating natural convection and heat transfer in cylindrical annular geometries, providing a reliable foundation for advanced entropy generation analyses [21].

Despite the extensive literature on natural convection in annular spaces, studies that simultaneously address vertical eccentric annular configurations and detailed entropy generation analysis remain relatively limited, particularly within open-source numerical frameworks [14]. This gap motivates the present chapter, which provides a systematic numerical investigation of heat flow and entropy generation in vertical eccentric annular convection using OpenFOAM.

The objectives of this chapter are threefold: (i) to examine the influence of Rayleigh number and geometric eccentricity on flow structure and thermal behavior; (ii) to quantify heat transfer performance using local and average Nusselt numbers; and (iii) to evaluate thermodynamic irreversibilities through local and global entropy generation rates and the Bejan number. The insights gained from this study aim to support the optimal thermal design of annular systems by balancing enhanced heat transfer with minimized entropy production.

The remainder of this chapter is organized as follows. Section 4.2 describes the physical model and geometric configuration of the vertical eccentric annulus. Section 4.3 presents the governing equations and relevant dimensionless parameters. The entropy generation formulation is introduced in Section 4.4. Section 4.5 outlines the numerical methodology, including the OpenFOAM solver, mesh strategy, boundary conditions, and validation. The results and physical discussion are presented in Section 4.6, followed by a parametric analysis in Section 4.7. Finally, Section 4.8 summarizes the main conclusions and suggests directions for future research.

4.2 PHYSICAL MODEL AND PROBLEM DESCRIPTION

The present study considers steady, laminar natural convection of an incompressible Newtonian fluid confined within a two-dimensional vertical eccentric annular enclosure. The physical configuration and computational domain are illustrated in Figure 4.1. The enclosure is formed by two long circular cylinders of radii R_i and R_o, representing the inner and outer boundaries, respectively. The axes of the cylinders are offset, resulting in an eccentric annular geometry characterized by the eccentricity ε, defined as the distance between the centers of the inner and outer cylinders. A Cartesian coordinate system (x, y) is adopted with the origin located at the center of the outer cylinder, as shown in Figure 4.1(a). The gravitational acceleration g acts in the negative y-direction, corresponding to a vertically oriented enclosure. The geometric relationship between the two cylinders is further described using the angular position φ, measured from the horizontal axis, and the eccentricity constraint $\varepsilon \leq L$, where $L = R_o - R_i$ denotes the annular gap width.

The inner and outer cylinders are maintained at uniform and constant temperatures, with the inner cylinder heated to $T_i = 298\ K$ and the outer cylinder cooled to $T_o = 288\ K$, such that $T_i > T_o$. Consequently, the inner and outer boundaries act as the hot and cold walls, respectively. No-slip velocity conditions are imposed on both cylinder surfaces. The imposed temperature difference generates density variations within the fluid, which, under the Boussinesq approximation, give rise to buoyancy forces and induce natural convection within the annular region. For the numerical simulations, the annular gap ratio is fixed at $L / R_i = 1.6$, consistent with configurations commonly examined in the literature for eccentric annular convection. The working fluid completely fills the annular space, and thermophysical properties are assumed to be constant except for the density variation appearing in the buoyancy

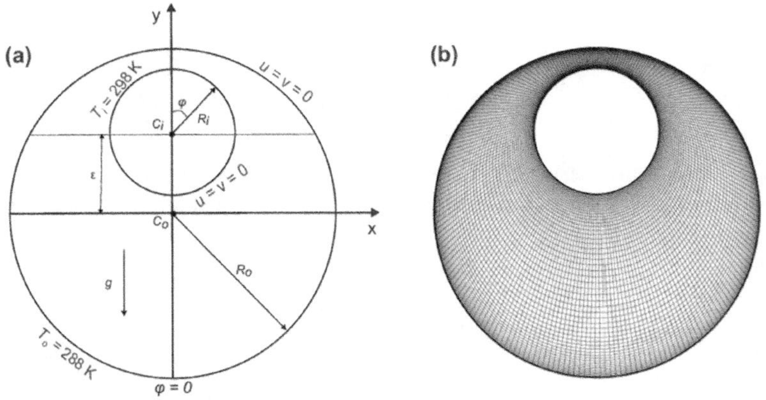

FIGURE 4.1 Schematic representation of the vertical eccentric annular enclosure considered in the present study: (a) two-dimensional physical configuration showing the inner and outer cylinders, eccentricity, coordinate system, gravitational direction, and thermal boundary conditions; and (b) structured quadrilateral computational mesh employed for the OpenFOAM simulations.

term. Figure 4.1(b) shows the structured quadrilateral mesh employed to discretize the computational domain. The mesh is refined near both the inner and outer cylinder walls to accurately resolve velocity and thermal boundary layers, which are critical for predicting heat transfer and entropy generation characteristics. This discretization strategy ensures adequate numerical accuracy while maintaining computational efficiency in the simulations.

4.3 MATHEMATICAL FORMULATION OF HEAT TRANSFER

4.3.1 Governing Equations

The buoyancy-driven flow inside the vertical eccentric annulus is modeled as a steady, two-dimensional, laminar motion of an incompressible Newtonian fluid. Density variations are considered only in the body-force term through the Boussinesq approximation, which is appropriate for the moderate temperature differences imposed between the inner and outer cylinders. All other thermophysical properties of the fluid are assumed to remain constant.

The conservation equations governing mass, momentum, and thermal energy are formulated in Cartesian coordinates (x, y). The continuity equation enforcing incompressibility is expressed as

$$\frac{\partial u}{\partial x} + \frac{\partial v}{\partial y} = 0. \tag{4.1}$$

The momentum transport in the horizontal and vertical directions is governed by

$$u\frac{\partial u}{\partial x} + v\frac{\partial u}{\partial y} = -\frac{1}{\rho}\frac{\partial p}{\partial x} + \nu\left(\frac{\partial^2 u}{\partial x^2} + \frac{\partial^2 u}{\partial y^2}\right), \tag{4.2}$$

$$u\frac{\partial v}{\partial x} + v\frac{\partial v}{\partial y} = -\frac{1}{\rho}\frac{\partial p}{\partial y} + \nu\left(\frac{\partial^2 v}{\partial x^2} + \frac{\partial^2 v}{\partial y^2}\right) + g\beta\left(T - T_c\right), \tag{4.3}$$

where the last term on the right-hand side of Equation (4.3) accounts for buoyancy effects induced by temperature differences. The thermal field evolves according to the energy equation,

$$u\frac{\partial T}{\partial x} + v\frac{\partial T}{\partial y} = \alpha\left(\frac{\partial^2 T}{\partial x^2} + \frac{\partial^2 T}{\partial y^2}\right). \tag{4.4}$$

In the above expressions, (u, v) denote the velocity components, p is the pressure, T is the temperature, ρ is the density, ν is the kinematic viscosity, α is the thermal diffusivity, β is the coefficient of thermal expansion, and g represents gravitational acceleration acting in the downward vertical direction.

The boundary conditions are prescribed in accordance with the physical configuration shown in Figure 4.1. Both the inner and outer cylindrical surfaces satisfy no-slip conditions. The inner cylinder is maintained at a constant temperature T_h, while the outer cylinder is held at a lower uniform temperature T_c, such that

$$u = v = 0, \quad T = T_h \quad \text{at } r = R_i, \tag{4.5}$$

$$u = v = 0, \quad T = T_c \quad \text{at } r = R_o. \tag{4.6}$$

To facilitate parametric analysis and reduce the governing equations to a generalized form, the following dimensionless variables are introduced:

$$X = \frac{x}{L}, \quad Y = \frac{y}{L}, \quad U = \frac{uL}{\alpha}, \quad V = \frac{vL}{\alpha}, \quad \theta = \frac{T - T_c}{T_h - T_c}, \tag{4.7}$$

$$P = \frac{pL^2}{\rho\alpha^2}, \quad \Pr = \frac{\nu}{\alpha}, \quad Ra = \frac{g\beta\left(T_h - T_c\right)L^3\Pr}{\nu^2}, \tag{4.8}$$

where $L = R_o - R_i$ denotes the characteristic length defined by the annular gap width, while Pr and Ra represent the Prandtl and Rayleigh numbers, respectively.

Substitution of these variables into Equations (4.1)–(4.4) yields the dimensionless governing equations

$$\frac{\partial U}{\partial X} + \frac{\partial V}{\partial Y} = 0, \tag{4.9}$$

$$U\frac{\partial U}{\partial X} + V\frac{\partial U}{\partial Y} = -\frac{\partial P}{\partial X} + \Pr\left(\frac{\partial^2 U}{\partial X^2} + \frac{\partial^2 U}{\partial Y^2}\right), \tag{4.10}$$

$$U\frac{\partial V}{\partial X} + V\frac{\partial V}{\partial Y} = -\frac{\partial P}{\partial Y} + \Pr\left(\frac{\partial^2 V}{\partial X^2} + \frac{\partial^2 V}{\partial Y^2}\right) + Ra \Pr \theta, \tag{4.11}$$

$$U\frac{\partial \theta}{\partial X} + V\frac{\partial \theta}{\partial Y} = \frac{\partial^2 \theta}{\partial X^2} + \frac{\partial^2 \theta}{\partial Y^2}. \tag{4.12}$$

The corresponding dimensionless boundary conditions are given by

$$U = V = 0, \quad \theta = 1 \quad \text{at } r = R_i, \tag{4.13}$$

$$U = V = 0, \quad \theta = 0 \quad \text{at } r = R_o. \tag{4.14}$$

The resulting formulation demonstrates that the coupled flow and heat transfer behavior within the eccentric annular enclosure is primarily governed by the Rayleigh number, the Prandtl number, and the geometric eccentricity. These parameters collectively determine the strength of buoyancy forces, thermal transport, and flow asymmetry within the system.

4.3.2 Thermal Transport Parameters

4.3.2.1 Streamfunction

To visualize and analyze the flow structure within the eccentric annular enclosure, a streamfunction, denoted by ψ, is introduced. For two-dimensional incompressible flows, the streamfunction provides a convenient scalar representation of the velocity field, allowing the identification of circulation cells and flow patterns without explicitly plotting velocity vectors.

In the dimensionless Cartesian coordinate system (X,Y), the velocity components (U,V) are related to the streamfunction through

$$U = \frac{\partial \psi}{\partial Y}, \qquad V = -\frac{\partial \psi}{\partial X}. \tag{4.15}$$

These relations automatically satisfy the continuity equation for incompressible flow. By combining the velocity definitions in Equation (4.15), the streamfunction is governed by the following Poisson-type equation:

$$\frac{\partial^2 \psi}{\partial X^2} + \frac{\partial^2 \psi}{\partial Y^2} = \frac{\partial U}{\partial Y} - \frac{\partial V}{\partial X}. \tag{4.16}$$

The sign and magnitude of the streamfunction provide physical insight into the rotational behavior of the flow. Positive values of ψ correspond to counterclockwise circulation, whereas negative values indicate clockwise motion.

4.3.2.2 Nusselt Numbers

The transport of thermal energy within the eccentric annular enclosure is quantified using the Nusselt number, which provides a measure of convective heat transfer relative to pure conduction. Once the velocity field and temperature distribution are obtained from the numerical simulations, the local and average heat transfer characteristics at the cylinder surfaces can be evaluated.

At any point along the inner and outer walls, the local Nusselt number is defined based on the temperature gradient normal to the surface. For the inner and outer cylinders, the local Nusselt numbers are expressed as

$$Nu_\ell^{\text{in}} = -\left(\frac{\partial T}{\partial S_n}\right)_i \frac{D_i}{T_h - T_c}, \qquad Nu_\ell^{\text{out}} = -\left(\frac{\partial T}{\partial S_n}\right)_o \frac{D_o}{T_h - T_c}, \tag{4.17}$$

where D_i and D_o denote the diameters of the inner and outer cylinders, respectively, and S_n represents the coordinate normal to the wall surface. The negative sign reflects the direction of heat transfer from the hot inner cylinder toward the colder outer boundary.

4.3.3 Entropy Generation

Irreversibilities in natural convection arise from heat transfer across finite temperature gradients and viscous dissipation associated with fluid motion. To quantify these effects, local entropy generation is evaluated under the assumptions of local thermodynamic equilibrium and linear transport theory.

The dimensionless local entropy generation due to heat transfer in Cartesian coordinates (X,Y) is defined as

$$S_{\theta,l} = \left(\frac{\partial\theta}{\partial X}\right)^2 + \left(\frac{\partial\theta}{\partial Y}\right)^2, \tag{4.18}$$

whereas the contribution from fluid friction is given by

$$S_{\psi,l} = \phi\left\{2\left[\left(\frac{\partial U}{\partial X}\right)^2 + \left(\frac{\partial U}{\partial Y}\right)^2\right] + \left(\frac{\partial U}{\partial Y} + \frac{\partial V}{\partial X}\right)^2\right\}. \tag{4.19}$$

Here, ϕ denotes the irreversibility distribution ratio, defined as

$$\phi = \frac{\mu T_0}{\kappa}\left(\frac{U_0}{\Delta T}\right)^2, \tag{4.20}$$

with $T_0 = (T_h + T_c)/2$. In the present study, ϕ is fixed at 10^{-3}, which is representative of liquid natural convection under moderate thermal gradients. The resulting entropy generation fields are used to assess the dominant sources of irreversibility within the eccentric annular enclosure.

In order to quantify the relative contribution of heat-transfer-induced and fluid-friction-induced irreversibilities, the Bejan number is introduced. The local Bejan number is defined as the ratio of entropy generation due to heat transfer to the total entropy generation and is given by

$$Be = \frac{S_{\theta,l}}{S_{\theta,l} + S_{\psi,l}}. \tag{4.21}$$

The Bejan number varies between zero and unity. Values of $Be \to 1$ indicate that entropy generation is dominated by heat transfer irreversibility, whereas $Be \to 0$ signifies dominance of fluid-friction-induced irreversibility. Intermediate values reflect a competitive contribution from both mechanisms.

4.4 NUMERICAL METHODOLOGY

The numerical simulations are conducted using the open-source CFD framework OpenFOAM [19], employing a modified buoyantBoussinesqSimpleFoam solver to model steady, two-dimensional laminar natural convection under the Boussinesq approximation in a vertical eccentric annulus. The governing equations are discretized using the finite volume method on body-fitted structured meshes, with second-order accurate schemes applied to convective and diffusive terms and pressure–velocity coupling handled via the SIMPLE algorithm. Mesh refinement is applied near the inner and outer cylinder walls to resolve thermal and hydrodynamic boundary layers, and grid independence is verified. No-slip velocity conditions and constant temperatures are imposed on the hot inner and cold outer cylinders, respectively, while fluid properties are assumed constant except for density variations in the buoyancy term. Solution convergence is ensured by reducing residuals of velocity, pressure, and temperature below 10^{-6}, and post-processing of the converged fields is used to compute streamlines, isotherms, Nusselt numbers, and entropy generation rates to assess heat transfer and irreversibility characteristics.

The numerical methodology and solver implementation are validated by comparing the present results with benchmark data reported for natural convection in eccentric annular enclosures. In particular, flow structures, Nusselt numbers, and entropy generation trends are found to be in good agreement with the results of Singh et al. [20, 21], thereby establishing the accuracy and reliability of the present simulations.

4.5 RESULTS AND DISCUSSION

In this section, the numerical results are presented and discussed to examine the combined effects of buoyancy strength and geometric eccentricity on natural convection, heat transfer, and entropy generation in a vertical annular enclosure. All simulations are conducted at a fixed Prandtl number of $\mathrm{Pr} = 10$, representative of liquid-phase natural convection, while the Rayleigh number is varied from $\mathrm{Ra} = 10^3$ to 10^4 in order to capture the transition from weak to moderately strong buoyancy-driven flow regimes. The impact of geometric asymmetry is investigated by considering eccentricity values of $\varepsilon = 0.8, 0.4, 0, -0.4, -0.8$. Based on a mesh independence study reported in [20], the numerical solutions were found to be insensitive to further grid refinement for a structured quadrilateral mesh comprising approximately 9,700 elements and 9,900 nodes, with a maximum skewness of 0.0399. Consequently, this mesh is employed for all simulations, as it provides an effective balance between numerical accuracy and computational efficiency.

Figure 4.2 illustrates the influence of geometric eccentricity on the thermal and flow structures in a vertical eccentric annular enclosure at $\mathrm{Pr} = 10$, $\mathrm{Ra} = 10^3$, and angular orientation $\varphi = 0$. The upper row presents contours of dimensionless temperature (isotherms), while the lower row shows the corresponding streamlines for eccentricity values ranging from $\varepsilon = 0.8$ to $\varepsilon = -0.8$. For the concentric configuration ($\varepsilon = 0$), the isotherms are nearly symmetric about the vertical axis, indicating a conduction-dominated regime with weak buoyancy-driven motion. The streamlines exhibit a pair of symmetric counter-rotating circulation cells, which is characteristic of low-Rayleigh-number natural convection in annular geometries. As the eccentricity magnitude increases ($|\varepsilon| = 0.4$ and 0.8), the symmetry of both temperature and velocity fields is progressively broken. The isotherms become increasingly distorted and cluster near the reduced annular gap, indicating enhanced local temperature

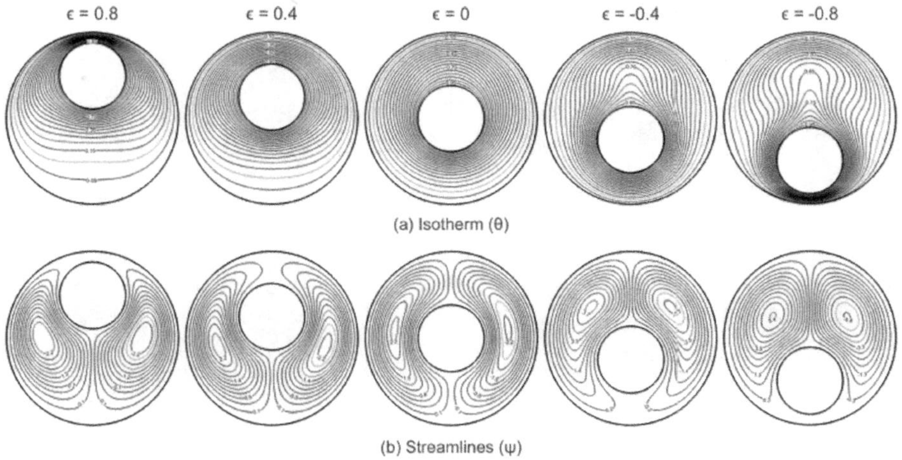

FIGURE 4.2 Contours of (a) isotherms (θ) and (b) streamline (ψ) for natural convection in a vertical eccentric annular enclosure at $Pr = 10$, $Ra = 10^3$ and $\varphi = 0$, shown for different eccentricity values $\varepsilon = 0.8, 0.4, 0, -0.4, -0.8$. The contours highlight the effect of eccentricity on thermal stratification and flow circulation, with increasing asymmetry in both temperature and velocity fields as ε increases.

gradients and intensified heat transfer in that region. Simultaneously, the circulation cells shift toward the narrow gap and exhibit unequal strengths, reflecting the influence of geometric asymmetry on buoyancy-driven flow. The sign of eccentricity determines the direction of the asymmetry. Positive eccentricity shifts the inner cylinder upward, strengthening thermal plumes and circulation near the upper narrow gap, whereas negative eccentricity produces analogous effects in the lower portion of the annulus. These observations demonstrate that even at relatively low Rayleigh numbers, eccentricity significantly alters thermal stratification and flow circulation.

Figure 4.3 illustrates the influence of increasing buoyancy strength on the thermal and flow structures in a vertical eccentric annular enclosure at $Pr = 10$ and $\varphi = 0$, when the Rayleigh number is increased from $Ra = 10^3$ (Figure 4.2) to $Ra = 10^4$. Compared with the lower Rayleigh number case, the temperature and velocity fields exhibit significantly stronger convection effects across all eccentricity values. The isotherm contours in the upper row show pronounced distortion relative to the nearly stratified patterns observed at $Ra = 10^3$. As Ra increases, thermal boundary layers near the cylinder surfaces become thinner and temperature gradients intensify, particularly in regions where the annular gap is reduced. This behavior indicates a transition from conduction-dominated transport toward convection-dominated heat transfer. The streamlines in the lower row demonstrate a corresponding increase in flow intensity. The circulation cells become stronger, more elongated, and increasingly asymmetric as the Rayleigh number increases. In contrast to the weak, nearly symmetric vortical structures at $Ra = 10^3$, the higher Rayleigh number promotes vigorous buoyancy-driven motion, leading to enhanced plume formation and stronger

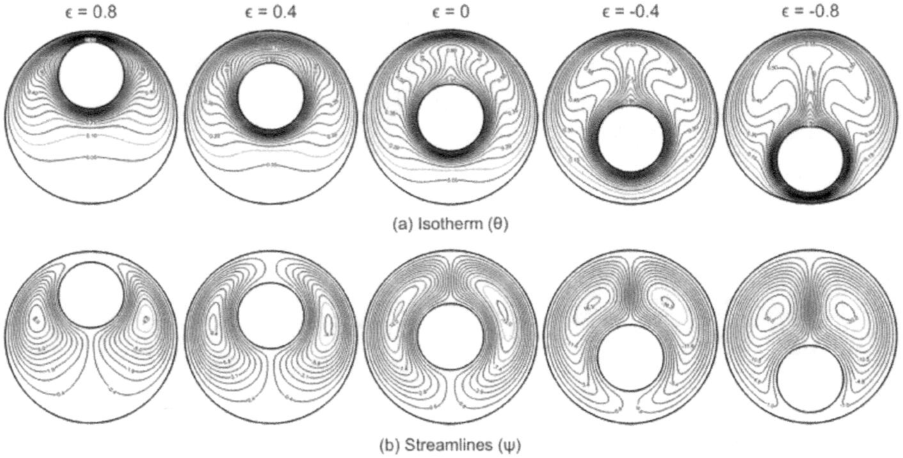

FIGURE 4.3 Contours of (a) dimensionless temperature (θ) and (b) streamlines (ψ) illustrating natural convection in a vertical eccentric annular enclosure at Pr = 10, Ra = 10^4 and $\varphi = 0$. Results are shown for different eccentricity values $\varepsilon = 0.8, 0.4, 0, -0.4, -0.8$. Increasing eccentricity intensifies flow asymmetry, alters plume development, and modifies the structure and strength of circulation cells, leading to pronounced variations in thermal and hydrodynamic fields.

recirculation zones. The effect of eccentricity is amplified at $\text{Ra} = 10^4$. For larger values of $|\varepsilon|$, the circulation cells migrate further toward the narrow annular gap and their strengths become increasingly unequal. Although positive and negative eccentricities produce mirror-image flow patterns, in both cases the interaction between geometric asymmetry and buoyancy leads to highly non-uniform thermal and hydrodynamic fields.

Figure 4.4 presents the spatial distribution of thermodynamic irreversibilities in a vertical eccentric annular enclosure at $\text{Pr} = 10$, $\text{Ra} = 10^3$, and angular orientation $\varphi = 0$, for different eccentricity values $\varepsilon = 0.8, 0.4, 0, -0.4, -0.8$. The first row illustrates contours of local entropy generation due to heat transfer $S_{\theta,l}$, the second row shows local entropy generation due to fluid friction $S_{\psi,l}$, and the third row depicts the corresponding Bejan number distributions. For the concentric configuration ($\varepsilon = 0$), entropy generation due to heat transfer is primarily concentrated near the inner and outer cylinder walls where temperature gradients are the strongest, while viscous entropy production remains relatively weak and symmetrically distributed. The Bejan number contours indicate that heat-transfer-induced irreversibility dominates most of the annular domain, consistent with the conduction-dominated regime at low Ra. As the magnitude of eccentricity increases, the entropy generation fields become increasingly asymmetric. For $|\varepsilon| = 0.4$ and 0.8, regions of elevated $S_{\theta,l}$ shift toward the narrow annular gap, reflecting intensified local temperature gradients caused by geometric confinement and enhanced buoyancy-driven transport. In contrast, entropy generation due to fluid friction $S_{\psi,l}$ increases in regions of stronger

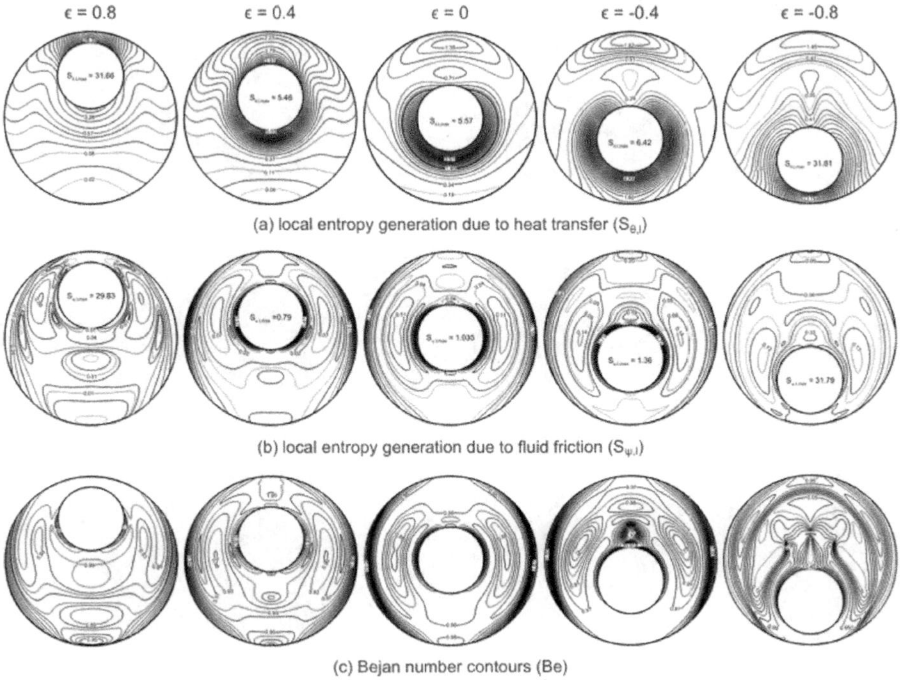

FIGURE 4.4 Contours of (a) local entropy generation due to heat transfer $S_{\theta,l}$, (b) local entropy generation due to fluid friction $S_{\psi,l}$ and (c) Bejan number (Be) for natural convection in a vertical eccentric annular enclosure at Pr = 10, Ra=10^3 and angular orientation $\varepsilon = 0.8, 0.4, 0, -0.4, -0.8$. Results are shown for different eccentricity values $\varphi = 0$. The contours illustrate how geometric eccentricity redistributes thermal and viscous irreversibilities, leading to pronounced asymmetry in entropy production and dominance of heat-transfer- or friction-induced irreversibility depending on the local flow structure.

circulation and shear, particularly near the boundaries of dominant convection cells. The Bejan number distributions clearly capture the competition between thermal and viscous irreversibilities. Regions with Bejan number approaching unity indicate dominance of heat-transfer-induced entropy generation, whereas lower Bejan number values identify zones where fluid-friction-induced irreversibility becomes significant. With increasing eccentricity, the spatial extent of friction-dominated regions expands, especially near the narrow annular gap where velocity gradients are amplified.

Figure 4.5 illustrates the combined effects of increased buoyancy strength and geometric eccentricity on entropy generation and irreversibility distribution in a vertical eccentric annular enclosure at $Pr = 10$, $Ra = 10^4$, and $\varphi = 0$. Compared with the lower Ra case ($Ra = 10^3$), the higher Ra significantly intensifies natural convection, leading to stronger velocity fields, thinner thermal boundary layers, and enhanced entropy production throughout the annulus. The contours of local entropy generation due to heat transfer $S_{\theta,l}$ (top row) indicate that thermal irreversibility is strongly

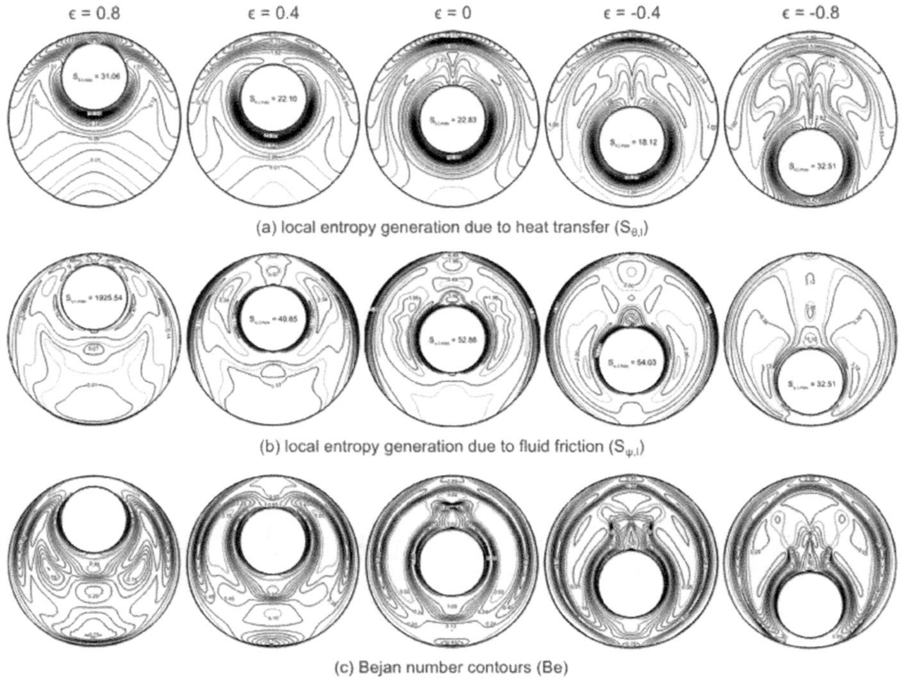

FIGURE 4.5 Contours of (a) local entropy generation due to heat transfer $S\theta,l$, (b) local entropy generation due to fluid friction $S\psi,l$, and (c) Bejan number for natural convection in a vertical eccentric annular enclosure at Pr = 10, Ra = 104, and angular orientation $\varphi = 0$. Results are shown for eccentricity values $\varepsilon = 0.8, 0.4, 0, -0.4, -0.8$. At the higher Rayleigh number, buoyancy-driven flow intensifies thermal and velocity gradients, leading to stronger entropy production, enhanced asymmetry, and a pronounced redistribution of heat-transfer- and friction-dominated irreversibilities across the annular domain.

localized near the cylinder surfaces, particularly in regions where the annular gap is reduced. With increasing eccentricity, these high-entropy zones migrate toward the narrow gap, reflecting steep temperature gradients associated with plume impingement and intensified thermal stratification. At $Ra = 10^4$, the distortion of isotherms observed earlier manifests as sharply localized peaks in thermal entropy generation, confirming a convection-dominated heat transfer regime. The second row presents the local entropy generation due to fluid friction $S_{\psi,l}$. At this elevated Rayleigh number, viscous dissipation becomes increasingly significant, especially within regions of intense circulation and high shear. Elevated values of $S_{\psi,l}$ are observed along the boundaries of dominant convection cells and in zones of strong flow acceleration and deceleration. Increasing eccentricity further enhances these effects by strengthening asymmetric circulation and amplifying velocity gradients near the narrow annular gap. The Bejan number contours (bottom row) highlight the competition between thermal and viscous irreversibilities. Regions where the Bejan number approaches unity correspond to dominance of heat-transfer-induced entropy generation, typically

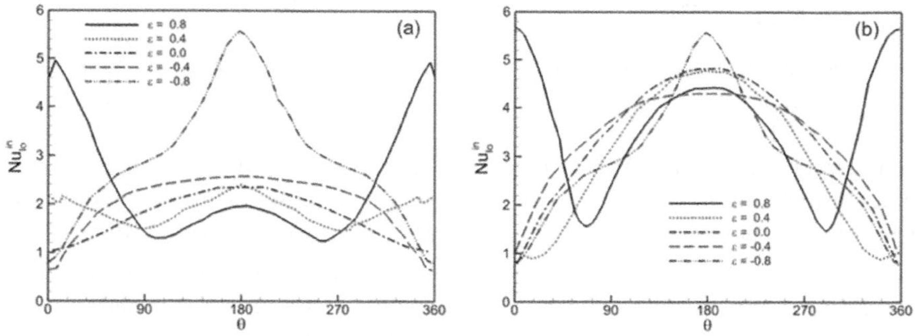

FIGURE 4.6 Angular variation of the local Nusselt number at the inner cylinder of a vertical eccentric annular enclosure for different eccentricity values ε = 0.8, 0.4, 0, −0.4, −0.8 at Pr = 10 and φ = 0: (a) Ra = 10^3 and (b) Ra = 10^4. The distributions highlight the influence of eccentricity and buoyancy strength on the spatial heat transfer characteristics along the heated inner wall.

near the heated and cooled walls. In contrast, lower Bejan number values indicate regions where fluid-friction-induced irreversibility becomes dominant, particularly within the core flow and near strong shear layers. As both Rayleigh number and eccentricity increase, the Bejan number distribution becomes highly non-uniform, revealing a complex spatial redistribution of dominant irreversibility mechanisms.

Finally, Figure 4.6 illustrates the angular variation of the local Nusselt number along the heated inner cylinder for different eccentricity values at $Pr = 10$ and $\varphi = 0$, comparing two buoyancy strengths: (a) $Ra = 10^3$ and (b) $Ra = 10^4$. The angular coordinate θ is measured along the circumference of the inner cylinder, enabling direct assessment of spatial non-uniformity in heat transfer induced by eccentricity and natural convection. At the lower Rayleigh number ($Ra = 10^3$), the local Nusselt number distributions exhibit relatively smooth variations, indicating a weakly convective and near-conduction-dominated regime. For the concentric configuration ($\varepsilon = 0$), the distribution is nearly symmetric about the vertical axis, with modest peaks occurring near the upper portion of the cylinder where buoyancy-driven plumes originate. As eccentricity increases, this symmetry is progressively broken, and higher local Nusselt numbers appear near angular locations corresponding to the narrow annular gap, reflecting enhanced temperature gradients caused by geometric confinement. When the Rayleigh number is increased to $Ra = 10^4$, natural convection becomes significantly stronger, leading to a substantial increase in the magnitude of the local Nusselt number for all eccentricity values. The angular distributions develop sharper and more pronounced peaks, which correspond to regions of plume impingement and intensified convective transport. For larger values of $|\varepsilon|$, the locations of maximum heat transfer shift toward the reduced-gap region, where flow acceleration and thinning of the thermal boundary layer occur.

4.6 CONCLUSIONS

This study numerically investigated laminar natural convection, heat transfer, and entropy generation in a vertical eccentric annular enclosure using the OpenFOAM framework. The results demonstrate that geometric eccentricity strongly influences flow symmetry, thermal stratification, and heat transfer distribution by shifting buoyancy-driven circulation toward the narrow annular gap, thereby intensifying local temperature gradients and producing highly non-uniform Nusselt number distributions along the heated inner cylinder. Increasing the Rayleigh number from 10^3 to 10^4 significantly strengthens convective motion, resulting in thinner thermal boundary layers, enhanced heat transfer rates, and sharper angular variations in local heat transfer. Entropy generation analysis reveals that heat-transfer-induced irreversibility dominates at lower Rayleigh numbers, whereas fluid-friction-induced irreversibility becomes increasingly important at higher buoyancy strengths and larger eccentricities, as reflected by the Bejan number distributions. The study highlights the strong coupling between buoyancy intensity and geometric asymmetry in governing both thermal performance and thermodynamic irreversibility, providing useful insights for the design and optimization of eccentric annular thermal systems.

ACKNOWLEDGMENTS

S.S. acknowledges funding through the German Research Foundation within the research unit DFG–FOR5409.

REFERENCES

1. F. P. Incropera, D. P. DeWitt, T. L. Bergman, and A. S. Lavine, *Fundamentals of Heat and Mass Transfer*, 7th ed., John Wiley & Sons, New York, 2011.
2. A. Bejan, *Convection Heat Transfer*, 4th ed., John Wiley & Sons, New York, 2013.
3. I. Catton, Natural convection in enclosures, *J. Heat Transfer*, 100 (1978), pp. 379–389.
4. T. H. Kuehn and R. J. Goldstein, An experimental study of natural convection heat transfer in concentric and eccentric horizontal cylindrical annuli, *J. Fluid Mech.*, 74 (1976), pp. 695–719.
5. G. D. Raithby and K. G. T. Hollands, Natural convection, in *Handbook of Heat Transfer*, W. M. Rohsenow et al., eds., McGraw–Hill, New York, 1976.
6. S. M. El-Sherbiny, G. D. Raithby, and K. G. T. Hollands, Heat transfer by natural convection across vertical and inclined air layers, *J. Heat Transfer*, 104 (1982), pp. 96–102.
7. A. Mojtabi and M. Charrier-Mojtabi, Natural convection in an eccentric annulus, *Int. J. Heat Mass Transfer*, 37 (1994), pp. 1063–1073.
8. M. Charrier-Mojtabi, A. Mojtabi, and J. C. Legros, Numerical study of natural convection in eccentric annular cavities, *Int. J. Heat Mass Transfer*, 35 (1992), pp. 2179–2190.
9. J. S. Yoo, Natural convection in a vertical eccentric annulus, *Int. J. Heat Fluid Flow*, 20 (1999), pp. 162–169.
10. G. Barakos, E. Mitsoulis, and D. Assimacopoulos, Natural convection flow in a square cavity revisited: Laminar and turbulent models with wall functions, *Int. J. Numer. Methods Fluids*, 18 (1994), pp. 695–719.

11. Y. Li and Q. Wang, Numerical investigation of natural convection heat transfer in vertical annular enclosures, *Appl. Therm. Eng.*, 98 (2016), pp. 1040–1052.
12. A. Bejan, *Entropy Generation Minimization: The Method of Thermodynamic Optimization of Finite-Size Systems and Finite-Time Processes*, CRC Press, Boca Raton, 1996.
13. S. Mahmud and R. A. Fraser, The second law analysis in fundamental convective heat transfer problems, *Int. J. Heat Mass Transfer*, 46 (2003), pp. 3957–3969.
14. G. Ibáñez, J. Pantoja, and J. L. Cuevas, Entropy generation analysis of natural convection in enclosures, *Energy*, 51 (2013), pp. 123–133.
15. S. V. Patankar, *Numerical Heat Transfer and Fluid Flow*, Hemisphere Publishing Corporation, Washington, DC, 1980.
16. H. K. Versteeg and W. Malalasekera, *An Introduction to Computational Fluid Dynamics: The Finite Volume Method*, 2nd ed., Pearson Education, London, 2007.
17. H. G. Weller, G. Tabor, H. Jasak, and C. Fureby, A tensorial approach to computational continuum mechanics using object-oriented techniques, *Comput. Phys.*, 12 (1998), pp. 620–631.
18. H. Jasak, *OpenFOAM: Open Source CFD in Research and Industry*, Ph.D. Thesis, Imperial College London, 2007.
19. OpenFOAM Foundation, *OpenFOAM User Guide*, Version 10, 2023.
20. S. Singh, B. Sengupta, and S. Rana, Analysis of thermal mixing and entropy generation during natural convection flows in arbitrary eccentric annulus, *Axioms*, 13 (2024), p. 233.
21. S. Rana, B. Sengupta, and S. Singh, Natural convection study in cylindrical annulus through OpenFOAM, in *Computational Fluid Flow and Heat Transfer*, CRC Press, Boca Raton, 2024, pp. 33–45.

5 Flow in Porous Media
Study of Hybrid Nanofluids

Sham Bansal

5.1 INTRODUCTION

Contemporary thermal engineering faces a critical bottleneck in expanding the efficiency of modern industrial systems, ranging from nuclear reactor cooling to microelectromechanical systems. Conventional heat transfer fluids, such as ethylene glycol and water, often suffer from inherently low thermal conductivity, which limits their cooling potential. To dismantle this barrier, research has pivoted from simple nanofluids to tri-hybrid nanofluids, which leverage the synergistic thermal potentials of three distinct nanoparticles suspended in a base medium. This investigation specifically targets the dynamics of these advanced ternary fluids flowing over a non-linear stretching sheet, a scenario ubiquitous in polymer processing and metallurgy. By incorporating the Darcy–Forchheimer relation, the analysis extends beyond standard Darcian limits to account for inertial drag within porous media, while simultaneously examining the regulating influence of magnetohydrodynamics (MHD). When coupled with an internal heat source, this configuration offers a rich, multi-physical framework for understanding how to maximize thermal efficiency in high-temperature industrial systems. Numerical computations in this work were carried out using the robust Keller-box finite difference implicit (KBFDI) method.

The transport of fluids through porous structures is a cornerstone of many industrial and geological processes, ranging from geothermal energy extraction to the filtration of chemical effluents. Analyzing the flow of these complex fluids through a Darcy–Forchheimer porous medium presents a unique challenge. The inertial drag forces inherent in high-velocity flows through porous boundaries interact dynamically with the enhanced viscosity of the tri-hybrid fluid. It allows for an accurate prediction of the resistance offered by the porous matrix against the fluid's motion. Memon et al. (2023) scrutinized Darcy–Forchheimer hybrid nanofluid flow; their research revealed the existence of dual solution branches. Stability analysis confirmed that the first branch represents a physically realizable and stable flow regime, whereas the second solution is unstable and physically invalid. Manoharkumar et al. (2024) utilized the RKF-NR computational method to analyze the combined impacts of viscous dissipation and Forchheimer drag on three-dimensional nanofluid flow over a non-linearly stretching sheet. An investigation into the rotating 3D flow of $Ag-CuO/H_2O$ hybrid nanofluids over a stretching sheet, incorporating radiative absorption, was conducted by Ramana et al. (2024). Their analysis demonstrated that

 DOI: 10.1201/9781003664345-5

the inertial drag introduced by the Forchheimer number effectively suppresses the fluid velocity components along both principal axes.

Conventional heat transfer fluids, such as water, ethylene glycol, and engine oil, have long been the standard in industrial engineering, yet their intrinsic low thermal conductivity places a hard ceiling on heat exchange efficiency. The pioneering introduction of mono-nanofluids marked a significant leap forward, but modern thermal loads require even greater capabilities. Consequently, recent research has pivoted toward ternary hybrid nanofluids. Unlike their predecessors, which suspend a single type of nanoparticle, or hybrid fluids which combine two, tri-hybrid systems suspend three distinct nanoparticle types, like metal, metal oxide, and carbon allotrope within a base fluid. This progression is not merely additive; it represents a strategic engineering of fluid properties to overcome the limitations of single-material suspensions. By synthesizing three distinct types of nanoparticles into a single host fluid, ternary nanofluids exploit the synergistic effects of various materials, yielding thermophysical properties that are significantly superior to their predecessors. While a hybrid fluid might balance thermal conductivity with stability, a ternary mixture allows for a more precise triangulation of properties. For instance, one particle may be selected for its magnetic susceptibility, another for its exceptional thermal conductivity, and a third for its ability to reduce agglomeration and enhance stability. Das et al. (2021) investigated the dynamics of hybrid nanofluid, incorporating Hall current and non-uniform magnetic field effects under zero mass-flux conditions. Ramireddy and Jayalakshmi (2025) utilized the KBFDI method to analyze the flow of a silver-gold/blood hybrid nanofluid, incorporating nonlinear thermal radiation and stratification effects. The findings offer significant insights into thermal conductivity enhancement and flow dynamics applicable to the treatment of vascular disorders. Laxmia et al. (2025) explored the combined influence of melting and Soret–Dufour effects on flow dynamics. The analysis elucidated how specific alumina volume fractions significantly modify velocity and temperature fields, offering critical insights for optimizing heat transfer efficiency. The interplay of magnetic forces and suction in modulating couple stress hybrid nanofluid flow was addressed by Khashi'ie et al. (2024), particularly noting the regulatory effect of convective boundary conditions on transport phenomena. Nabwey et al. (2024) explored the dynamics of Williamson hybrid nanofluid flow past an inclined stretching cylinder, reporting superior heat transfer compared to conventional nanofluids. Swain and Kar (2024) investigated the dynamics of a ternary hybrid nanofluid, considering slip boundaries and Soret diffusion. They observed that the Soret effect notably improves fluid motion. Alnahdi and Gul (2024) examined the hybrid nanofluid flow, accounting for magnetic fields and thermal radiation. The research underscores the critical role of external heat sources and dissipation mechanisms in modifying thermal profiles.

Considerable attention has been devoted to the dynamics of MHD flows due to their profound impact on transport phenomena in porous media. The central principle relies on the interaction between an applied magnetic field and the electrical currents induced within the fluid, which generates a retarding body force. This interaction is not merely theoretical; it provides a practical method for managing the cooling rates of electronic devices and improving the efficiency of MHD power generators. The integration of MHD with nanofluids has gained traction, as the magnetic field offers a way to manipulate the suspended nanoparticles, thereby optimizing thermal conductivity and stability in high-temperature systems. Adel et al. (2025) investigated the transport phenomena of MHD nanofluids over a porous stretching sheet subjected to convective heating. The study evaluated the combined influence of thermal radiation and viscous dissipation on the flow characteristics. Pennelli et al. (2025) researched the MHD viscoelastic nanofluid flow via an inclined stretching sheet, incorporating multiple slip boundaries and chemical reaction effects. Reddy et al. (2024) analyzed the MHD flow of a tangent hyperbolic fluid, with a focus on slip boundary effects. The findings emphasize the significant influence of slip conditions on optimizing heat transport and fluid velocity profiles in practical engineering systems. Pal and Mandal (2015) evaluated the impact of induced magnetic fields on buoyancy-driven mixed convection in nanofluids. The study focused on the flow dynamics near a stagnation point along a stretching surface. Bansal (2024) analyzed non-Darcy MHD nanofluid flow over an inclined non-linear stretching surface, contributing to a deeper understanding of transport mechanisms in chemically reacting flows. Prashanth and Rao (2024) highlighted the influence of Dufour and radiative effects on the MHD flow of nanofluids. The findings elucidate how thermophoresis and Brownian motion significantly modulate the concentration profiles. Bafe et al. (2023) numerically evaluated the dynamics of Carreau nanofluid MHD flow using the continuation method, integrating nonlinear thermal radiation and Joule heating. Athal et al. (2023) highlighted the dynamics of MHD hyperbolic tangent nanofluids over a stretching sheet, specifically examining the roles of thermal radiation, chemical reactions, and nanoparticle mechanisms. Their findings, derived through similarity transformations, offer valuable data on shear stress and heat transfer rates relevant to the design of chemical reactors and heat exchangers.

The analysis of boundary layer flow generated by moving solid surfaces is a cornerstone of contemporary fluid mechanics, primarily due to its vast applicability in industrial manufacturing. Processes such as the extrusion of polymer sheets, glass fiber production, hot rolling, and wire drawing rely heavily on the interaction between a cooling fluid and a stretching boundary. During actual polymer extrusion, the sheet often stretches at a rate that varies non-uniformly due to factors like varying tension or material viscosity. Consequently, the rigid assumption of linear velocity fails to predict the flow characteristics accurately. To address this, researchers have pivoted toward the non-linear stretching sheet model. In this framework, the velocity of the sheet is not proportional to the distance from the slit but is instead governed by a power-law index. Focusing on the rheological advantages of the Casson model, Jaffrullah et al. (2024) scrutinized nanofluid flow via a non-linearly stretching sheet, incorporating Joule heating and chemical reaction effects. The analysis aimed to overcome the thermal conductivity limitations of conventional fluids, offering optimized heat transfer solutions for metallurgical and industrial cooling systems. To enhance the design of cooling devices, Kallem et al. (2025) investigated the influence of absorbing radiation and heat generation over a permeable, moving plate. These nanofluids are vital for effective thermal management in electronic devices, making their flow analysis essential. Bansal, Kumar et al. (2024) researched the impact of slip boundary conditions on MHD nanofluid flow over a non-linear stretching sheet. The analysis elucidates how permeability and slip factors modify heat and mass transfer rates, offering critical insights for optimizing transport mechanisms in engineering systems. To optimize industrial thermal processes, Triveni et al. (2023) scrutinized the behavior of MHD Casson nanofluids over a nonlinear stretching sheet using the Runge–Kutta–Fehlberg simulation method. The analysis integrates the simultaneous effects of non-uniform heat sources, chemical reactions, and viscous dissipation. A key focus is placed on quantifying the impact of Brownian motion and thermophoresis on temperature and velocity profiles, offering significant insights for designing efficient heat transfer systems.

Since the governing equations for such a system are highly non-linear and coupled, analytical solutions are often intractable. Consequently, this study employs the KBFDI method. This numerical scheme is chosen specifically for its unconditional stability and second-order accuracy, ensuring robust convergence even when handling the complex boundary layer approximations required for ternary hybrid nanofluid flows. Aly et al. (2025) employed the KBFDI method to study the behavior of hybrid Casson nanofluids; their findings provide critical correlations for skin friction and Nusselt numbers, highlighting the role of entropy optimization in engineering flow systems. Utilizing the KBFDI method, Kiyani et al. (2025) models nanofluid flow over a nonlinear surface to evaluate the combined effects of inertial forces and diffusion mechanisms on thermal reflection. Ramya et al. (2018) employed the KBFDI method to investigate nanofluid flow, integrating both velocity and thermal slip boundary conditions. The analysis highlights how various governing parameters influence the flow field. Kouki et al. (2025) utilized the KBFDI method to simulate the dynamics of magnetized micropolar nanofluid flow over an exponentially contracting surface, incorporating cross-diffusion effects. Their analysis indicates that the combined influence of buoyancy forces and the Soret effect leads to enhanced energy and momentum fields in both assisting and opposing flow regimes. Bansal, Pal et al. (2024) utilized the KBFDI method to investigate water-based nanofluids over an inclined, permeable stretching surface. Their analysis reveals that while intensifying the velocity slip parameter improves heat transfer efficiency. Employing the KBFDI method, Al-Zubaidi et al. (2023) researched the MHD nanofluid flow through a vertical expanding sheet, accounting for viscous dissipation and Soret–Dufour effects.

Despite the extensive literature, the complex interaction of ternary nanofluids comprising copper (Cu) aluminum oxide (Al_2O_3) and ferrous ferric oxide (Fe_3O_4) flowing over a nonlinearly stretching surface within a Darcy–Forchheimer porous medium with heat source and chemical reactions remains underexplored. This gap highlights a need for rigorous mathematical modeling that accounts for these compounded physical effects.

5.2 MATHEMATICAL FORMULATION

This study models the flow behavior of a steady, incompressible tri-hybrid nanofluid within a Darcy–Forchheimer porous medium, characterized by homogeneity and isotropy, in the presence of a magnetic field and a heat source. The flow is induced by a non-linearly stretching sheet $\left(u_w = ax^n\right)$ maintained at fixed thermal $\left(T_W\right)$ and solutal $\left(C_W\right)$ conditions relative to the ambient fluid $\left(T_\infty, C_\infty\right)$. As illustrated in Figure 5.1, the geometrical framework incorporates the effects of a variable magnetic field defined by $B(x) = B_0 x^{\frac{n-1}{2}}$, with the governing conservation equations derived and adapted from standard theoretical models (Devi & Devi, 2016; Ramya et al., 2018; Swain & Kar, 2024)).

The governing flow models and boundary limits utilized here are those stipulated by Ramya et al. (2018).

$$\frac{\partial u}{\partial x} + \frac{\partial v}{\partial y} = 0 \tag{5.1}$$

$$u\frac{\partial u}{\partial x} + v\frac{\partial u}{\partial y} = \nu_{thnf}\frac{\partial^2 u}{\partial y^2} - \frac{\sigma_{thnf}\left(B(x)\right)^2}{\rho_{thnf}}u - \nu_{thnf}\frac{1}{K_d}u - \frac{Fu^2}{\rho_{thnf}} \tag{5.2}$$

$$u\frac{\partial T}{\partial x} + v\frac{\partial T}{\partial y} = \alpha_{thnf}\frac{\partial^2 T}{\partial y^2} + \frac{\left(\rho C_p\right)_{np}}{\left(\rho C_p\right)_{thnf}}\left[D_B\frac{\partial T}{\partial y}\frac{\partial C}{\partial y} + \frac{D_T}{T_\infty}\left(\frac{\partial T}{\partial y}\right)^2\right] + \frac{Q_0\left(T - T_\infty\right)}{\left(\rho C_p\right)_{thnf}} \tag{5.3}$$

$$u\frac{\partial C}{\partial x} + v\frac{\partial C}{\partial y} = D_B\frac{\partial^2 C}{\partial y^2} + \frac{D_T}{T_\infty}\frac{\partial^2 T}{\partial y^2} \tag{5.4}$$

$$\left.\begin{array}{l} u = u_w = ax^n,\ v = v_w,\ T = T_w = T_\infty + ax^{2n} \text{ at } y = 0 \\ \text{and } v = 0,\ \ u = 0,\ \ C \to C_\infty \ \ \&\ \ T \to T_\infty \text{ as } y \to \infty \end{array}\right\} \tag{5.5}$$

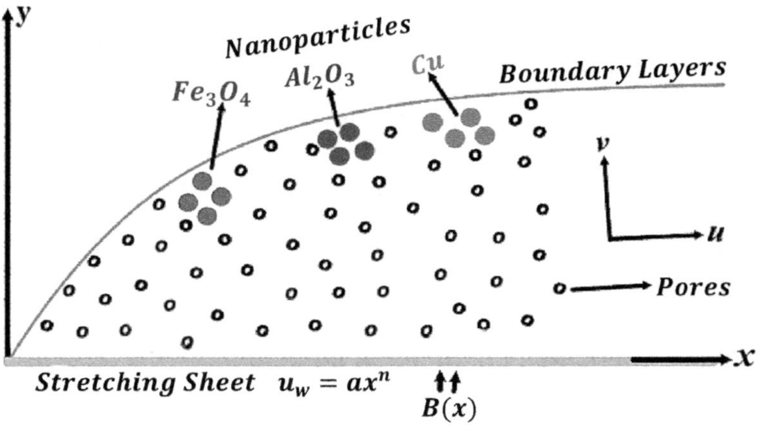

FIGURE 5.1 Physical model of present research.

Adopting the theoretical frameworks presented in Devi and Devi (2016) and Swain and Kar (2024), the constitutive equations for the ternary nanofluid's kinematic viscosity ν_{thnf}, effective thermal diffusivity α_{thnf}, and effective viscosity μ_{thnf} are expressed as follows:

$$\left.\begin{aligned}&\nu_{thnf}=\frac{\mu_{thnf}}{\rho_{thnf}},\alpha_{thnf}=\frac{\kappa_{thnf}}{(\rho C_p)_{thnf}},\mu_{thnf}=\frac{\mu_f}{(1-\phi_1)^{2.5}(1-\phi_2)^{2.5}(1-\phi_3)^{2.5}},\\&\text{where, }\rho_{thnf}=(1-\phi_1)\left[(1-\phi_2)\left[(1-\phi_3)\rho_f+\phi_3\rho_3\right]+\phi_2\rho_2\right]+\phi_1\rho_1\text{ and}\\&(\rho C_p)_{thnf}=(1-\phi_1)\left[(1-\phi_2)\left[(1-\phi_3)(\rho C_p)_f+\phi_3(\rho C_p)_3\right]+\phi_2(\rho C_p)_2\right]+\phi_1(\rho C_p)_1\end{aligned}\right\}$$

$$\left.\begin{aligned}&\text{Also, }\frac{\kappa_{thnf}}{\kappa_{hnf}}=\frac{\kappa_1+2\kappa_{hnf}-2\phi_1(\kappa_{hnf}-\kappa_1)}{\kappa_1+2\kappa_{hnf}+\phi_1(\kappa_{hnf}-\kappa_1)},\frac{\kappa_{hnf}}{\kappa_{nf}}=\frac{\kappa_2+2\kappa_{nf}-2\phi_2(\kappa_{nf}-\kappa_2)}{\kappa_2+2\kappa_{nf}+\phi_2(\kappa_{nf}-\kappa_2)},\\&\frac{\kappa_{nf}}{\kappa_f}=\frac{\kappa_3+2\kappa_f-2\phi_3(\kappa_f-\kappa_3)}{\kappa_3+2\kappa_f+\phi_3(\kappa_f-\kappa_3)},\frac{\sigma_{thnf}}{\sigma_{hnf}}=\frac{\sigma_1+2\sigma_{hnf}-2\phi_1(\sigma_{hnf}-\sigma_1)}{\sigma_1+2\sigma_{hnf}+\phi_1(\sigma_{hnf}-\sigma_1)},\\&\frac{\sigma_{hnf}}{\sigma_{nf}}=\frac{\sigma_2+2\sigma_{nf}-2\phi_2(\sigma_{nf}-\sigma_2)}{\sigma_2+2\sigma_{nf}+\phi_2(\sigma_{nf}-\sigma_2)}\text{ and }\frac{\sigma_{nf}}{\sigma_f}=\frac{\sigma_3+2\sigma_f-2\phi_3(\sigma_f-\sigma_3)}{\sigma_3+2\sigma_f+\phi_3(\sigma_f-\sigma_3)}\end{aligned}\right\}\tag{5.6}$$

Similarity transformations are described as:

$$\left.\begin{aligned}&\eta=yx^{\frac{n-1}{2}}\left(\frac{a(n+1)}{2\nu_f}\right)^{\frac{1}{2}},u=ax^nf'(\eta),\xi(\eta)=\frac{C-C_\infty}{C_w-C_\infty}\\&\theta(\eta)=\frac{T-T_\infty}{T_w-T_\infty}\text{ and }v=-x^{\frac{n-1}{2}}\left(\frac{a(n+1)\nu_f}{2}\right)^{\frac{1}{2}}f(\eta)\end{aligned}\right\}\tag{5.7}$$

Presenting the stream function $\psi(x,y)$, that fulfills the continuity equation (5.1) stated as:

$$u=\frac{\partial\psi}{\partial y}\text{ and }v=-\frac{\partial\psi}{\partial x}\tag{5.8}$$

From Equations (5.2)–(5.4) using Equations (5.6)–(5.8), it can be written as

$$f'''(\eta)+\frac{S_2}{S_1}f(\eta)f''(\eta)-\frac{2}{n+1}\left[\frac{S_2}{S_1}(n+Fr)(f'(\eta))^2+\frac{S_4}{S_1}Mf'(\eta)+\delta f'(\eta)\right]=0\tag{5.9}$$

$$\theta''(\eta)+\frac{Pr}{S_5}\left[S_3\theta'(\eta)f(\eta)+Nb\xi'(\eta)\theta'(\eta)+Nt\left(\theta'(\eta)\right)^2+S_3\left(\frac{2}{n+1}\right)Q_1\theta\right]=0 \quad (5.10)$$

$$\xi''(\eta)+Sc\,f(\eta)\xi'(\eta)+\frac{Nt}{Nb}\theta''(\eta)=0 \quad (5.11)$$

The physical quantities in these equations are defined as:

$$\left.\begin{aligned}
&S_1=\frac{\mu_{thnf}}{\mu_f}, S_2=\frac{\rho_{thnf}}{\rho_f}, S_3=\frac{(\rho C_p)_{thnf}}{(\rho C_p)_f}, S_4=\frac{\sigma_{thnf}}{\sigma_f}, S_5=\frac{\kappa_{thnf}}{\kappa_f},\\
&\Pr=\frac{v_f}{\alpha_f}, Fr=\frac{C_b}{\sqrt{k_d}}, M=\frac{\sigma_f B_0^2}{T_\infty \rho_f}, Nb=\frac{(\rho C_p)_{np}(C_f-C_\infty)D_B}{(\rho C_p)_f v_f}\\
&\delta=\frac{v_f}{aK_d}, Nt=\frac{(\rho C_p)_{np}(T_f-T_\infty)D_T}{(\rho C_p)_f v_f T_\infty}, Sc=\frac{v_f}{D_B} \text{ and } Q_1=\frac{xQ_0}{\rho C_p}
\end{aligned}\right\} \quad (5.12)$$

The specific boundary conditions in Equation (5.5) result directly from utilizing the transformations set forth in Equation (5.7).

$$u=u_w=ax^n \Rightarrow ax^n f'(\eta)=ax^n \Rightarrow f'(\eta)=1,$$

$$C=C_w \Rightarrow C_w-C_\infty=\xi(\eta)(C_w-C_\infty)\Rightarrow \xi(\eta)=1$$

$$v=v_w \Rightarrow -x^{\frac{n-1}{2}}\left(\frac{a(n+1)v_f}{2}\right)^{\frac{1}{2}} f(\eta)=v_w$$

$$\Rightarrow f(\eta)=\frac{-v_w}{x^{\frac{n-1}{2}}\left(\frac{a(n+1)v_f}{2}\right)^{\frac{1}{2}} f(\eta)}=S,$$

and $T=T_w \Rightarrow T_w-T_\infty=\theta(\eta)(T_w-T_\infty)\Rightarrow \theta(\eta)=1.$

Therefore, the boundary conditions in the transformed domain take the following form:

$$\left.\begin{aligned}
&f'(\eta)=\theta(\eta)=\xi(\eta)=1, f(\eta)=S \text{ at } \eta=0\\
&\text{and } f'(\eta), \xi(\eta) \text{ and } \theta(\eta)\to 0 \text{ as } \eta\to\infty
\end{aligned}\right\} \quad (5.13)$$

TABLE 5.1
Specialized Properties of Nanoparticles

Physical Properties	$C_p\left(\mathrm{J\,Kg^{-1}K^{-1}}\right)$	$\rho\left(Kg\,\mathrm{m^{-3}}\right)$	$\sigma\left(S\,\mathrm{m^{-1}}\right)$	$\kappa\left(W\,\mathrm{m^{-1}K^{-1}}\right)$
Al_2O_3	765	3970	3.5×10^{7}	40
Cu	385	8933	5.96×10^{7}	400
Fe_3O_4	670	5180	2.5×10^{4}	9.7
$Water\left(H_2O\right)$	4180	997	5.5×10^{-6}	0.6071

5.2.1 Quantification of Surface Friction and Transport Phenomena

The key physical quantities, including Nusselt/Sherwood numbers and skin friction, are expressed through the following relations:

$$Nu = \frac{xq_w}{\kappa_f\left(T_w - T_\infty\right)},\ Sh = \frac{xq_m}{D_m\left(T_w - T_\infty\right)} \text{ and } C_f = \frac{\tau_w}{\rho_f u_w^2}$$

The flux and shear stress rates at the wall $(y=0)$ are defined using the relations $q_w = -\kappa_{thnf}\dfrac{\partial T}{\partial y},\ q_m = -D_m\dfrac{\partial C}{\partial y}$ and $\tau_w = \mu_{thnf}\dfrac{\partial u}{\partial y}$.

Based on the analysis, the dimensionless skin friction coefficient $\left(f''(0)\right)$ and rates of mass and heat transfer $\left(-\xi'(0), -\theta'(0)\right)$ can be expressed as follows:

$$\left.\begin{aligned} &\mathcal{S}_1 f''(0) = C_f\left(\frac{2Re_x}{n+1}\right)^{1/2}, -\xi'(0) = Sh\left(\frac{(n+1)Re_x}{2}\right)^{-1/2} \\ &\text{and} - \mathcal{S}_5\theta'(0) = Nu\left(\frac{(n+1)Re_x}{2}\right)^{-1/2}, \text{ where } Re_x = \frac{xu_w}{\nu_f} \end{aligned}\right\} \tag{5.14}$$

5.2.2 Specialized Properties of Nanoparticles

Table 5.1 outlines the specific thermophysical properties of the nanoparticles used in this study.

5.3 RESULTS AND DISCUSSION

The current analysis employs the KBFDI technique to model the flow of Darcy–Forchheimer MHD tri-hybrid nanofluids $\left(Al_2O_3 / Cu / Fe_3O_4\right)$ driven by a non-linearly stretching sheet with a heat source. By numerically resolving the system of ordinary differential equations (ODEs) $(5.9-5.11)$ and boundary conditions 5.13, the study elucidates the sensitivity of the flow to various physical parameters.

Specifically, we assessed the interplay between the nanoparticle volume fractions (ϕ_1, ϕ_2, ϕ_3) and porous media factors like, porosity (δ), and the Forchheimer coefficient (Fr). The results also highlight the determining roles of the suction factor (S), stretching parameter (n), Prandtl number (Pr), Schmidt number (Sc), and heat generation (Q_1) in shaping the fluid's dynamic behavior.

5.3.1 Present Research Validation

Establishing the reliability of numerical results is a fundamental prerequisite for their application in future engineering designs. To verify the precision of the current algorithm, the problem was simplified to a limiting case for cross-validation with Devi and Devi (2016). The general model was reduced to align with the referenced study by neglecting specific effects, namely, setting $Sc = Fr = \delta = Nb = Nt = Q_1 = \phi_1 = \phi_3 = 0$, while fixing the parameters at $Pr = 6.135$ and $\phi_2 = 0.04$. As evidenced in Table *5.2*, the computed outcomes for the skin friction coefficient $-f''(0)$ and the rate of heat transfer $-\theta'(0)$ across distinct values of suction parameter S and magnetic factor M exhibit a strong correlation with the established literature, thereby validating the robustness of our numerical scheme.

5.3.2 Present Research Results

This research offers a comprehensive analysis of water-based nanofluids, ranging from single-particle suspensions to hybrid and ternary hybrid configurations involving Al_2O_3, Cu and Fe_3O_4.

Through visual plots and tabulated metrics, we explore the complexities of fluid dynamics with a specific aim toward industrial relevance. Unless specified

TABLE 5.2

Results Comparison of $-\theta'(0)$ and $-f''(0)$ for Distinct Values of Suction Parameter S and Magnetic Factor M, When

$Sc = Fr = \delta = Nb = Nt = Q_1 = \phi_1 = \phi_3 = 0, Pr = 6.135$ and $\phi_2 = 0.04$.

S	*M*	$-\theta'(0)$ (Devi and Devi, 2016)	Present	$-f''(0)$ (Devi and Devi, 2016)	Present
0	2	1.531486	1.5314858	2.538415	2.5384147
0.5	2	3.985379	3.9853789	2.889322	2.8893222
1	2	6.750319	6.7503190	3.281808	3.2818083
1.5	2	9.638723	9.6387232	3.712996	3.7129964
0.5	0	4.132241	4.1322409	1.582096	1.5820960
0.5	1	4.081784	4.0817844	2.007264	2.0072638
0.5	3	3.888601	3.8886008	3.904730	3.9047397

otherwise, the baseline constants provided in Table 5.3 serve as the input for all numerical calculations.

Figure 5.2 illustrates the impact of non-Darcy flow, where rising values of the Forchheimer number Fr introduce significant inertial drag, thereby impeding the momentum boundary layer and dampening the fluid velocity $f'(\eta)$. In contrast, Figure 5.3 shows that this flow retardation facilitates energy accumulation, resulting in an elevated thermal profile $\theta(\eta)$ for higher Fr values. When analyzing fluid variations, the mono-nanofluid (Fe_3O_4) exhibits the least resistance and maintains the highest velocity. Notably, the Ternary hybrid composite $(Al_2O_3 + Cu + Fe_3O_4)$ retains a higher velocity than the binary hybrid $(Cu + Fe_3O_4)$. From a thermal perspective, the ternary hybrid leverages the synergistic conductivity of its three constituents to achieve the highest temperatures, outperforming the hybrid and mono-fluids. Finally, Figure 5.4 highlights a concentration crossover in $\xi(\eta)$; while the mono-nanofluid

TABLE 5.3
Default Parameter Values Employed in the Numerical Computations

Parameter Symbols	Default Value
S, Nb, Nt	0.1
ϕ_1, ϕ_2, ϕ_3	0.05
Fr, Sc	0.5
M	1
Q_1	0.2
n	1.5
Pr	7
δ	0.6

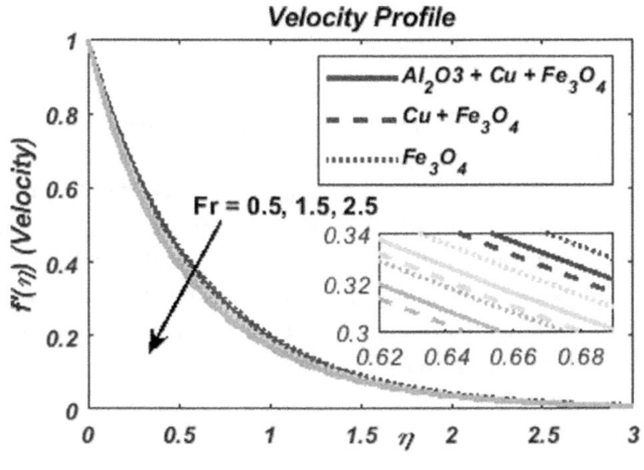

FIGURE 5.2 Influence of Fr on velocity.

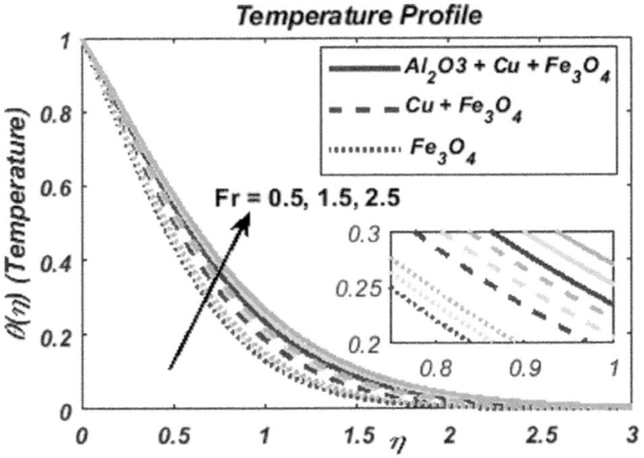

FIGURE 5.3 Influence of *Fr* on temperature.

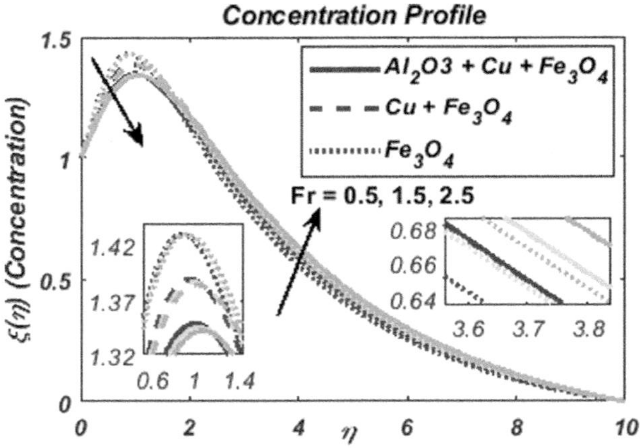

FIGURE 5.4 Influence of *Fr* on concentration.

dominates particle concentration near the wall $(\eta = 0)$, it depletes more rapidly than the composites in the free stream region $(\eta > 2)$.

Figure 5.5 highlights the suppressive influence of the Schmidt number (Sc) on the nanoparticle concentration $\xi(\eta)$. This decay is attributed to the weaker mass diffusion associated with higher values of Sc, which restricts the spread of nanoparticles away from the surface. Additionally, the fluid composition plays a complex role: near the surface $(\eta = 0)$, the mono-nanofluid records the highest concentration levels compared to the hybrid and ternary variants, respectively. However, this behavior is non-uniform, as a distinct crossover occurs deeper into the fluid domain $(\eta > 2)$, where the ternary nanofluids mixture eventually exhibits a reversed trend.

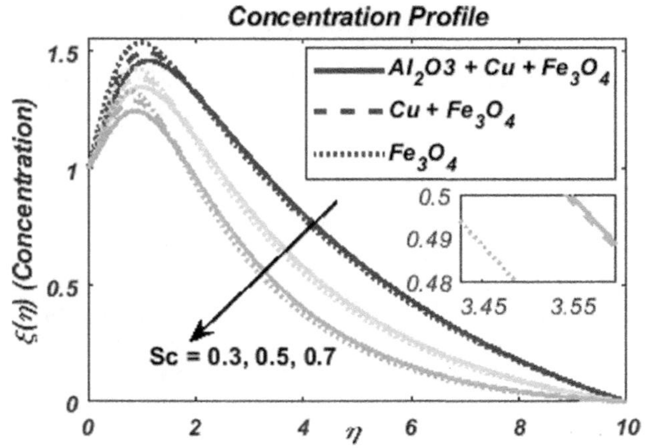

FIGURE 5.5 Influence of *Sc* on concentration.

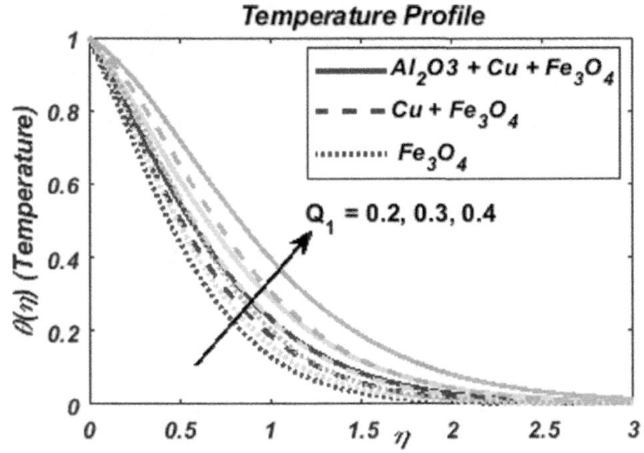

FIGURE 5.6 Influence of Q_l on temperature.

The influence of internal heat generation, represented by the parameter Q_1, is critical to the flow's thermal characteristics. As illustrated in Figure 5.6, intensifying the heat source supplies additional caloric energy to the system and energizes the fluid particles. This energy injection thickens the thermal boundary layer, leading to a substantial enhancement in the temperature profile $\theta(\eta)$. Among the fluids tested, the ternary mixture exhibits the most pronounced thermal response, outperforming both the hybrid and mono-nanofluids due to its enhanced synergistic thermal properties. Finally, Figure 5.7 presents a distinctive intersection in the concentration field $\xi(\eta)$, mirroring the trends observed with the Forchheimer number, suggesting that the lighter mono-particles diffuse differently in the far-field compared to the heavier hybrid composites.

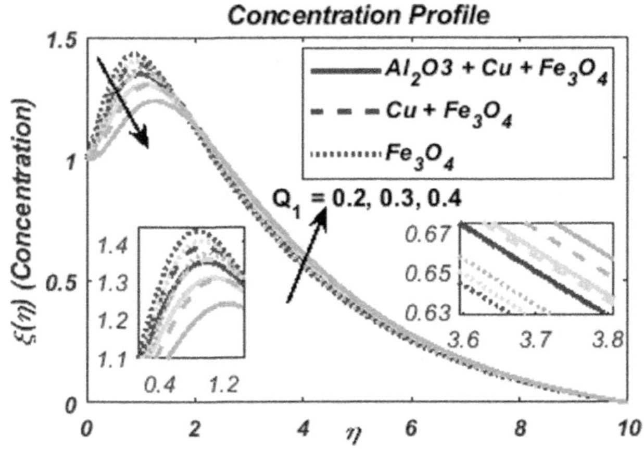

FIGURE 5.7 Influence of Q_1 on concentration.

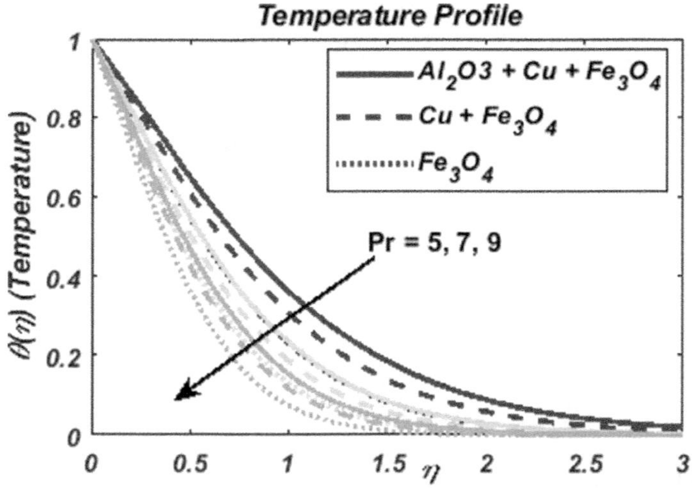

FIGURE 5.8 Influence of Pr on temperature.

Figure 5.8 elucidates the inverse relationship between the Prandtl number (Pr) and the thermal field. Higher Pr values indicate a reduction in thermal diffusivity, which thins the thermal boundary layer and effectively dampens the fluid temperature $\theta(\eta)$. Despite this suppression, the choice of nanoparticle composition remains critical: the Ternary hybrid $(Al_2O_3 + Cu + Fe_3O_4)$ consistently exhibits the highest thermal profile, outperforming both the hybrid $(Cu + Fe_3O_4)$ and the mono-nanofluid (Fe_3O_4) due to its superior composite thermal conductivity. Interestingly, the Prandtl number also triggers a dual-nature response in the concentration profile $\xi(\eta)$, as shown in Figure 5.9. While an increase in Pr yields a higher nanoparticle concentration in the near-wall region $(\eta = 0)$, this behavior inverts deeper into the fluid ($\eta > 2$), indicating a complex coupling between the thermal and solutal boundary layers.

Figure 5.10 depicts the hydrodynamic impact of the porosity parameter, where higher values correspond to increased Darcy resistance. This resistance creates a retarding force within the porous medium, significantly impeding the momentum boundary layer and suppressing fluid velocity. Consequently, as shown in Figure 5.11, this flow deceleration reduces the heat removal rate, allowing energy to accumulate and thickening the thermal boundary layer, which results in higher temperatures.

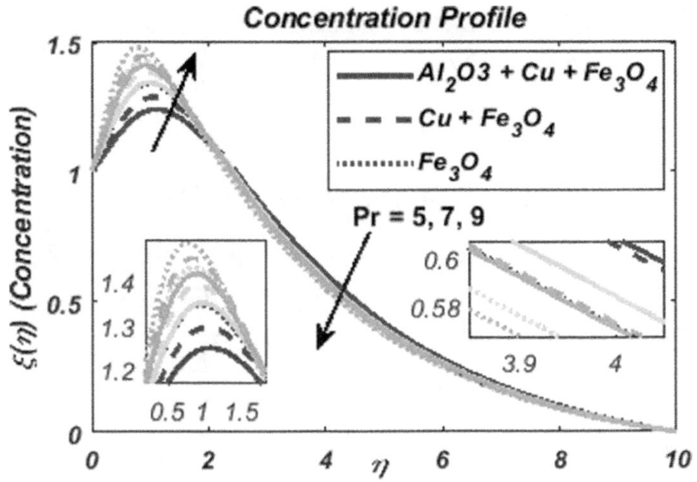

FIGURE 5.9 Influence of Pr on concentration.

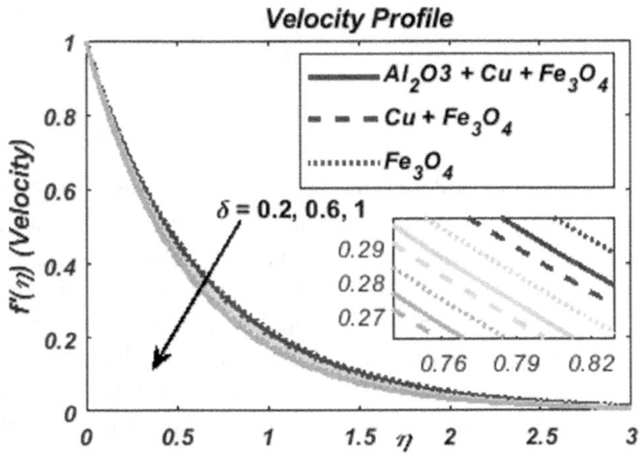

FIGURE 5.10 Influence of δ on velocity.

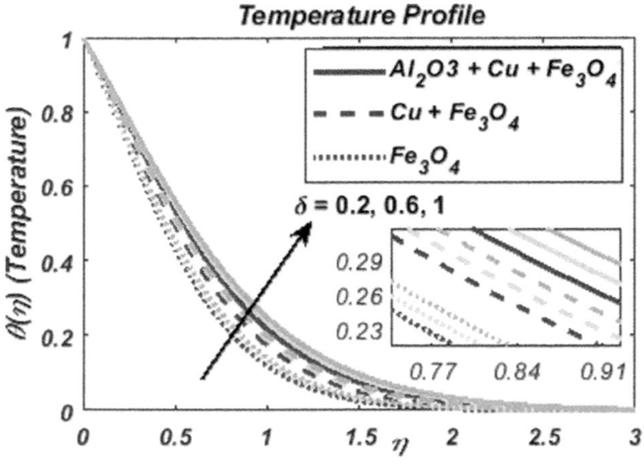

FIGURE 5.11 Influence of δ on temperature.

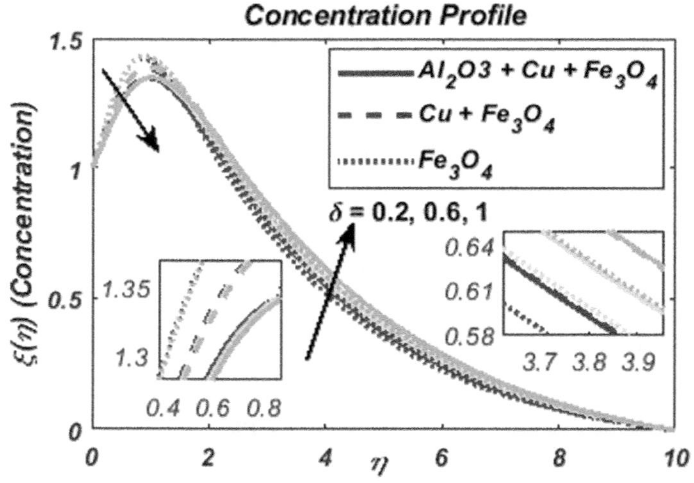

FIGURE 5.12 Influence of δ on concentration.

Paralleling the trends observed with the Forchheimer number, the fluid composition also dictates performance: the mono-nanofluid encounters the least hydraulic resistance, maintaining peak velocity, whereas the ternary hybrid unexpectedly retains higher momentum than the binary hybrid. Thermally, however, the ternary composite dominates, utilizing the combined thermal properties of its three nanoparticles to achieve maximum temperatures. Finally, Figure 5.12 illustrates a concentration crossover similar to inertial drag effects; the mono-nanofluid exhibits the highest concentration near the wall but is surpassed by the composite fluids in the free stream region.

The computed values for the skin friction coefficient $\left(-f''(0)\right)$, local Nusselt number $\left(-\theta'(0)\right)$, and Sherwood number $\left(-\xi'(0)\right)$ are summarized in Table 5.4, for hybrid and ternary systems. The computations indicate that non-linear stretching (n) and the Forchheimer number (Fr) act as a catalyst for heat and mass exchange, simultaneously reducing drag. Conversely, the rise in a magnetic field (M) hinders thermal and solutal transport while increasing wall friction due to Lorentz drag. When comparing fluid types, the hybrid nanofluid $\left(Cu + Fe_3O_4\right)$ unexpectedly outperforms the ternary mixture $\left(Al_2O_3 + Cu + Fe_3O_4\right)$ in all three metrics, implying that the inclusion of aluminum oxide acts as a limiting factor in this specific stretching regime.

Further analysis in Table 5.5 isolates the impact of porosity and volume fraction. Increasing the porosity parameter (δ) restricts fluid motion, thereby diminishing heat and mass transfer rates while raising skin friction. The sensitivity to volume fraction varies significantly by material: elevating ϕ_1 leads to a decline across all monitored outputs, whereas increasing ϕ_2 or ϕ_3 boosts skin friction but stifles transfer rates. Comparative metrics indicate that among these three nanoparticles, Al_2O_3 dominated fluids offer the best theoretical heat transfer potential, Fe_3O_4 maximizes mass transfer and Cu creates the highest frictional resistance.

5.4 CONCLUSION

The present computational analysis scrutinizes the dynamics of Darcy–Forchheimer MHD flow, isolating the impact of nanoparticle composition, ranging from mono (Fe_3O_4) to hybrid $(Cu + Fe_3O_4)$ and ternary $\left(Al_2O_3 + Cu + Fe_3O_4\right)$ systems, architectures over a non-linearly stretching surface with internal heat generation. The outcomes elucidate that the precise configuration of the nanoparticle matrix is a determinant factor in modulating flow velocity, thermal and solutal transport rates, as well as wall shear stress. These findings hold significant promise for industrial design, reinforcing the paradigm that hybrid and ternary nanofluids are superior candidates for optimizing advanced heat exchange systems compared to conventional fluids. The salient conclusions are summarized below:

1. **Inertial and Porous Resistance:** Increasing the Forchheimer number or the porosity parameter introduces significant drag forces. This inertial resistance suppresses fluid velocity but paradoxically enhances thermal accumulation, leading to a thicker thermal boundary layer. In velocity comparisons, the mono-nanofluid $\left(Fe_3O_4\right)$ proves the most agile, maintaining the highest flow speeds compared to the composite fluids.
2. **Thermal Dynamics:** The ternary hybrid $\left(Al_2O_3 + Cu + Fe_3O_4\right)$ consistently exhibits superior thermal performance, achieving the highest temperature profile compared to mono and hybrid variants regardless of variations in the heat source or Prandtl number. While a stronger heat source augments the

TABLE 5.4

Computations of $-\theta'(0), -\xi'(0)$ and $-f''(0)$ for Hybrid and Ternary Nanofluids across Various Ranges of n, M and Fr

			Hybrid $(Cu + Fe_3O_4)$			Ternary $(Al_2O_3 + Cu + Fe_3O_4)$		
n	M	Fr	$-\theta'(0)$	$-\xi'(0)$	$-f''(0)$	$-\theta'(0)$	$-\xi'(0)$	$-f''(0)$
0.5	1	0.5	0.4817173	0.1342576	1.9162200	0.3348809	0.0039642	1.8933602
1	1	0.5	0.8064063	0.4265780	1.8068027	0.7114550	0.3356126	1.7811171
1.5	1	0.5	0.9436818	0.5449889	1.7376959	0.8513782	0.4556254	1.7100905
2	1	0.5	1.0235292	0.6115946	1.6899636	0.9309600	0.5213536	1.6609629
1.5	2	0.5	0.8689712	0.5070227	1.9599982	0.7705984	0.4124242	1.9343259
1.5	3	0.5	0.7912106	0.4557801	2.1593247	0.6843323	0.3531770	2.1349364
1.5	4	0.5	0.7068730	0.391708	2.3416466	0.5849707	0.2745692	2.3181572
1.5	1	1	0.9244615	0.5333493	1.8285938	0.8324215	0.4441131	1.7976585
1.5	1	1.5	0.9059171	0.5217414	1.9153716	0.8140854	0.4326007	1.8813297
1.5	1	2	0.8879532	0.5101616	1.9985310	0.7962761	0.4210818	1.9615709

TABLE 5.5

Computations of $-\theta'(0), -\xi'(0)$ and $-f''(0)$ for Various Ranges of ϕ_1, ϕ_2, ϕ_3 and δ.

			$-\theta'(0)$		$-\xi'(0)$		$-f''(0)$	
ϕ_1	ϕ_2	ϕ_3	$\delta = 0.2$	$\delta = 0.6$	$\delta = 0.2$	$\delta = 0.6$	$\delta = 0.2$	$\delta = 0.6$
0	0.03	0.03	1.0608627	1.0337426	0.6326986	0.6250461	1.6133880	1.7100695
0.03	0.03	0.03	0.9999408	0.9715151	0.5748441	0.5660226	1.6055027	1.7026244
0.05	0.03	0.03	0.9617136	0.9324054	0.5382264	0.5286785	1.5979188	1.6954692
0.03	0	0.03	1.0791956	1.0521347	0.6405607	0.6345012	1.5574065	1.6573129
0.03	0.05	0.03	0.9538930	0.9246069	0.5359771	0.5256168	1.6298935	1.7256612
0.03	0.03	0	1.0613882	1.0342212	0.6310104	0.6236640	1.5999216	1.6973601
0.03	0.03	0.05	0.9614976	0.9322239	0.5392739	0.5295537	1.6057735	1.7028800
0.05	0.05	0.05	0.8824743	0.8513782	0.4674566	0.4556254	1.6134093	1.7100905

TABLE 5.6
Nomenclature

$B_0, B(x)$	The uniform and variable magnetic field strengths	**Greek Letters**	
a	Stretching rate	κ	Thermal conductivity
C	Nanofluid concentration	$\theta(\eta)$	Non-dimensional temperature
C_f	Skin-friction	β_T	Thermal expansion coefficient
C_p	Nanoparticle specific heat	σ	Electric conductivity
N_b, N_t	Brownian motion and thermophoresis parameters	η	Non-dimensional similarity variable
D_T	Thermal diffusion coefficient	ρ	Density
T	Nanofluid temperature	μ	Viscosity
$f(\eta)$	Dimensionless stream function	τ	Wall shear stress
$f'(\eta)$	Non-dimensional velocity	$-\theta'(0)$	Local Nusselt number
$-f''(0)$	Skin friction coefficient	α	Thermal diffusivity
δ	Non-dimensional porosity parameter	ϕ_1, ϕ_2, ϕ_3	Volume fractions of nanoparticles Al_2O_3, Cu and Fe_3O_4 respectively
K_d	Dimensional permeability parameter	β_C	Concentration expansion coefficient
Sc	Schmidt number	$\xi(\eta)$	Non-dimensional concentration
S	Suction/injection factor	v	Kinematic viscosity
Pr	Prandtl number	ρC_p	Heat capacitance
Fr	Forchheimer Number	$-\xi'(0)$	Local Sherwood number
x, y	Cartesian coordinate axis	ψ	Stream function
T_w, T_∞	Surface and free stream temperature respectively	**Subscripts**	
n	Non-linearly stretching parameter	$f, nf, hnf, thnf$	Base fluid, nanofluid, hybrid, and ternary nanofluids, respectively
q_w	Wall heat flux	w	Boundary conditions at surface

(Continued)

TABLE 5.6 (CONTINUED)
Nomenclature

Re	Reynolds number	np	Nanoparticles
Q_1	Heat source	1,2,3	First, second, and third nanoparticles
D_B	Brownian diffusion coefficient	∞	Free stream boundary conditions
v,u	Velocity component parallel to the y and x directions	**Superscripts**	
Nu	Nusselt number	$'$	Differentiation with respect to η

temperature field, an increasing Prandtl number effectively dampens it by reducing thermal diffusivity.

3. **Concentration Crossover Phenomenon:** A distinct "crossover" trend is observed in the concentration profile $(\xi(\eta))$ under the influence of Forchheimer, heat source, and Prandtl parameters. The mono-nanofluid dominates particle concentration in the near-wall region $(\eta \approx 0)$ but depletes rapidly in the far-field $(\eta > 2)$, where it is surpassed by the ternary and hybrid mixtures. Furthermore, the Schmidt number acts as a strong suppressant for concentration due to weaker mass diffusivity.
4. **Optimization of Transfer Rates:** Non-linear stretching factor and Forchheimer inertial forces act as catalysts for improving Nusselt and Sherwood numbers while reducing skin friction. Conversely, magnetic and porosity factors hinder these transport rates.
5. **Component Efficiency:** Although the ternary fluid is thermally dominant in the boundary layer, the binary hybrid $(Cu + Fe_3O_4)$ yields higher wall transfer rates (Nusselt/Sherwood numbers). When isolating specific nanoparticles, Al_2O_3 offers the highest theoretical heat transfer potential, Fe_3O_4 maximizes mass transfer efficacy, and Cu creates the highest frictional resistance.

REFERENCES

Adel, M., Khader, M. M., Babatin, M. M., Alraddadi, I., Alaidrous, A., & Ismail, G. M. (2025). Heat transfer and magnetohydrodynamic nanofluid flow caused by a stretching sheet heated convectively: Numerical investigation. *European Journal of Pure and Applied Mathematics*, *18*(1), 5502.

Alnahdi, A. S., & Gul, T. (2024). MHD hybrid nanofluid flow over a stretched surface with convective boundary conditions: Applications of heat transfer. *Modern Physics Letters B*, *38*(08), 2450046.

Aly, E. S., Alamri, A. A., Sedki, A. M., Aly, E. S., Alamri, A. A., & Sedki, A. M. (2025). Nonlinear slip flow of Forchheimer MHD hybrid Casson nanofluid over stretching permeable surface with thermal radiation and entropy generation. *ZAMM - Journal of Applied Mathematics and Mechanics / Zeitschrift Für Angewandte Mathematik und Mechanik*, *105*(9), e70203.

Al-Zubaidi, A., Abutuqayqah, H., Ahmad, B., Bibi, S., Abbas, T., & Saleem, S. (2023). Analysis of slip condition in MHD nanofluid flow over stretching sheet in presence of viscous dissipation: Keller box simulations. *Alexandria Engineering Journal*, *82*, 26–34.

Athal, I., Haewon, B., Sasikala, A., Reddy, B. N., Govindan, V., Maddileti, P., Saritha, K., Reddy, B. S., Rajakumari, S., Tawade, J. V., Tamam, N., Abdullaeva, B. S., Chohan, J., & Mishra, R. (2024). Combined impact of radiation and chemical reaction on MHD hyperbolic tangent nanofluid boundary layer flow past a stretching sheet. *Modern Physics Letters B*, *38*(16), 2341010.

Bafe, E. E., Firdi, M. D., & Enyadene, L. G. (2023). Numerical investigation of MHD Carreau nanofluid flow with nonlinear thermal radiation and Joule heating by employing Darcy–Forchheimer effect over a stretching porous medium. *International Journal of Differential Equations*, *2023*, 1–17.

Bansal, S. (2024). Impacts of thermophoresis and Brownian motion on MHD nanofluid flow over an inclined Non-Linear stretching surface with chemical reaction. *Nanotechnology Perceptions*, *20*(3), 379–392.

Bansal, S., Kumar, A., Pal, J., Goyal, I., & Negi, A. S. (2024). Influence of thermal wall and velocity slips on non-darcy MHD boundary layer flow of a nanofluid over a non-linear stretching sheet. *Journal of Physics Conference Series*, *2844*(1), 012018.

Bansal, S., Pal, J., Bisht, M. S., & Fartyal, P. (2024). Influence of nanofluids on boundary layer flow over an inclined stretching sheet in a porous media along with magnetic field. *International Journal of Mathematical Engineering and Management Sciences*, *9*(2), 267–282.

Das, K., Giri, S. S., & Kundu, P. K. (2021). Influence of Hall current effect on hybrid nanofluid flow over a slender stretching sheet with zero nanoparticle flux. *Heat Transfer*, *50*(7), 7232–7250.

Devi, S. P. A., & Devi, S. S. U. (2016). Numerical investigation of hydromagnetic hybrid Cu – Al2O3/Water nanofluid flow over a permeable stretching sheet with suction. *International Journal of Nonlinear Sciences and Numerical Simulation*, *17*(5), 249–257.

Jaffrullah, S., Sridhar, W., & Ganesh, G. R. (2024). Analysis of MHD casson nanofluid flow over a nonlinearly stretching surface with joule heating, radiation and suction effects. *International Journal of Heat and Technology*, *42*(2), 414–426.

Kallem, S. R., Sheri, S. R., Shankar, G., Prathiba, A., & Sunitha, M. (2025). Numerical investigation of the effects of radiation and heat source on the MHD flow of hyperbolic tangent fluid over a non-linear stretching sheet embedded in a porous. *CFD Letters*, *17*(10), 1–16.

Khashi'ie, N. S., Hamzah, K. B., Waini, I., Zainal, N. A., Arifin, N. M., & Pop, I. (2024). Magnetohydrodynamic couple stress hybrid nanofluid flow over a convectively heated stretching surface. *Magnetohydrodynamics*, *60*(3/4), 197–206.

Kiyani, M. Z., Aeman, A., Abbasi, A., Khan, S. U., Adnan, N., Eladeb, A., Kolsi, L., & Mebarek-Oudina, F. (2025). Analysis of nanofluid through porous media with darcy–forchheimer and mass suction effects: Keller box numerical simulations. *Journal of Nanofluids*, *14*(3), 438–446.

Kouki, M., Shukat, S., Ullah, I., Alam, M. M., & Ali, A. H. (2025). Keller-box based computational investigation of magnetized gravity-driven Micropolar nanofluid flow past an exponentially contracting surface with cross diffusion effect and engineering applications. *Alexandria Engineering Journal, 111*, 160–170.

Laxmia, N. B., Chand, N. D. K., & Thakur, P. (2025). Hydrodynamic Casson hybrid nanofluid flow across a stretching sheet in the regime of velocity slip and temperature jump, including viscous dissipation, melting, Soret and Dufour effects. *Journal of the Nigerian Society of Physical Sciences, 7*(4), 2936.

Manoharkumar, K. N., Patil, S. M., Dinesh, P. A., & Suma, K. V. (2024). Combined effect of viscous dissipation and forchheimer drag on 3D nanofluid flow over nonlinear stretching sheet. *Journal of Mines Metals and Fuels*, 453–461. https://doi.org/10.18311/jmmf/2023/47296

Memon, M. A., Jacob, K., Lanjwani, H. B., Khan, U., Sherif, E. M., & Pop, I. (2023). Thermal analysis for hydromagnetic flow of Darcy-Forchheimer hybrid nanofluid with velocity and temperature slip effects: Scrutinization of stability and dual solutions. *Advances in Mechanical Engineering, 15*(11).

Nabwey, H. A., El-Hakiem, A., Khan, W. A., Abdelrahman, Z. M., Rashad, A., & Hawsah, M. A. (2024). Magnetic Williamson hybrid nanofluid flow around an inclined stretching cylinder with joule heating in a porous medium. *Chemical Engineering Journal Advances, 18*, 100604.

Pal, D., & Mandal, G. (2015). MHD convective stagnation-point flow of nanofluids over a non-isothermal stretching sheet with induced magnetic field. *Meccanica, 50*(8), 2023–2035.

Pennelli, S., Shaik, M., & Prathi, V. (2025). The MHD viscoelastic nanofluid-flow upon an inclined stretching sheet with radiation, chemical reaction and multiple slip effects-homotopy analysis method. *Thermal Science, 29*(4 Part B), 3293–3306.

Prashanth, M., & Rao, V. S. (2024). The impact of Soret, Dufour, and chemical reaction on MHD nanofluid over a stretching sheet. *Partial Differential Equations in Applied Mathematics, 10*, 100674.

Ramana, R. M., Maheswari, C., Shaw, S. M., Dharmaiah, G., Fernandez-Gamiz, U., & Noeiaghdam, S. (2024). Numerical investigation of 3-D rotating hybrid nanofluid Forchheimer flow with radiation absorption over a stretching sheet. *Results in Engineering, 22*, 102019.

Ramireddy, P., & Jayalakshmi, K. (2025). The impression of thermal stratification and joule heating on hybrid nanofluid over a porous stretching cylinder. *JP Journal of Heat and Mass Transfer, 38*(1), 47–76.

Ramya, D., Raju, R. S., Rao, J. A., & Chamkha, A. (2018). Effects of velocity and thermal wall slip on magnetohydrodynamics (MHD) boundary layer viscous flow and heat transfer of a nanofluid over a non-linearly-stretching sheet: A numerical study. *Propulsion and Power Research, 7*(2), 182–195.

Reddy, B. N., Dodda, R., & Reddy, B. S. (2024). A numerical investigation on boundary layer flow of MHD tangent hyperbolic fluid flow over a stretching sheet with slip boundary conditions. *Journal of Advanced Research in Fluid Mechanics and Thermal Sciences, 120*(1), 122–139.

Swain, B. K., & Kar, S. (2024). Dissipative MHD flow of ternary-hybrid nanofluid over an inclined stretching sheet with heat source and mass flux due to temperature gradient. *ZAMM - Journal of Applied Mathematics and Mechanics / Zeitschrift Für Angewandte Mathematik und Mechanik, 104*(4), e202200360.

Triveni, B., Rao, M. V. S., Gangadhar, K., & Chamkha, A. J. (2023). Heat transfer analysis of MHD Casson nanofluid flow over a nonlinear stretching sheet in the presence of non-uniform heat source. *Numerical Heat Transfer Part A Applications, 85*(13), 2145–2164.

6 Particle-Laden Flows

Modelling Particle Transport in Non-Newtonian Flows: A Finite Difference Solution

Subrata Ghosh and Sahin Ahmed

6.1 INTRODUCTION

In today's world, the study of Casson fluid, nanofluid, boundary layer flow through an extending sheet has a critical role in solving many manufacturing and engineering problems. Understanding flow behaviours allows researchers or engineers to control heat and mass transfer smoothly, enhancing manufacturing and the overall efficiency of the fluid product. Over the past few years, nanofluid has become a revolutionary class of engineered fluid; it is basically a traditional fluid (like water, kerosene, oil, etc.) containing a suspension of nanosized particles, namely metal, metal oxides, or carbon-based oxides, having a diameter in the range of 1–100 nm. Nanofluids have excellent thermal and transport properties compared to conventional fluids. Their characteristics, like increased surface area and improved thermal stability, attract a lot of researchers in this field and make them a great choice for modern appliances like cooling systems and solar energy devices, etc. Due to its enhanced thermophysical properties and numerous applications across different fields, many researchers are attracted to nanotechnology. Heat conductivity of a fluid is improved by mixing nanoparticles in a traditional fluid, which is called nanofluids (Choi *et al.* 1995). After that, a detailed study on convective transport phenomena of nanofluids was also conducted (Buongiorno 2006). A detailed study has been done using the Adams-Bashforth Predictor-Corrector method that examines the impacts of thermal radiation, variable heat flux, and magnetic field on CuO-Fe_2O_3/H_2O hybrid nanofluid flow past a stretchable cylinder (Ali *et al.* 2021). In recent years, a study explored the steady flow of Powell-Eyring nanofluids with MHD effect past an elastic cylinder (Hayat *et al.* 2016). They conducted a comprehensive study on how prime parameters affect the thermal, momentum, and mass diffusion profiles and presented them graphically. The hybrid nanofluid flow was mathematically and numerically analysed in magnetically influenced stretching and shrinking surfaces under unsteady, incompressible conditions. Key parameters and heat transport were illustrated using graphs (Naqvi *et al.* 2024). A computational analysis of hybrid nanofluid flow using

DOI: 10.1201/9781003664345-6

the finite element method under three-dimensional conditions and magnetic field past a porous stretchable sheet was jointly conducted. The results revealed that increasing the magnetic field parameter thickens the thermal boundary layer and reduces fluid velocity (Kabra and Sharma 2025). In recent years, the hydromagnetic flow through a stretchable and shrinkable sheet with the effects of variable heat conductivity was numerically investigated. Prime physical parameters influencing thermal diffusion and flow velocity were presented graphically (Kaye and Ahmed 2024). The nanofluid flow containing Cu, Ag, and Fe_2O_3 nanoparticle suspended in water past a permeable, magnetically influenced extending sheet within a porous medium was numerically examined, including several key parameters. The study reveals that the velocity of Fe_2O_3-water nanofluid is more than Cu-H_2O and Ag-H_2O nanofluids, while the trend is reversed for the thermal profile (Hazarika and Ahmed 2021). Recently few authors used the MATLAB bvp4c solver to study various aspects of fluids behaviour along an extending sheet under a magnetic field (Das and Ahmed 2025; Ahmed and Hussain 2023).

Non-Newtonian fluids refer to those which do not obey Newton's law of viscosity. Casson fluid is a special class of fluid that possesses yield stress behaviour. In 1995, Casson developed the model of Casson fluid for viscoelastic fluid flow. Blood and paint are some examples of Casson fluid. Understanding how non-Newtonian fluids behave past an elastic sheet has become a crucial topic in engineering research. In earlier work, the flow of Casson nanofluid under magnetic and heat radiation effects was explored. The effects of prime parameters on momentum, heat, and concentration profiles were also analyzed and the results are presented graphically (Reddy *et al.* 2018). In another investigation, the effect of thermal energy radiation and suction on steady Casson nanofluid flow under a magnetic field past anelastic and shrinking cylinder was mathematically analyzed. The investigation confirms the existence of mass suction values and the stability test showed that the initial solution is linearly stable; however, it becomes linearly unstable in the second solution (Soomro *et al.* 2024). In recent investigations, a detailed study on Casson non-Newtonian fluid flow through an extending cylinder under MHD was done (Dang *et al.* 2024). The model contains parameters for Brownian diffusion Nb and thermophoretic parameter Nt and the results shows that as Nb and Nt increases, the concentration of nanoparticle decreases with greater heat parameter values. A detailed comparative study of Casson fluid in mixed convection boundary layer flow under the effects of MHD through a cylindrical and flat surface was also conducted. Additionally, the impacts of several prime parameters such as Prandtl number and mixed convection were studied and demonstrated graphically (Rehman *et al.* 2017). In later investigations, tangent hyperbolic fluid flow under magnetic field and heat radiation effects through an extending cylinder was numerically studied. The investigation showed that with higher velocity slip parameters, the fluid velocity diminishes (Reddy *et al.* 2017). Again, through the Keller box approach, the tangent hyperbolic fluid flow under MHD through a stretchable sheet was carried out (Malik *et al.* 2015). In another study, Eyring-Powell fluid flow under the impacts of MHD through a stretchable sheet was conducted numerically using the finite difference scheme. This study reveals that for higher magnetic field parameters, it slows down the fluid flow (Akbar *et al.* 2015).

The aim of this research is to investigate the velocity, thermal, and concentration profiles of a non-Newtonian MHD fluid under the combined effects of thermophoretic force and Brownian diffusion inside a stretchable cylinder with slip velocity. The objective involve modelling and using MATLAB's bvp4c to solve the governing non-linear equations, studying the influence of slip, magnetic field, and nanoparticle effects on momentum, energy and concentration profiles, and validating the numerical results for accuracy and stability. The study aims to give a better understanding of thermal and mass transmission behaviour in non-Newtonian magnetic nanofluids for practical industrial and biomedical applications.

The present study is driven by the following research questions:

- How do magnetic field strength and porous medium permeability together influence the momentum, thermal, and concentration fields in hybrid nanofluid flow?
- How does surface curvature modify warmth and molar transmission characteristics, and what roles do the Prandtl and Schmidt numbers play in this behaviour?
- In what way do thermophoretic and Brownian diffusion effects control nanoparticle movement and temperature distribution inside a stretchable cylinder?
- How do non-Newtonian effects (Casson parameter) and viscous dissipation (Eckert number) interact to alter flow resistance and boundary-layer heating?
- Can the developed finite difference model accurately predict flow and thermal behaviour compared to existing benchmark results?

6.2 MATHEMATICAL FORMULATION

The steady-state laminar 2-D motion of a non-Newtonian incompressible liquid in a stretchable cylinder under slip velocity is taken into account in this investigation. In this formulation, the surface velocity is modelled by $U_w(x) = ax/l$, with 'a' being a constant greater than zero and where l is the characteristic length and the component x acts along the direction of the stretchable cylinder and r = radial direction.

The study is based on the following assumptions

1. A constant transverse magnetic field $B = B_0$ is imposed normal to the fluid flow.
2. The flow is steady, laminar, and satisfies the Boussinesq approximation.
3. The kinematic viscosity and heat diffusivity of the fluid are considered constant throughout the flow field.
4. The surface of the cylinder follows velocity slip and convective boundary conditions.

5. The impacts of thermal radiation, Brownian motion (Nb) and thermophoretic forces (Nt) are included in the mathematical model.
6. The flow is supposed to be axisymmetric, and the influence of induced magnetic fields and Ohmic heating is negligible (Figure 6.1).

The fluid flow is modelled using the following equations by Kabra and Sharma (2025)

$$\frac{\partial(ru)}{\partial x}+\frac{\partial(rv)}{\partial r}=0, \tag{6.1}$$

$$u\frac{\partial u}{\partial x}+v\frac{\partial u}{\partial r}=\nu_f\left(1+\frac{1}{\beta}\right)\frac{\partial^2 u}{\partial r^2}-\frac{\sigma B_o^2}{\rho_f}u-\frac{\nu}{k}U, \tag{6.2}$$

$$\left(u\frac{\partial T}{\partial x}+v\frac{\partial T}{\partial r}\right)=\left[\begin{array}{l}\frac{\alpha}{r}\frac{\partial}{\partial r}\left(r\frac{\partial T}{\partial r}\right)+\tau\left\{D_B\frac{\partial C}{\partial r}\frac{\partial T}{\partial r}+\frac{D_T}{T_\infty}\left(\frac{\partial T}{\partial r}\right)^2\right\}\\ +\frac{1}{\left(\rho C_p\right)_f}\left\{-\frac{1}{r}\frac{\partial}{\partial r}(rq_r)+\mu\left(\frac{\partial u}{\partial y}\right)^2+\sigma B^2u^2\right\}\end{array}\right], \tag{6.3}$$

$$\left(u\frac{\partial C}{\partial x}+v\frac{\partial C}{\partial r}\right)=\frac{D_B}{r}\frac{\partial}{\partial r}\left(r\frac{\partial C}{\partial r}\right)+\frac{D_T}{T_\infty}\left(\frac{\partial^2 T}{\partial r^2}+\frac{1}{r}\frac{\partial T}{\partial r}\right), \tag{6.4}$$

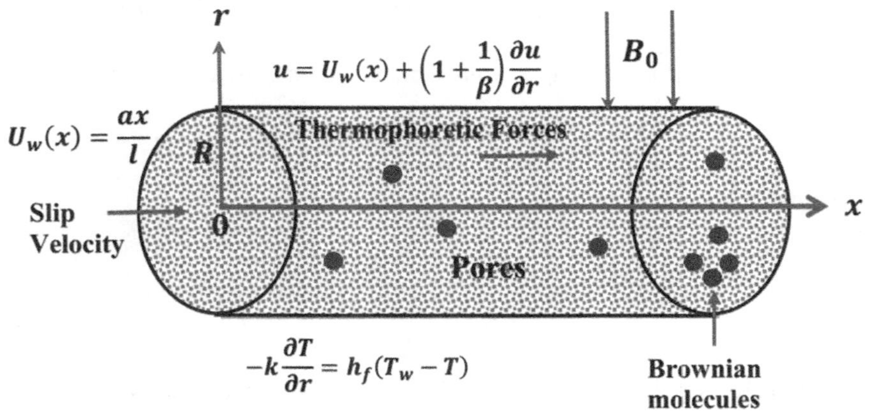

FIGURE 6.1 Flow geometry of stretchable cylinder.

where u, v are the fluid velocity along $x, r-$ direction. $\tau = \frac{\rho C_p}{(\rho C_p)_f}$, denotes the ratio effective nanoparticle heat capacity over base fluid heat capacity.

The boundary conditions relevant to the flow are given by Kabra and Sharma (2025):

$$\left\{\begin{array}{c} u = U_w(x) + \left(1 + \frac{1}{\beta}\right)\frac{\partial u}{\partial r}, \quad v = 0, \quad -k\frac{\partial T}{\partial r} = h_f(T_w - T), \\ -D_m\frac{\partial C}{\partial r} = k_m(C_W - C) \; at \; r = R, \quad u \to 0, \quad T \to T_\infty, \quad C \to C_\infty \; as, \; r \to \infty \end{array}\right\} \quad (6.5)$$

For isotropic and incompressible non-Newtonian fluid, the extra stress tensor equation is as follows:

$$\tau_{ij} = \begin{cases} 2\left(\mu + P_y/\sqrt{2\pi}\right)e_{ij}, & \pi > \pi_c \\ 2\left(\mu + P_y/\sqrt{2\pi_c}\right)e_{ij}, & \pi < \pi_c \end{cases} \quad (6.6)$$

Here, P_y denotes the fluid's yield stress, π equals the product of deformation rate with themselves, $\pi = e_{ij}, e_{ij}$, where e_{ij} is the (i, j)th component of deformation rate and π_c is defined as the critical value of this product in the non-Newtonian model.

Radiative heat flux through Roseland approximation is given by

$$q_r = -\frac{4\sigma_s}{3k_e}\frac{\partial T^4}{\partial r}, \quad (6.7)$$

where σ_s and k_e are the Stefan–Boltzmann constant and the mean absorption coefficient.

After using Taylor series approximation and deleting the terms of higher order, we obtain

$$T^4 \cong 4T_\infty^3 T - 3T_\infty^3 \quad (6.8)$$

Now

$$\frac{\partial q_r}{\partial r} = -\frac{16\sigma_s T_\infty^3}{3k_e}\frac{\partial^2 T}{\partial r^2} \quad (6.9)$$

Using Equations (6.7) and (6.8), we get the energy equation as follows:

$$\left(u\frac{\partial T}{\partial x}+v\frac{\partial T}{\partial r}\right)=\left[\begin{array}{c}\left\{\alpha+\frac{16\sigma_s T_\infty^3}{3k_e\left(\rho C_p\right)_f}\right\}\frac{\partial^2 T}{\partial r^2}+\frac{1}{r}\left\{\alpha+\frac{16\sigma_s T_\infty^3}{3k_e\left(\rho C_p\right)_f}\right\}\frac{\partial T}{\partial r}+\\ \tau\left\{D_B\frac{\partial C}{\partial r}\frac{\partial T}{\partial r}+\frac{D_T}{T_\infty}\left(\frac{\partial T}{\partial r}\right)^2\right\}+\frac{1}{\left(\rho C_p\right)_f}\left\{\mu\left(\frac{\partial u}{\partial y}\right)^2+\sigma B^2u^2\right\}\end{array}\right] \quad (6.10)$$

Let $\Psi(x,r)$ be a stream function in the flow field. Here

$$u=\frac{1}{r}\left(\frac{\partial\Psi}{\partial r}\right),\quad v=-\frac{1}{r}\left(\frac{\partial\Psi}{\partial x}\right)$$

The similarity transformations to solve the governing equations are considered as:

$$\left\{\begin{array}{c}\eta=\sqrt{\frac{a}{lv_f}}\left(\frac{r^2-R^2}{2R}\right),\quad \Psi=\sqrt{\frac{av_f}{l}}xRf(\eta)\\ \theta(\eta)=\frac{T-T_\infty}{T_w-T_\infty},\quad \phi(\eta)=\frac{C-C_\infty}{C_w-C_\infty},\end{array}\right\} \quad (6.11)$$

Using the above similarity transformations, the governing equations reduce to

$$\left[\begin{array}{c}\left(f'(\eta)\right)^2-f''(\eta)f(\eta)-\left(1+1/\beta\right)nf''(\eta)-\left(1+1/\beta\right)(2n\eta+1)f'''(\eta)\\ +Mf'(\eta)+\frac{1}{Da}f'(\eta)\end{array}\right]=0, \quad (6.12)$$

$$\left[\begin{array}{c}Pr\left\{\begin{array}{c}-MEc\left(f'(\eta)\right)^2-Ec(2n\eta+1)\left(f''(\eta)\right)^2-f(\eta)\theta'(\eta)\\ -(2n\eta+1)\left(Nb\theta'(\eta)\phi'(\eta)+Nt\left(\theta'(\eta)\right)^2\right)\end{array}\right\}\\ -\left(1+\frac{4}{3}\right)Rd\left\{(2n\eta+1)\theta''(\eta)+2n\theta'(\eta)\right\}\end{array}\right]=0, \quad (6.13)$$

$$\left[\begin{array}{c}-Scf(\eta)\phi'(\eta)-2n\phi'(\eta)-(2n\eta+1)\phi''(\eta)\\ -\frac{Nt}{Nb}\left\{2n\theta'(\eta)+(2n\eta+1)\theta''(\eta)\right\}\end{array}\right]=0, \quad (6.14)$$

Now the boundary conditions reduce to:

$$\left\{\begin{array}{c} f(0)=0,\quad f'(0)=1+\alpha_1\left(1+\frac{1}{\beta}\right)f''(0),\quad \theta'(0)=-Bi_1\left(1-\theta(0)\right), \\ \phi'(0)=-Bi_2\left(1-\phi(0)\right)\quad at\ \eta=0 \\ f'(\eta)\to 0,\quad \theta(\eta)\to 0, \phi(\eta)\to 0,\ as\ \eta\to\infty \end{array}\right\} \tag{6.15}$$

The non-dimensional numbers are

$$\left\{\begin{array}{c} M=\frac{\sigma B_0^2 l}{\left(\rho_f a\right)}, Rd=\frac{4\sigma_s T_\infty^3}{k_e k}, \Pr=\frac{\nu_f}{D_B}; n\left(1/R\right)\sqrt{\frac{l\nu_f}{a}}, \\ Da=\frac{ak}{\nu l}, Nt=\frac{\tau D_T\left(T_w-T_\infty\right)}{T_\infty \nu_f}, Nb=\frac{\tau D_B\left(C_w-C_\infty\right)}{\nu_f}, \\ \alpha_1=\frac{r}{R\sqrt{\frac{a}{\left(l\nu_f\right)}}}, Bi_1=\left(\frac{h_f}{k}\right)\left(\frac{R}{r}\right)\sqrt{\frac{l\nu_f}{a}}, Bi_2=\left(\frac{k_m}{D_m}\right)\left(\frac{R}{r}\right)\sqrt{\frac{l\nu_f}{a}} \end{array}\right\}$$

6.3 NUMERICAL SIMULATIONS

The transformed ordinary differential equations (6.12)–(6.14) are extremely nonlinear in nature, making it tough to obtain analytical solutions. Therefore, we numerically solved the equations and the boundary conditions (6.15), through MATLAB's **bvp4c**.

The nonlinear governing equations are first transformed into a system of first-order ordinary differential equations (ODEs) by defining the following set of new variables:

$$\left\{\begin{array}{c} y(1)=f,\ y(2)=f',\ y(3)=f'',\ y(4)=\theta, \\ y(5)=\theta',\ y(6)=\phi,\ y(7)=\phi' \end{array}\right\} \tag{6.16}$$

Using Equation (6.16), the transformed equations (6.12)–(6.14) become

$$\begin{Bmatrix} y(1)' \\ y(2)' \\ y(3)' \\ y(4)' \\ y(5)' \\ y(6)' \\ y(7)' \end{Bmatrix} = \begin{Bmatrix} y(2) \\ y(3) \\ \left[\left\{ \dfrac{1}{\left(1+\frac{1}{\beta}\right)(2n\eta+1)} \right\} \left\{ \begin{array}{l} y(2)^2 - y(3)\,y(1) + My(2) + \\ \left(\frac{v}{k}\right)\left(\frac{l}{a}\right) y(2) - \left(1+\frac{1}{\beta}\right) ny(3) \end{array} \right\} \right] \\ y(5) \\ \left[\left\{ \dfrac{1}{\left(1+\frac{1}{\beta}\right)(2n\eta+1)\,Rd} \right\} \left\{ \begin{array}{l} -\left(1+\frac{4}{3}\right) Rd2ny(5) + \\ \Pr \left\{ \begin{array}{l} -MEcy(2)^2 - \\ Ec(2n\eta+1)\,y(3)^2 \\ -y(1)\,y(5) - (2n\eta+1) \\ \left\{ Nby(5)\,y(7) - Nty(5)^2 \right\} \end{array} \right\} \end{array} \right\} \right] \\ y(7) \\ \left[\left\{ \dfrac{1}{(2n\eta+1)\left(1+\frac{Nt}{Nb}\right)} \right\} \left(-Scy(1)\,y(7) - 2ny(7) - \left(\frac{Nt}{Nb}\right) 2ny(5) \right) \right] \end{Bmatrix} \tag{6.17}$$

Along with the boundary conditions

$$\left\{ \begin{array}{c} y0(1)-0,\, y0(2)-1-\alpha_1\left(1+\dfrac{1}{\beta}\right) y0(3),\, y0(5)+Bi_1\left(1-y0(4)\right), \\ y0(7)+Bi_2\left(1-y0(6)\right),\, y1(2)-0,\, y1(4)-0,\, y1(6)-0 \end{array} \right\} \tag{6.18}$$

6.4 VALIDITY AND ACCURACY

Comparative study of present velocity $\left(f'(\eta)\right)$ and thermal $\left(\theta(\eta)\right)$ profiles with those from the previous study Reddy et al. (2018) confirms the reliability of the current numerical model. As shown in Table 6.1, the results display similar qualitative behaviour—both velocity and temperature decrease steadily with increasing η, which aligns with the boundary layer theory. The inclusion of a higher magnetic parameter, porosity resistance, and Prandtl number causes slight variations in

TABLE 6.1
Comparison Study of Present and Previous Research Reddy et al. (2018) on the Velocity and Temperature When $\beta = 0.5$, $n = 0.04$, $Da = 0.2$, $M = 2$, $Rd = 0.3$, $Ec = 0.01$, $Nb = 2$, $Sc = 0.27$, $Pr = 7$, $\alpha_1 = 0.05$, $Bi_1 = 13$, $Bi_2 = 12$

Nt	η	$f'(\eta)$-Present	$f'(\eta)$-Previous	$\theta(\eta)$-Present	$\theta(\eta)$-Previous
0.1	0	0.78841	0.63376	0.99893	0.94225
0.1	0.2	0.55124	0.51365	0.98886	0.81023
0.1	0.4	0.38524	0.4163	0.96369	0.68961
0.1	0.6	0.26948	0.3374	0.92344	0.5782
0.1	0.8	0.18842	0.27345	0.86395	0.47481
0.1	1	0.13174	0.22163	0.80364	0.3796
0.3	0	0.78841	0.63376	1.001	0.94864
0.3	0.2	0.55124	0.51365	0.99878	0.83147
0.3	0.4	0.38524	0.4163	0.9869	0.72505
0.3	0.6	0.26948	0.3374	0.9659	0.62721
0.3	0.8	0.18842	0.27345	0.93641	0.53601
0.3	1	0.13174	0.22163	0.89916	0.45001

magnitude, but the overall trend remains consistent. The effects of the thermophoresis parameter are also clear; an increase in Nt causes the temperature profile to slightly rise without having a major impact on velocity. This behaviour demonstrates how thermophoresis enhances heat transfer through nanoparticles. Therefore, the validity and accuracy of the current computational approach for upcoming nanofluid and hybrid nanofluid analyses are confirmed by the close agreement with prior findings and accurate parameter responses.

6.5 GRID STABILITY TEST

Table 6.2 indicates that the computed temperature $\theta(\eta)$ remains unchanged for all grid sizes (h = 0.1, 0.01, 0.001) and across different values of η. This consistency signifies that the numerical solution is stable and grid independent, confirms good match and dependability of the discretization method used in the simulation. The results show very little variation with decreasing grid spacing, ensuring confidence in the computed temperature profiles.

6.6 RESULTS AND DISCUSSION

This investigation examines how prime physical parameters influence the fluid's momentum, energy, and concentration profiles using graphical analysis.

The motion of the fluid reduces as the curvature parameter $\boldsymbol{n}$ increases, as seen in Figure 6.2. This happens as a result of increased surface stretching caused by greater

TABLE 6.2
Stability Analysis at Various Grid Sizes for Temperature $\theta(\eta)$

η	$h = 0.1$	$h = 0.01$	$h = 0.00$
0	0.61759	0.61759	0.61759
0.2	0.59374	0.59374	0.59374
0.4	0.56946	0.56946	0.56946
0.6	0.54548	0.54548	0.54548
0.8	0.52196	0.52196	0.52196
1	0.49898	0.49898	0.49898
1.2	0.47657	0.47657	0.47657
1.4	0.45475	0.45475	0.45475
1.6	0.43356	0.43356	0.43356

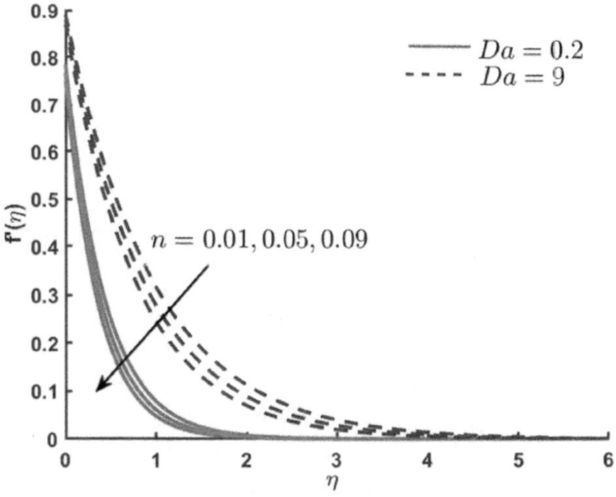

FIGURE 6.2 Effect of n on $f'(\eta)$ for different Da.

curvature, which raises the boundary layer's resistance to fluid motion. Thus, the flow slows down near the wall, producing lower velocity profiles. Similarly, a smaller Darcy number ($\boldsymbol{Da}$) indicates lower permeability of the porous medium, which results in higher drag forces that oppose the flow. Therefore, velocity decreases across the boundary layer results from both enhanced curvature and decreased permeability.

Figure 6.3 illustrates that as the Casson parameter (β) increases, fluid velocity decreases. A larger β indicates stronger non-Newtonian behaviour, which increases the fluid's effective viscosity. This higher viscosity resists deformation and slows the flow, resulting in reduced velocity profiles. Similarly, fluid motion decreases for higher curvature parameter. Due to the fact that a greater curvature enhances surface stretching and increases resistance near the boundary layer. Hence, the fluid motion within the flow region is slowed down by both higher β and **n** values.

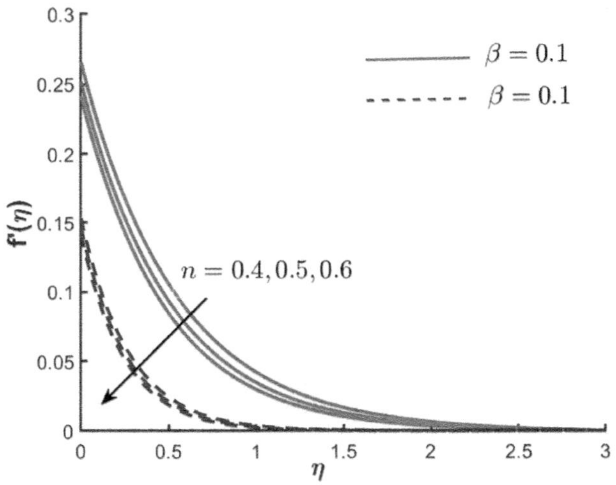

FIGURE 6.3 Effect of n on $f'(\eta)$ for different β.

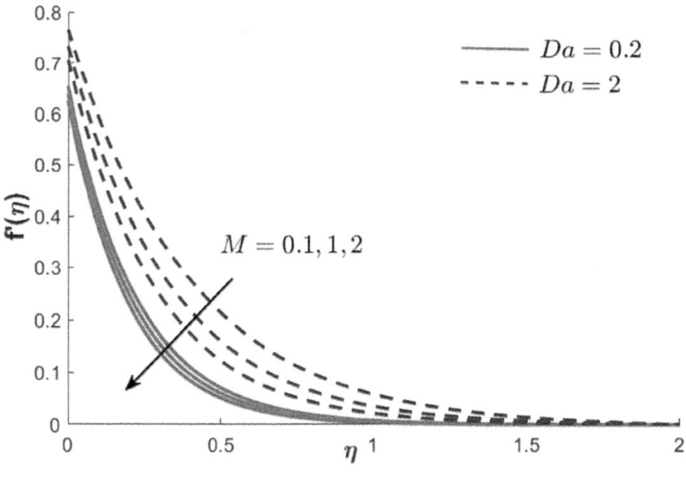

FIGURE 6.4 Effect of M on $f'(\eta)$ for different Da.

Figure 6.4 signifies that as the value of the magnetic parameter $\boldsymbol{M}$ rises, the speed of the fluid reduces. This occurs because the applied magnetic field induces Lorentz forces that act opposite to the direction of flow, creating magnetic drag and slowing down the flow of the electrically conductive fluid. Conversely, when the Darcy number $\boldsymbol{Da}$ increases, it gives rise to fluid velocity. This is because higher $\boldsymbol{Da}$ means a more permeable porous medium with less resistance to flow. Hence, the fluid moves more easily over the porous structure.

As thermophoretic parameter Nt rises, it enhances the fluid's temperature profile, as shown in Figure 6.5. Due to the fact that thermophoresis causes nanoparticles to move from hotter to cooler regions, increasing energy transfer and reducing

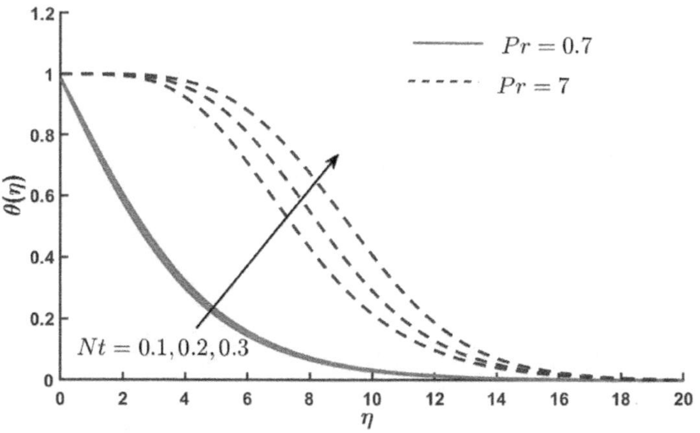

FIGURE 6.5 Effect of Nt on θ for different Pr.

heat loss from the fluid. This energy accumulation enhances the overall temperature. Consequently, the fluid's temperature increases when we increase the value of Prandtl number $\boldsymbol{Pr}$. Th reason is for a greater Prandtl number $\boldsymbol{Pr}$, it implies weaker heat diffusivity compared to momentum diffusivity, which means thermal energy spreads more slowly within the fluid. This low heat conduction retains thermal energy within the fluid, which leads to the enhancement of the thermal profile.

When the Eckert number $\boldsymbol{Ec}$ increases, it enhances the fluid's temperature profile, as demonstrated in Figure 6.6. Because a greater Eckert number implies that more kinetic energy from fluid motion is transformed into internal energy through viscous dissipation, which heats the fluid. Consequently, fluid temperature rises when the value of Prandtl number $\boldsymbol{Pr}$ increases, as it implies lower thermal diffusivity compared to momentum diffusivity. As a result, heat transfers more slowly, allowing thermal energy to accumulate within the fluid. Therefore, a greater $\boldsymbol{Ec}$ increases temperature through viscous heating, a higher $\boldsymbol{Pr}$ decreases heat loss, and the combined effects result in an overall increase in fluid temperature.

Figure 6.7 signifies that for greater values of magnetic parameter $\boldsymbol{M}$, the fluid temperature rises. The magnetic field creates Lorentz force that oppose the fluid velocity, increase viscous dissipation, and decrease convective heat transport. As a result, more energy remains within the fluid, which leads to the increase in temperatures. On the other hand, for a higher Darcy number $\boldsymbol{Da}$ the fluid temperature decreases, since a higher Darcy number means the porous media is more permeable, which makes it easier for fluid flow, which also improves convective heat transfer and promotes better cooling.

Figure 6.8 illustrates how the curvature parameter $\boldsymbol{n}$ affects fluid concentration depending on the Schmidt number $\boldsymbol{Sc}$. At a higher $\boldsymbol{Sc} = 1.5$, mass diffusivity is lower, so species diffuse slowly. In this case, increasing curvature strengthens surface stretching and limits convective mass transport away from the boundary, raising concentration near the surface. In contrast, at a lower $\boldsymbol{Sc} = 0.2$, mass diffusivity is

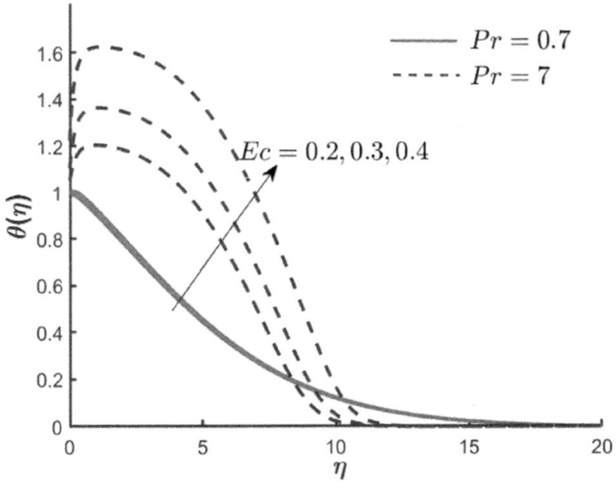

FIGURE 6.6 Effect of Ec on θ for different Pr.

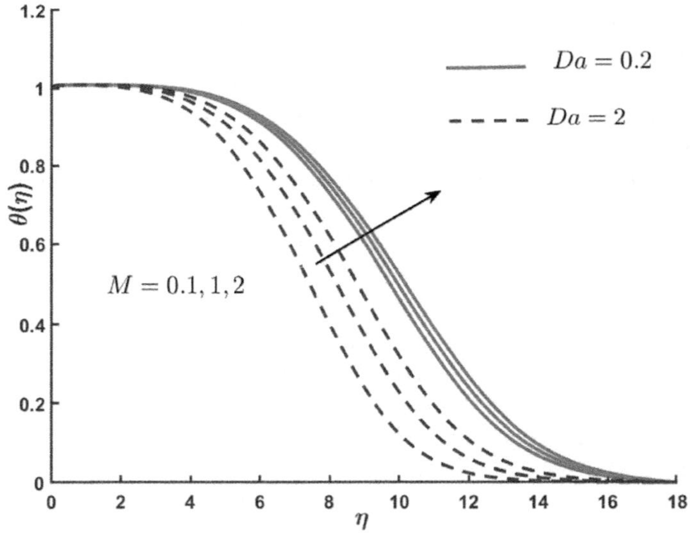

FIGURE 6.7 Effect of Da on θ for different M.

higher, allowing species to spread more easily. Here, greater curvature enhances the species transport away from the surface, reducing concentration. Therefore, the collective influence of curvature-driven convection and mass diffusion explains why the same increase in $\boldsymbol{n}$ can produce opposite concentration trends at different Schmidt numbers.

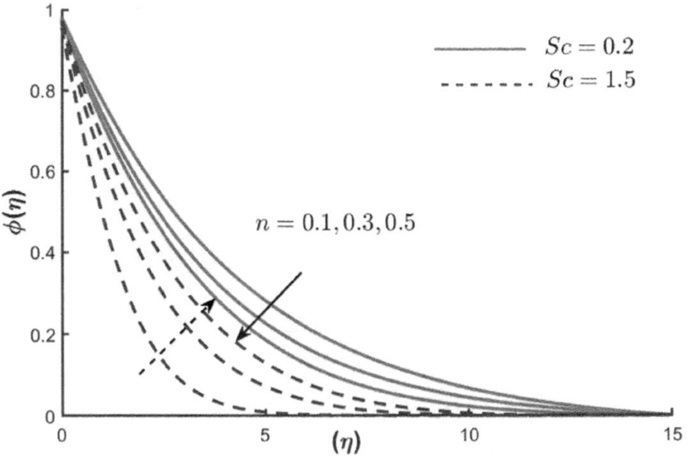

FIGURE 6.8 Effect of Sc on ϕ for different n.

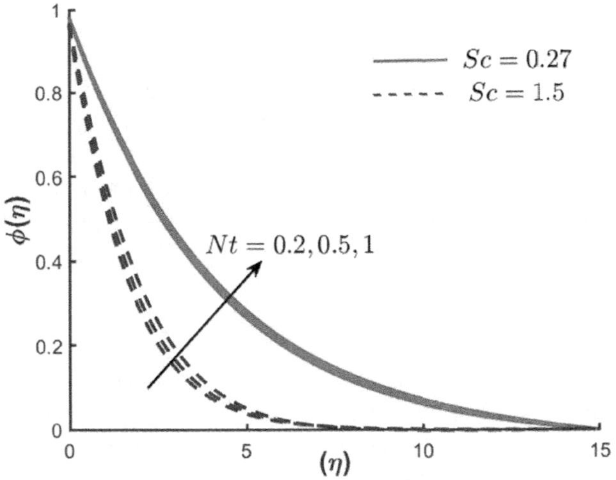

FIGURE 6.9 Effect of Sc on ϕ for different Nt.

Figure 6.9 illustrates that the fluid concentration increases with the thermophoretic parameter $\boldsymbol{Nt}$, as thermophoresis drives nanoparticles or solute particles from warmer regions to cooler regions, which leads to local buildup near the surface. Conversely, increasing the Schmidt number $\boldsymbol{Sc}$ reduces concentration because when $\boldsymbol{Sc}$ increases it indicates that mass diffusion is weaker relative to momentum diffusion, slowing down the spread of species and limiting their accumulation near the surface. In essence, thermophoresis enhances mass buildup through directed particle motion, while a larger Schmidt number restricts it by reducing diffusion efficiency.

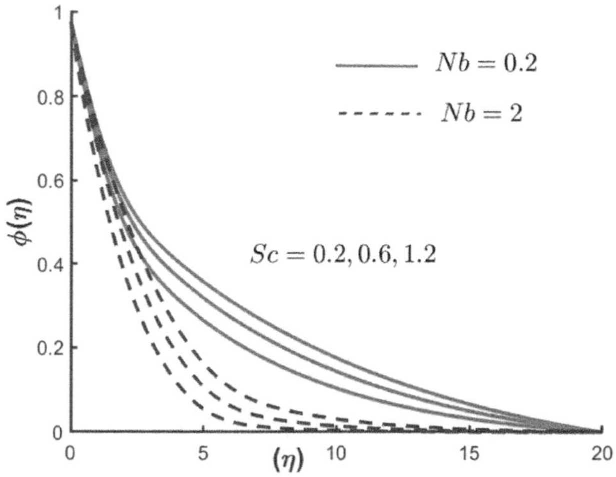

FIGURE 6.10 Effect of Sc on ϕ for different Nb.

Figure 6.10 shows that the fluid concentration reduces when we increase the Brownian motion parameter $\boldsymbol{Nb}$. Stronger Brownian motion improves the arbitrary nanoparticle movement, causing them to disperse away from high-concentration regions and lowering the local concentration near the surface. Similarly, increasing the Schmidt number $\boldsymbol{Sc}$ also reduces concentration, as a larger $\boldsymbol{Sc}$ indicates that mass diffusivity is lower relative to momentum diffusivity, limiting the ability of species to diffuse and accumulate. In summary, both improved Brownian motion and higher Schmidt number decrease fluid concentration by promoting particle dispersion and limiting mass transport.

6.7 CONCLUSION

This study examined the impacts of magnetic fields, surface curvature, porous media, and nanoparticles on mass transfer, thermal energy transfer, and fluid flow. The results show that crucial roles of prime parameters like Casson number, Darcy number, Prandtl number, Schmidt number, thermophoresis, Brownian motion, and Eckert number in determining momentum, energy and concentration profiles, and the key significant conclusions are drawn as follows:

- The combined presence of non-Newtonian characteristics (Casson parameter) and viscous dissipation (Eckert number) raises flow resistance and promotes additional heating within the boundary layer.
- Thermophoresis raises both temperature and concentration by pushing nanoparticles toward cooler regions.
- Through enhanced convective heat removal, a higher Darcy number increases velocity and reduces fluid.

- Higher Brownian motion spreads nanoparticles evenly, which lowers the wall concentration.
- Thermal and mass transfer are strongly affected by the surface curvature, where higher curvature thickens the thermal boundary layer. It shows greater influence for different Prandtl and Schmidt numbers.
- The investigation shows that in Brownian fluid-based thermal systems, fluid velocity, thermal energy, and mass transfer can be effectively managed by adjusting these parameters.
- The velocity slip conditions weaken the wall shear stress and moderate fluid velocity, which provides flow control in microfluidic and biomedical applications.
- Accurate and stable solutions are obtained using the finite difference method, validated against already published studies.
- This investigation reveals that energy and mass transfer in devices like cooling, polymer processing, and drug delivery can be regulated by adjusting thermophoretic force, Brownian motion, and magnetic effects.

Further work can look at hybrid nanofluid flow in different geometries, turbulent and unstable flows with changing magnetic fields. Uses in advanced cooling systems, microfluidic, and biomedical devices can be extended by combining chemical reactions, multi-phase flows, or bio-convection effects.

NOMENCLATURE

Symbols	Description (SI unit)	Symbols	Description (SI unit)
M	Hartmann number [–].	α	Thermal diffusivity (m^2s^{-1}).
Bi_1	Thermal Biot number [–].	ρC_p	heat capacitance of fluid (J/m^3K).
Bi_2	Concentration Biot number [–].	$(\rho C_p)_f$	Heat capacity of the base fluid (J/m^3K).
α_1	First order velocity slip parameter (m).	D_B	Brownian motion coefficient (m^2/s).
n	Curvature parameter [–].	D_T	Thermophoretic diffusion coefficient (m^2/s).
Sc	Schmidt number [–].	κ	Thermal conductivity (W/m.K).
Rd	Radiation parameter [–].	τ	Stress tensor (Pa).
Pr	Prandtl number [–].	P_y	Yield stress (Pa).
Nt	Thermophoresis parameter [–].	C_p	Specific heat [–].
Nb	Brownian motion parameter [–].	η	Similarity variable [–].
σ_s	Stephen Boltzmann constant ($W.m^{-2}K^{-4}$).	β	Non-Newtonian fluid parameter [–].

k_e	Mean absorption coefficient (m^{-1}).	Ec	Eckert number [–].
q_r	Radiative heat flux ($W.m^{-2}$).	U_w	Stretching velocity (ms^{-1}).
a	Fixed number [–].	U_∞	Free stream velocity (ms^{-1}).
l	Characteristic length (m).	T_∞	Constant temperature (K).
(u, v)	Velocity component of the fluid along x and r direction (ms^{-1}).	C_∞	Constant concentration (kg/m^3).
B_0	Magnetic field strength (T).	CuO	Copper oxide [–].
ν_f	Kinematic viscosity of the fluid (m^2s^{-1}).	Fe_2O_3	Iron oxide [–].
σ	Electric conductivity (S / m).	Cu	Copper [–].
μ_f	Dynamic viscosity of the fluid (Pa.s or $kg.m^{-1}s^{-1}$).	Ag	Silver [–].
ρ_f	Fluid density (kgm^{-3}).	C_w	Concentration of the wall ($molm^{-3}$).
T	Surface temperature parameter (K).	T_w	Wall temperature (K).

REFERENCES

Akbar, N. S., Ebaid, A., & Khan, Z. H. (2015). Numerical analysis of magnetic field effects on Eyring-Powell fluid flow towards a stretching sheet. *Journal of magnetism and Magnetic Materials, 382*, 355–358.

Ali, A., Kanwal, T., Awais, M., Shah, Z., Kumam, P., & Thounthong, P. (2021). Impact of thermal radiation and non-uniform heat flux on MHD hybrid nanofluid along a stretching cylinder. *Scientific Reports, 11*(1), 20262.

Buongiorno, J. (2006). *Convective transport in nanofluids. ASME Journal of Heat and Mass Transfer.*

Choi, S. U. S. (1995). *Enhancing thermal conductivity of fluids with nanoparticle.* The Proceedings of the 1995 ASME International Mechanical Engineering Congress and Exposition, San Francisco, pp. 99–105.

Dang, K., Makkar, V., & Sharma, N. (2024). Radiative MHD casson non-Newtonian nanofluid slip flow induced by stretching cylinder: A numerical approach. *Indian Journal of Science and Technology, 17*(38), 3993–4004.

Das, B., & Ahmed, S. (2025). Numerical Modelling in Heat Transfer Analysis of SWCNT/EG Nanofluid in a Stretching Sheet: A Darcy–Forchheimer Model. In *Advances in topology, dynamical systems and interdisciplinary applications* (pp. 211–224). Singapore: Springer Nature Singapore.

Gnaneswara Reddy, M., Manjula, J., & Padma, P. (2017). Mass transfer flow of MHD radiative tangent hyperbolic fluid over a cylinder: A numerical study. *International Journal of Applied and Computational Mathematics, 3*(Suppl 1), 447–472.

Hayat, T., Gull, N., Farooq, M., & Ahmad, B. (2016). Thermal radiation effect in MHD flow of Powell—Eyring nanofluid induced by a stretching cylinder. *Journal of Aerospace Engineering, 29*(1), 04015011.

Hazarika, S., Ahmed, S., & Chamkha, A. J. (2021). Investigation of nanoparticles Cu, Ag and Fe_3O_4 on thermophoresis and viscous dissipation of MHD nanofluid over a stretching sheet in a porous regime: a numerical modeling. *Mathematics and Computers in Simulation*, *182*, 819–837.

Hussain, M. A., & Ahmed, S. (2023). Dual solution of heat transfer in hydromagnetic micropolar fluid flow over a stretching sheet in a porous media: A numerical approach. *Heat Transfer*, *52*(2), 1215–1231.

Kabra, Y., & Sharma, V. K. (2025). MHD boundary layer flow of hybrid nanofluids over a porous stretching sheet with thermal radiation: A three-dimensional perspective. *Dandao Xuebao / Journal of Ballistics*, 37(1), 164–175.

Kaye, R., & Ahmed, S. (2024). Melting heat transfer and variable viscosity of a hydromagnetic flow over a stretching/shrinking sheet: numerical approach. *JP Journal of Heat and Mass Transfer*, *37*(2), 185–199.

Malik, M. Y., Salahuddin, T., Hussain, A., & Bilal, S. (2015). MHD flow of tangent hyperbolic fluid over a stretching cylinder: using Keller box method. *Journal of Magnetism and Magnetic Materials*, *395*, 271–276.

Muhammad Raza Shah Naqvi, S., Manzoor, U., Waqas, H., Liu, D., Naeem, H., Eldin, S. M., & Muhammad, T. (2024). Numerical investigation of thermal radiation with entropy generation effects in hybrid nanofluid flow over a shrinking/stretching sheet. *Nanotechnology Reviews*, *13*(1), 20230171.

Reddy, M. G., Kumari, P. V., & Padma, P. (2018). Effect of thermal radiation on MHD casson nano fluid over a cylinder. *Journal of Nanofluids*, *7*(3), 428–438.

Rehman, K. U., Qaiser, A., Malik, M. Y., & Ali, U. (2017). Numerical communication for MHD thermally stratified dual convection flow of Casson fluid yields by stretching cylinder. *Chinese Journal of Physics*, *55*(4), 1605–1614.

Soomro, A. M., Fadhel, M. A., Lund, L. A., Shah, Z., Alshehri, M. H., & Vrinceanu, N. (2024). Dual solutions of magnetized radiative flow of Casson Nanofluid over a stretching/shrinking cylinder: Stability analysis. *Heliyon*, *10*(8), e29696.

7 Non-Newtonian Fluid Dynamics

Satish Kumar, Satya Deo, Biswajit Mallick, and Pawan Kumar Patel

7.1 INTRODUCTION

The theory of non-Newtonian fluids has evolved gradually and intensively, driven by scientific findings, mathematical modeling, and technological advances, beginning in the 17th century after the elucidation of the law of viscosity. The law states that there exists a linear proportionality constant between shear stress and strain rate – which was defined by Sir Isaac Newton (Batchelor, 2000) and named them Newtonian fluids. However, various real-world fluids like blood, toothpaste, paints, starch pastes, molten plastics, and slurries revealed behaviors incompatible with Newton's assumptions for fluids, such as shear-thinning, yield stress, and time-dependent viscosity during the late 19th and early 20th centuries. Eugene C. Bingham reacted by explaining the behavior of some non-Newtonian fluids (Bingham, 1917) and developing a model to characterize materials that flow only above a specific stress, known as yield stress. Bingham and Reiner (Nae, 1993) introduced the term *Rheology*, a combination of Greek words "Rheo" (*to flow*) and "logos" (*study*), to describe the flow of materials due to deformation under the influence of applied forces.

The development of various constitutive models in the mid-20th century stimulated theoretical refinement. Ostwald and De Waele (Richardson, 2011) developed the power-law fluid model to describe fluids with shear-dependent viscosity, while Casson (Khan *et al.*, 2021) proposed the Casson model in 1959 to explain yield stress in substances like blood and printer inks. Rivlin and Ericksen (1997) developed a class of tensors known as Rivlin–Ericksen tensors to analyze the kinematic behavior of real-world fluids. These tensors provide a mathematical framework for explaining memory effects, elastic behavior, and nonlinear flow in complex fluids. The Rivlin–Ericksen tensors, a sequence of symmetric tensors used in continuum mechanics, play a crucial role in describing non-Newtonian and viscoelastic fluids, like second-grade and third-grade fluids. In 1960, Noll and Coleman (Indiana University, 1952) formalized second-grade fluids within a thermodynamic framework in which the stress tensor not only depends on the rate of strain but also on its time derivatives via Rivlin–Ericksen tensors, effectively capturing normal stress differences, elastic recoil, and other viscoelastic behaviors in polymeric and biological fluids. Pierre Carreau (1972) devised a fluid model in 1972 with shear-thinning behavior across a

DOI: 10.1201/9781003664345-7

wide range of shear rates, which became known as the Carreau fluid model, which is widely used in biofluid modeling and polymer science.

Alongside these advancements, viscoelastic models like Maxwell, Oldroyd-B, and Jeffery were developed to include material memory, stress relaxation, and retardation effects – providing the foundation for modern non-Newtonian fluid mechanics. Eringen (1966) introduced micropolar fluids in 1966, which have a linear relation between stress and rate of strain but are regarded as non-Newtonian due to their own microstructure effects such as microrotation, couple stresses, and stress asymmetry – especially in the modeling of complex fluids such as suspensions and biological flows. Stokes (1984) established the couple stress theory to study microscopic effects such as rotational interactions and couple stress in complex structured fluids; such fluids are called couple stress fluids. The simulation of complex flows has become possible due to the development of computational rheology during the last two decades of the 20th century. Recently, several contemporary models have been constructed to define the behavior of biofluids, nanofluids, and fractional-order viscoelasticity influenced by electromagnetic fields and non-local effects, initially by Rajagopal and Srinivas (2000) and later by Chhabra and Richardson (2008). Thus, non-Newtonian fluid theory continues to evolve at the interface of mathematics, physics, and engineering.

Porosity, magnetic fields, heat transport, and multi-phase interactions are some of the phenomena that researchers have incorporated into classical fluid models in recent years to investigate more realistic scenarios. For instance, Maurya *et al.* (2024) analyzed hydromagnetic behavior of Casson fluid across a stretching sheet through a porous medium. Patel *et al.* (2025) investigated hydromagnetic couple stress flows through porous spheres in layered porous media. The authors of the book chapter have made a substantial contribution to this research field: Kumar *et al.* (2022) examined MHD flow of a second-grade fluid surrounded by Newtonian layers in a porous medium; Kumar and Deo (2025) provided an exact solution for heat transfer in magnetohydrodynamic Jeffery fluid flow employing the traveling wave methodology; and Kumar *et al.* (2025) examined magnetohydrodynamic transport of non-miscible non-Newtonian fluids in a porous cylindrical annulus. Mallick (2024) conducted a comprehensive numerical investigation on electrokinetic energy conversion in microfluidic systems, incorporating the couple stress fluid model to capture the microstructural effects of biomolecules and electrolytes. The results highlighted that the non-Newtonian couple stress effects significantly influence flow characteristics, streaming potential generation, and overall energy conversion efficiency. In another study, Mallick and Misra (2019) examined the electromagnetic effects on peristaltic transport of Powell-Eyring fluid, employing a perturbation method for the analytical formulation. Extending the classical Fourier heat flux model, Misra *et al.* (2018) numerically analyzed the influence of Cattaneo-Christov heat flux on the electrothermofluidic transport of third-order fluids in the presence of a magnetic field. Using a local thermal equilibrium (LTE) model, Mallick *et al.* (2023) mathematically analyzed steric induced electrothermal flow of micropolar fluid. By performing entropy generation analysis, the study demonstrated that the global entropy of the microfluidic system can be effectively minimized through the regulation of the micropolarity parameter, thereby offering potential applications in the development of thermofluidic micropumps and biomicrofluidic diagnostic devices. Collectively, these studies highlight the significance of MHD effects, electroosmosis, porosity, multiphase interactions, and non-Newtonian rheology, along with the associated heat transfer phenomena, as key directions in contemporary fluid dynamics research.

7.2 CLASSIFICATION OF NON-NEWTONIAN FLUIDS

Non-Newtonian fluids can be grouped based on their predominant material properties. Broadly, these fluid models fall into three main categories:

1. Time-independent fluids
2. Time-dependent fluids
3. Viscoelastic fluids

7.2.1 Time-Independent Fluids

This group may further be divided into two subgroups, viz.

1. **Purely viscous fluids:** A purely viscous fluid is a fluid whose shear stress is directly proportional to the shear rate at every point of the fluid. These fluids exhibit no elastic behavior and respond instantly to applied forces. In simple shear flow, the constitutive equations of purely viscous fluids are

$$\tau = \mu(\dot{\gamma})\dot{\gamma}, \tag{7.1}$$

where τ is the stress tensor, $\mu(\dot{\gamma})$ the apparent viscosity, $\dot{\gamma} = \nabla\vec{v} + (\nabla\vec{v})^T$ the rate of deformation tensor, and $\dot{\gamma} = \sqrt{\frac{1}{2}\dot{\gamma} : \dot{\gamma}}$ the scalar shear rate. Non-Newtonian purely viscous fluids can be categorized into two classes, viz.

(a) **Shear-Thinning or Pseudoplastic Fluids:** A purely viscous fluid is classified as a shear-thinning or pseudoplastic fluid if its apparent viscosity decreases with increasing shear rate. Most real non-Newtonian fluids, such as polymer melts, polymer solutions, biological fluids, and mayonnaise, exhibit shear-thinning behavior. The term "**pseudoplastic**" reflects the similarity between the viscosity function of a shear-thinning fluid and that of viscoplastic fluid models. Although shear-thinning fluids do not possess a true yield stress, their viscosity response mimics that of viscoplastic materials in the sense that both exhibit a nonlinear, decreasing stress-strain rate relationship.

(b) **Shear-Thickening Fluids or Dilatant Fluids:** For a very few real fluids, the apparent viscosity increases with increasing shear rate. Such fluids are known as shear-thickening fluids, or dilatant fluids. The term "**dilatant**" reflects the fact that these fluids often expand in volume when subjected to shear stresses.

Table 7.1 provides a comprehensive overview of various mathematical models for purely viscous fluids and their constitutive relations.

2. **Viscoplastic Fluids:** Viscoplastic materials are a special type of materials that behave like solids under small magnitudes of force and like fluids under large magnitudes of force. This unique behavior is governed by a property called the yield shear stress, denoted by τ_0. When the applied shear stress τ is less than the yield stress τ_0, the material behaves like a rigid solid and does not flow. In this state, it may undergo slight elastic deformation. On the other hand, when the applied shear stress exceeds the yield value $(\tau > \tau_0)$, the material begins to flow like a fluid. This transition from solid-like to fluid-like behavior is characteristic of viscoplastic fluids. Table 7.2 presents a summary of the different mathematical models describing viscoplastic fluids along with their corresponding constitutive equations.

7.2.2 Time-Dependent Fluids

In general, the apparent viscosity of a fluid is influenced not only by the instantaneous shear rate but also by the period over which the fluid has been exposed to shear. More precisely, fluids whose apparent viscosity decreases or increases over time under applied shear along with strain rate fall into this category. Time-dependent fluids are further sub-divided into two groups:

1. **Thixotropic Fluids**: Fluids, such as paint, yogurt, ketchup, drilling mud, and printing inks, have a time-dependent shear thinning property, called thixotropic property. Fluids with thixotropic properties show a reduction in viscosity over time when subjected to constant shear and become less viscous over time. Once the shear is removed, the viscosity slowly recovers to its original state.
2. **Rheopectic fluids**: Rheopectic fluids behave in a manner opposite to that of thixotropic fluids; that is, such fluids exhibit a time-dependent shear thickening property, called rheopecty property. The viscosity of rheopectic fluids increases gradually with time when subjected to a constant shear stress or deformation. A rheopectic fluid becomes thicker over time upon constant shearing. Gypsum pastes, polymer suspensions, and lubricants are some well-known examples of rheopectic fluids.

The constitutive equations for this type of fluid are of a similar form as Equation (7.1) with the addition of a structure parameter λ_s in the apparent viscosity as follows:

TABLE 7.1
Various Mathematical Models for Purely Viscous Fluids and Their Constitutive Relations

Model	Apparent Viscosity	Key Characteristics
Power law (Ostwald–de Waele) (Shit *et al.*, 2016)	$\mu(\dot{\gamma}) = K\lvert\dot{\gamma}\rvert^{n-1}$; K : Flow consistency index, n : Flow behavior index	Shear-thinning when $n<1$; Newtonian at $n=1$; Shear-thickening when $n>1$
Carreau/Bird–Carreau (Ullah and Ashraf, 2024)	$\mu(\dot{\gamma}) = \mu_\infty + \frac{(\mu_0 - \mu_\infty)}{\left[1+(\lambda\dot{\gamma})^2\right]^{(1-n)/2}}$; μ_0 : Finite viscosity at zero shear rate, n: Power index, μ_∞ : Viscosity at infinite shear rate, λ: Characteristic time	Effectively captures a shear-rate-dependent viscosity exhibiting Newtonian behavior in both the low- and high-shear-rate limits, with a shear-thinning (or thickening) region in between.
Carreau–Yasuda (Yasuda, 1979)	$\mu(\dot{\gamma}) = \mu_\infty + \frac{(\mu_0 - \mu_\infty)}{\left[1+(\lambda\dot{\gamma})^a\right]^{(1-n)/a}}$; a : Controlling parameter for transition region between the low-shear-rate Newtonian plateau and the power-law region	Generalized five parameter Carreau model; effectively describe shear-thinning fluids; applicable in polymer processing, biomedical engineering
Cross (Cross, 1965)	$\mu(\dot{\gamma}) = \mu_\infty + \frac{\mu_0 - \mu_\infty}{1+(\lambda\dot{\gamma})^{1-n}}$	Flexible across the entire range of shear stress. When $n=0$, cross model reduces to **Williamson model** (Williamson, 1929).
Sisko (Khan *et al.*, 2010)	$\mu(\dot{\gamma}) = a + b\lvert\dot{\gamma}\rvert^{n-1}$; a : Newtonian viscosity parameter; b: consistency index; n: flow behavior index	Blends Newtonian and non-Newtonian behavior; smoothly transit to power-law-type behavior
Ellis (Pal *et al.*, 2025)	$\mu(\dot{\gamma}) = \frac{\mu_0}{1+(\tau/\tau_{1/2})^{n-1}}$; μ_0 : Viscosity at zero shear, τ: Shear stress; $\tau_{1/2}$: Shear stress at $\mu = \mu_0/2$, n: Power law index	A higher value of n indicates a more pronounced shear-thinning behavior of the fluid; predicts Newtonian fluid behavior in the limit $\tau_{1/2} \to \infty$

(*Continued*)

TABLE 7.1 (CONTINUED)
Various Mathematical Models for Purely Viscous Fluids and Their Constitutive Relations

Model	Apparent Viscosity	Key Characteristics
Prandtl (Soomro *et al.*, 2017)	$\mu_e(\dot{\gamma}) = \mu_0 \frac{\sin^{-1}\left(\lambda\dot{\gamma}\right)}{\lambda\dot{\gamma}}$; μ_0 : Finite viscosity at zero shear rate, λ: Time relaxation parameter	A nonlinear rheological model particularly well-suited to high-shear applications involving polymers and viscous industrial fluids
Eyring (Qi and Ng, 2017)	$\mu_e(\dot{\gamma}) = \mu_0 \frac{\sinh^{-1}(\lambda\dot{\gamma})}{\lambda\dot{\gamma}}$; μ_0 : Finite viscosity at zero shear rate; λ: Time relaxation parameter	Simple shear-thinning fluids based on molecular rearrangement
Eyring Powell (Yoon and Ghajar, 1987)	$\mu_e(\dot{\gamma}) = \mu_\infty + (\mu_0 - \mu_\infty)\frac{\sinh^{-1}(\lambda\dot{\gamma})}{\lambda\dot{\gamma}}$; μ_∞ : viscosity at infinite shear rate	Captures Newtonian and non-Newtonian transitions

$$\tau = \mu(\lambda_s, \dot{\gamma})\dot{\gamma}. \tag{7.2}$$

Here, λ_s reflects the internal structure of the material and varies with shear. The evolution of the structure parameter is frequently represented by a kinetic equation, such as Fredrickson's model (Fredrickson, 1970) (Table 7.3), which accounts for both the breakdown and the build-up of the material structure.

7.2.3 Viscoelastic Fluids

Viscoelastic fluids are those that exhibit dual behavior – viscous as a liquid, and elastic as a solid when subjected to a deformation. Viscoelastic fluids are capable of storing a part of energy as elastic energy during deformation. Viscoelastic fluids can deform as well as flow in response to stress, but also recover partially when the stress is removed (elastic behavior). The behavior of such fluids is time-dependent due to the memory effect. Mucus, gels, toothpaste, molten plastics, etc., are a few real-life examples of viscoelastic fluids. Some important mathematical models along with their constitutive relations are listed in Table 7.4.

7.2.4 Microstructured Fluids

Non-Newtonian fluids, which exhibit internal structures – such as suspended particles, polymers, or micelles – that substantially influence flow behavior, fall under the

TABLE 7.2
Various Mathematical Models for Viscoplastic Fluids and Their Constitutive Relations

Model	Constitutive Equation	Key Charactertistics
Bingham Plastic (Bingham, 1922)	$\tau = \left(\mu + \frac{\tau_0}{\lvert\dot{\gamma}\rvert}\right)\dot{\gamma};\ \text{if}\ \lvert\tau\rvert > \tau_0,$ $\dot{\gamma} = 0;\ \text{if}\ \lvert\tau\rvert \le \tau_0$ τ_0 : Yield stress, μ : Plastic viscosity	Requires a minimum shear stress τ_0 to start flow; apparent viscosity is shear-thinning type
Herschel–Bulkley (Herschel and Bulkley, 1926)	$\tau = \left(K\lvert\dot{\gamma}\rvert^{n-1} + \frac{\tau_0}{\lvert\dot{\gamma}\rvert}\right)\dot{\gamma};\ \text{if}\ \lvert\tau\rvert > \tau_0,$ $\dot{\gamma} = 0;\ \text{if}\ \lvert\tau\rvert \le \tau_0$ K : Consistency index, n : Power law index	Shear-thinning when $n < 1$ and shear-thickening when $n > 1$; reduces to Bingham plastic fluid when $n = 1$; reduces to power law fluid when $\tau_0 = 0$
Casson (Casson, 1959)	$\tau = \left(\sqrt{\mu\lvert\dot{\gamma}\rvert} + \sqrt{\tau_0}\right)^2 \frac{\dot{\gamma}}{\lvert\dot{\gamma}\rvert};\ \text{if}$ $\lvert\tau\rvert > \tau_0,$ $\dot{\gamma} = 0;\ \text{if}\ \lvert\tau\rvert \le \tau_0$ τ_0 : Yield stress, μ : Plastic viscosity	At very high shear rates, apparent viscosity approaches a constant, but at moderate rates it varies nonlinearly with shear rate.
Papanastasiou model (Papanastasiou, 1987)	$\tau = \left(\mu + \frac{\tau_0}{\lvert\dot{\gamma}\rvert}\left(1 - e^{-m\dot{\gamma}}\right)\right)\dot{\gamma};\ \text{if}$ $\lvert\tau\rvert > \tau_0,$ $\dot{\gamma} = 0;\ \text{if}\ \lvert\tau\rvert \le \tau_0$ m : Controlling parameter for the exponential growth of stress	Removes the sharp yield point of Bingham plastic model and offer a smooth, continuous stress–strain relationship valid for all shear rates
Robertson–Stiff model (Ferrás *et al.*, 2012)	$\tau = \left(K^{1/n}\lvert\dot{\gamma}\rvert^{(n1)/n} + \left(\frac{\tau_0}{\lvert\dot{\gamma}\rvert}\right)^{1/n}\right)^n \dot{\gamma};$ $\text{if}\ \lvert\tau\rvert > \tau_0,$ $\dot{\gamma} = 0;\ \text{if}\ \lvert\tau\rvert \le \tau_0$ K : Consistency index, n : Power law index	Provides a good fit for fluids that display both yield stress and shear-thinning (pseudo-plastic) properties

category of microstructured fluids. Microstructured fluids are capable of explaining phenomena such as shear thinning, normal stress differences, or couple stresses that cannot be elucidated by means of Newtonian fluids. Eringen (1966) – micropolar fluid theory and Stokes (1984) – couple stress fluid theory have developed advanced

TABLE 7.3
Basic Mathematical Relations for Evolution of Structure Parameter for Time-Dependent Fluids

Model	Structure Parameter Equation	Main Features/Examples
Thixotropic (Larson, 2015)	$\frac{d\lambda_s}{dt} = -a\lambda_s\lvert\dot{\gamma}\rvert + b(1-\lambda_s)$; a : Constant rate for shear-induced breakdown, b : Constant rate for structural build-up	Apparent viscosity decreases over time; internal microstructure of the fluid breaks down with continuous stress; ketchup, inks
Rheopectic (Bautista *et al.*, 1999)	$\frac{d\lambda_s}{dt} = +a\lambda_s\lvert\dot{\gamma}\rvert - b(1-\lambda_s)$; a : Constant rate for shear-induced breakdown, b : Constant rate for structural build-up	Apparent viscosity increases over time; internal microstructure augments over time; Gypsum paste, lubricants

theories for capturing the flow behavior of microstructured fluids, which play a vital role in the study of biological fluid, polymer solutions, and nanofluids with prominent microstructural effects (Table 7.5).

7.3 GOVERNING EQUATIONS OF MOTION

Dynamics of fluids, whether Newtonian or non-Newtonian, is governed by the basic principles of continuum mechanics specifically, conservation of mass, linear momentum, angular momentum (if applicable), and energy. These principles do not create challenges during the modeling of non-Newtonian fluids. However, it is very challenging to incorporate how fluids respond to their deformation. Non-Newtonian fluids exhibit complex and frequently non-linear relationships, necessitating extensive constitutive modeling, whereas Newtonian fluids have a linear and simple relationship between stress and strain rate.

In this section, we shall address the general forms of the governing equations of motion for incompressible fluid flows. Further, we will discuss how these equations are modified for diverse non-Newtonian fluids using their constitutive models.

7.3.1 Continuity Equation

The equation of continuity for a fluid motion is obtained by utilizing the principle of conservation of mass. It is valid for all fluid types, regardless of their rheological behavior, because it ensures that mass cannot be produced nor destroyed within the

TABLE 7.4
Various Mathematical Models for Viscoelastic Fluids and Their Constitutive Relations

Model	Constitutive Equation	Key Characteristics/Examples
Maxwell (Salahuddin *et al.*, 2022)	$\tau + \lambda_1 \frac{D\tau}{Dt} = \mu\dot{\gamma}$; λ_1 : Relaxation time; μ : Viscosity of fluid; τ : Deviatoric stress tensor; $\frac{D}{Dt}$: Material derivative	Exhibit elastic resistance to quick deformations but viscous nature over longer timescales; polymer melts and biological fluids
Upper Convective Maxwell fluid (Mederos *et al.*, 2024)	$\tau + \lambda_1 \overset{\nabla}{\tau} = \mu\dot{\gamma}$; $\overset{\nabla}{\tau}$: Upper-convected time derivative of τ	Extends the classical Maxwell model to account for large deformation flows; polymer engineering, lubrication, and computational rheology
Linear Phan-Thein-Tanner (PTT) fluid (Walait *et al.*, 2024)	$\left(1 + \frac{\varepsilon\lambda_1}{\mu} tr(\tau)\right)\tau + \lambda_1 \overset{\nabla}{\tau} = \mu\dot{\gamma}$; $\overset{\nabla}{\tau}$: Upper-convected time derivative of τ ε: PTT fluid parameter	Modified stress relaxation model for stress magnitude, improved prediction accuracy at high deformation rates
Exponential Phan-Thein-Tanner (PTT) fluid (Walait *et al.*, 2024)	$\left(\exp\left(\frac{\varepsilon\lambda}{\mu}\right) tr(\tau)\right)\tau + \lambda_1 \overset{\nabla}{\tau} = \mu\dot{\gamma}$; $\overset{\nabla}{\tau}$: Upper-convected time derivative of τ ε: PTT fluid parameter	Introduces stronger nonlinearities and fits experimental data better at high deformations
Kelvin–Voigt (Bulıcek *et al.*, 2012)	$\tau = E\gamma + \mu\dot{\gamma}$; E : Elastic modulus; γ: Shear strain	Describes creep behavior
Jeffery (Bird *et al.*, 1987)	$\tau + \lambda_1 \frac{D\tau}{Dt} = \mu\left(\dot{\gamma} + \lambda_2 \frac{D\dot{\gamma}}{Dt}\right)$; λ_1 : Relaxation time; λ_2 : Retardation time	Exhibits both relaxation and retardation times, allowing it to capture creep and stress-relaxation behaviors, which makes it especially suitable for modeling biological and polymeric fluids.
Oldroyd-B (Bird *et al.*, 1987)	$\tau + \lambda_1 \overset{\nabla}{\tau} = \mu\left(\dot{\gamma} + \lambda_2 \overset{\nabla}{\dot{\gamma}}\right)$; λ_1 : Relaxation time, λ_2 : Retardation time	Viscoelastic, frame invariant, constant shear viscosity, nonzero first normal stress difference. It is especially significant in modeling polymer solutions and biofluids.

(Continued)

TABLE 7.4 (CONTINUED)
Various Mathematical Models for Viscoelastic Fluids and Their Constitutive Relations

Model	Constitutive Equation	Key Characteristics/Examples
Oldroyd 4-constant fluid (Bird *et al.*, 1987)	$\tau + \lambda_1 \overset{\nabla}{\tau} + \frac{\lambda_5}{2}\left(tr\tau\right)\dot{\gamma} = \mu\left(\dot{\gamma} + \lambda_2 \overset{\nabla}{\dot{\gamma}}\right)$; λ_5: Coupling parameter linking trace of stress and rate of strain	Extension of Oldroyd-B model by adding coupling parameter linking stress trace $\left(tr\tau\right)$ to the rate-of-strain tensor (important in compressible and nonlinear flows).
Oldroyd 8-constant fluid (Bird *et al.*, 1987)	$\tau + \lambda_1 \overset{\nabla}{\tau} + \frac{1}{2}\lambda_5\left(tr\,\tau\right)\dot{\gamma}$ $+\frac{1}{2}\lambda_3\left(\tau\cdot\dot{\gamma} + \dot{\gamma}\cdot\tau\right)$ $+\frac{1}{2}\lambda_7\left(\dot{\gamma}:\dot{\gamma}\right)\mathbf{I}$ $+\frac{1}{2}\lambda_6\left(\tau:\dot{\gamma}\right)\mathbf{I}$ $= \mu\left\{\dot{\gamma} + \lambda_2 \overset{\nabla}{\dot{\gamma}} + \lambda_4\left(\dot{\gamma}\cdot\dot{\gamma}\right)\right\}$; λ_3 : Non-linear coupling parameter between stress and rate of strain which controls anisotropic effect, λ_4 : Quadratic strain rate parameter which describes shear thinning effect and elongational thickening, λ_6 : Non-linear coupling parameter between stress and rate of strain which controls isotropic effect , λ_7 : Isotropic quadratic term parameter, contributes to normal stress differences and extensional thickening	Most general thermodynamically consistent viscoelastic fluid model: includes memory, retardation, trace effects, nonlinear stress–strain couplings, normal stress differences, shear-thinning, extensional thickening.
Second Grade Fluid (Kumar *et al.*, 2022)	$\tau = \mu\dot{\gamma} + \alpha_1\dot{\gamma}_{(1)} + \alpha_2\left(\dot{\gamma}\right)^2$; α_1 : Elasticity parameter, α_2 : Cross viscosity parameter, $\dot{\gamma}_{(1)} = \frac{D\dot{\gamma}}{Dt} + \dot{\gamma}\,\nabla\vec{v} + \left(\nabla\vec{v}\right)^T \dot{\gamma}$.	Normal stress effects, Reduces to **Reiner Rivlin fluid model** if $\alpha_1 = 0$.

(*Continued*)

TABLE 7.4 (CONTINUED)
Various Mathematical Models for Viscoelastic Fluids and Their Constitutive Relations

Model	Constitutive Equation	Key Characteristics/Examples
Fourth-grade fluid (Truesdell and Noll, 1992)	$\tau = \mu\dot{\gamma} + \alpha_1\dot{\gamma}_{(1)} + \alpha_2\dot{\gamma}^2 + \beta_1\dot{\gamma}_{(2)} + \beta_2\left(\dot{\gamma}\dot{\gamma}_{(1)} + \dot{\gamma}_{(1)}\dot{\gamma}\right) + \beta_3\dot{\gamma}^3 + \epsilon_1\dot{\gamma}_{(3)} + \epsilon_2\left(\dot{\gamma}\dot{\gamma}_{(2)} + \dot{\gamma}_{(2)}\dot{\gamma}\right) + \epsilon_3\dot{\gamma}_{(1)}^2 + \epsilon_4\left(\dot{\gamma}^2\dot{\gamma}_{(1)} + \dot{\gamma}_{(1)}\dot{\gamma}^2\right) + \epsilon_5\dot{\gamma}^4$, where $\dot{\gamma}_{(n+1)} = \frac{D\dot{\gamma}_{(n)}}{Dt} + \dot{\gamma}_{(n)}\nabla\vec{v} + \left(\nabla\vec{v}\right)^T\dot{\gamma}_{(n)}$; $n = 1, 2$. The symbols $\beta_i, i = 1, 2, 3$ and $\epsilon_j, j = 1, 2, 3, 4, 5$ are additional material constants.	Captures nonlinear viscoelastic effects such as shear thinning/thickening, normal stress differences, and memory effects. Reduces to **third-grade fluid** if $\epsilon_j = 0 \,\forall\, j = 1, 2, 3, 4, 5$.

control volume. For compressible fluids, the continuity equation is expressed in the following form:

$$\frac{\partial \rho}{\partial t} + \nabla \cdot \left(\rho\vec{v}\right) = 0. \tag{7.3}$$

The symbols $\rho, \vec{v}$, and $\nabla \cdot \left(\rho\vec{v}\right)$ are fluid density, velocity vector, and the divergence of the mass flux, respectively. Since ρ is constant for an incompressible homogeneous fluid, therefore the equation of continuity is reduced to the form

$$\nabla \cdot \vec{v} = 0. \tag{7.4}$$

7.3.2 Momentum Equation (Cauchy Equation)

The momentum equation expresses Newton's second law of motion as applied to a moving fluid control volume; the rate of change of momentum is equal to the sum

TABLE 7.5
Microstructured Fluids

Model	Constitutive Equation	Main Features/Examples
Couple stress fluid	$\tau = \mu\dot{\gamma} - \eta\nabla^2\dot{\gamma}$, η : Couple stress viscosity parameter	Includes couple stresses, nonlocal effects
Micropolar fluid	$\tau = (\mu + \kappa)\dot{\gamma} + \kappa\left(\nabla\vec{v} - (\nabla\vec{v})^T - 2e_{ijk}\,\vec{\omega}\right)$, Couple stress tensor $\mathbf{M} = \alpha(\nabla \cdot \vec{\omega})\boldsymbol{I} + \beta\nabla\vec{\omega} + \gamma(\nabla\vec{\omega})^T$; κ : Microrotational viscosity parameter, e_{ijk} : Levi-Civita tensor, $\vec{\omega}$: Microrotational vector, $\mathbf{M}$: Couple stress tensor, $\boldsymbol{I}$: Identity tensor, α, β, γ : Spin gradient viscosity coefficients.	Micro-rotation and asymmetric stress tensor

of forces acting on the fluid. For an incompressible, continuum fluid with no body torques, the general form is

$$\rho\left(\frac{\partial\vec{v}}{\partial t} + \vec{v}\cdot\nabla\vec{v}\right) = -\nabla p + \nabla\cdot\tau + \vec{f}, \tag{7.5}$$

where the left-side terms, $\frac{\partial\vec{v}}{\partial t}$ – local acceleration and $\vec{v}\cdot\nabla\vec{v}$ – convective acceleration, describe how the fluid's velocity alters with time (t) at a fixed point in space and the change in velocity due to spatial variation during fluid motion, respectively. The right-side terms $-\nabla p$, $\nabla\cdot\tau$, and $\vec{f}$ are pressure (p) gradient force, divergence of extra stress tensor (τ), and external body force like gravity, electromagnetic forces, or centrifugal effects.

7.3.3 Conservation of Angular Momentum

In classical fluid mechanics (Newtonian), when the stress tensor is symmetric then conservation of angular momentum is satisfied inherently. Angular momentum balancing must be realized in microstructured non-Newtonian fluids, such as micropolar fluids and couple stress fluids, due to the presence of asymmetric stress tensors and body couple.

Conservation of angular momentum principle expresses that the rate of change of the angular momentum of a fluid element is equal to the sum of the moments of the forces acting on it. The general form of the angular momentum conservation is

$$\rho\left(\frac{D\vec{L}}{Dt}\right)=\nabla\cdot\mathbf{M}+2\vec{T}_{axl}, \tag{7.6}$$

where the symbols $\vec{L}, \mathbf{M}$, and $\vec{T}_{axl}$ denote, respectively, angular momentum per unit mass, couple stress tensor, and axial vector corresponding to the anti-symmetric part of the Cauchy stress tensor. Also, $\frac{D}{Dt}$ denote the material derivative.

7.3.4 Energy Conservation Principle

Thermodynamical analysis associated with non-Newtonian fluid flow is crucial due to temperature effects, viscous heating, or to detect alterations in thermal properties. In these circumstances, the energy conservation principle – the total energy (including internal, kinetic, and potential energy) of a fluid element is conserved when the work done by pressure and viscous forces, as well as heat transfer, is taken into account – becomes extremely important. For an incompressible fluid, the general energy conservation principle can be written in the following form:

$$\rho c_p\left(\frac{\partial T}{\partial t}+\vec{v}\cdot\nabla T\right)=k\nabla^2 T+\Phi+Q, \tag{7.7}$$

where specific heat, temperature, thermal conductivity, viscous dissipation, and external heat source are denoted by symbols c_p, T, k, Φ and Q, respectively. Mathematically, $\Phi=\tau:\nabla\vec{v}$, representing the conversion of mechanical energy into heat due to internal friction in non-Newtonian fluids.

7.4 GOVERNING EQUATIONS OF MOTION FOR NON-NEWTONIAN FLUIDS

Non-Newtonian fluid dynamics does not have a linearity between stress tensor and the rate of strain, as in Newtonian fluids. The stress tensors of non-Newtonian fluids are extremely non-linear, time-dependent, or have higher-order derivatives depending on the rheological model chosen – power-law, viscoelastic, micropolar, etc. Thus, the solution of the momentum equations strongly depends on the choice of the constitutive relation, even if the structure of the momentum equation remains the same.

7.4.1 For Time-Independent Non-Newtonian Fluid Models

Time-independent fluids are the class of non-Newtonian fluids whose viscosity alters with instantaneous shear rate without being impacted by prior deformation. The motion of time-independent fluid models, either purely viscous models (Carreau, Cross, etc.) or viscoplastic models (Bingham plastic, Casson, etc.), is described by the fundamental conservation principles of mass, momentum, and energy (if applicable) of continuum mechanics, along with an appropriate constitutive relation between the shear stress and the rate of strain through a shear-rate–dependent viscosity. The

TABLE 7.6
Governing Equations of Motion for Time-Independent Fluids

Model	Equations of Motion
Purely viscous fluid models	Continuity equation $\nabla \cdot \vec{v} = 0$, Linear momentum equation $\rho\left(\frac{\partial \vec{v}}{\partial t} + \vec{v} \cdot \nabla \vec{v}\right) = -\nabla p + \nabla \cdot \left(\mu(\dot{\gamma})\dot{\gamma}\right) + \vec{f}$.
Viscoplastic models	Continuity equation $\nabla \cdot \vec{v} = 0$, Linear momentum equation $\rho\left(\frac{\partial \vec{v}}{\partial t} + \vec{v} \cdot \nabla \vec{v}\right) = -\nabla p + \nabla \cdot \tau + \vec{f}$; $\lvert\tau\rvert > \tau_0$.

general governing equations for such incompressible non-Newtonian fluids are outlined in Table 7.6.

7.4.2 For Time-Dependent Non-Newtonian Fluid Models

Time-dependent non-Newtonian fluids exhibit viscosities that are influenced by both the instantaneous shear rate, duration, and previous history of deformation. Thixotropic fluids have shear-thinning behavior with time at constant shear, whereas rheopectic fluids have shear-thickening behavior as a result of microstructural breakdown or formation, which is indicated by the time-evolving structure parameter (λ_s). The governing equations of motion for incompressible thixotropic or rheopectic fluids can be obtained by substituting the constitutive relation of thixotropic or rheopectic fluids in Equation (7.5) with the equation of structural evolution – a common approach adopted by Moore (Larson, 2015) are as follows:

Continuity equation

$$\nabla \cdot \vec{v} = 0, \tag{7.8}$$

Momentum equation

$$\rho\left(\frac{\partial \vec{v}}{\partial t} + \vec{v} \cdot \nabla \vec{v}\right) = -\nabla p + \nabla \cdot \left(2\mu(\lambda_s, \dot{\gamma})\dot{\gamma}\right) + \vec{f}, \tag{7.9}$$

Equation of structural evolution

$$\frac{d\lambda_s}{dt} = \underbrace{\frac{\lambda_{eq} - \lambda_s}{t_b}}_{\text{recovery/build-up}} - \underbrace{k_d \lambda_s (\dot{\gamma})^m}_{\text{breakdown under shear}}, \tag{7.10}$$

Here, equilibrium structure parameter (at rest), characteristic structural recovery time, structural breakdown rate constant, and shear-rate sensitivity are expressed by symbols λ_{eq}, t_b, k_d, and m, respectively.

- When the breakdown term dominates under shear stress ($k_d > 0$), then λ_s decreases which implies viscosity drops – **Thixotropic fluids**.
- When the build-up term dominates under shear stress ($k_d < 0$), then λ_s increases which implies viscosity rises – **Rheopectic fluids**.

7.4.3 For Viscoelastic Fluid Models

In order to describe the motion of viscoelastic fluids, it is essential to combine the fundamental balance laws of continuum mechanics with an appropriate constitutive equation that takes into consideration the material's memory and elastic effects. Different viscoelastic fluid models are distinguished by the constitutive relation, which links the extra stress tensor to the kinematics of deformation through material-specific parameters such as relaxation time, retardation time, and viscosity. The governing equations for viscoelastic fluid flow are determined by combining the incompressible Navier–Stokes equations with the appropriate constitutive relation of the viscoelastic fluid model. In the following, we present Table 7.7 which include the general form of flow equations for various incompressible viscoelastic fluids.

7.4.4 For Microstructured Fluid Models

Microstructured fluid theories, micropolar and couple stress fluids, are extensions of classical mechanics with the addition of microrotation, couple stresses, and micro-deformation. These theories are crucial in substances such as liquid crystals, liquids containing suspended particles, polymeric solutions, and biological fluids because particle-scale rotations and size-dependent stresses affect macroscopic flow behavior. The governing equations of motion for incompressible microstructured fluids are included in Table 7.8.

The symbols $\vec{\omega}, j, \kappa, \vec{c}$, and α, β are microrotation vector, microinertia per unit mass, vortex viscosity, body couple per unit mass, and spin gradient viscosity coefficients, respectively. Couple stress viscosity coefficient of couple stress fluid is denoted by η. All other symbols appearing in Table 7.8 carry their conventional meanings, as defined in the preceding sections.

7.5 CONCLUSION

The complexity of governing equations associated with various non-Newtonian fluid models demands a combination of analytical, numerical, and experimental strategies for their solution. Analytical solutions work only in very limited cases, typically with simple geometries in which the flow phenomena take place and low parameter conditions, such as small Deborah or Reynolds numbers, which are functions of structural behavior like shear-thinning, shear-thickening, yield stress, etc. However,

TABLE 7.7
Governing Equations of Motion for Viscoelastic Fluids

Model	Equations of Motion
Maxwell (Salahuddin *et al.*, 2022)	Continuity equation $\nabla \cdot \vec{v} = 0$, Linear momentum equation $\rho \frac{\partial \vec{v}}{\partial t} + \rho(\vec{v} \cdot \nabla)\vec{v} = -\nabla p + \nabla \cdot \tau + \vec{f}$, where $\tau + \lambda_1 \frac{D\tau}{Dt} = \mu \dot{\gamma}$.
Upper convective Maxwell fluid (Mederos *et al.*, 2024)	Continuity equation $\nabla \cdot \vec{v} = 0$, Linear momentum equation $\rho \frac{\partial \vec{v}}{\partial t} + \rho(\vec{v} \cdot \nabla)\vec{v} = -\nabla p + \nabla \cdot \tau + \vec{f}$, where $\tau + \lambda_1 \overset{\nabla}{\tau} = \mu \dot{\gamma}$ and $\overset{\nabla}{\tau} = \frac{\partial \tau}{\partial t} + \vec{v} \cdot \nabla \tau - (\nabla \vec{v})^T \cdot \tau - \tau \cdot (\nabla \vec{v})$.
Kelvin-Voigt (Bulıcek *et al.*, 2012)	Continuity equation $\nabla \cdot \vec{v} = 0$, Linear momentum equation $\rho \frac{\partial \vec{v}}{\partial t} + \rho(\vec{v} \cdot \nabla)\vec{v} = -\nabla p + \mu \nabla^2 \vec{v} + E \nabla \cdot \dot{\gamma} + \vec{f}$.
Oldroyd-B (Bird *et al.*, 1987)	Continuity equation $\nabla \cdot \vec{v} = 0$, Linear momentum equation $\rho \frac{\partial \vec{v}}{\partial t} + \rho(\vec{v} \cdot \nabla)\vec{v} = -\nabla p + \nabla \cdot \tau + \vec{f}$, where $\tau + \lambda_1 \overset{\nabla}{\tau} = \mu\left(\dot{\gamma} + \lambda_2 \overset{\nabla}{\dot{\gamma}}\right)$.
Oldroyd 4 – constant fluid	Continuity equation $\nabla \cdot \vec{v} = 0$, Linear momentum equation $\rho \frac{\partial \vec{v}}{\partial t} + \rho(\vec{v} \cdot \nabla)\vec{v} = -\nabla p + \nabla \cdot \tau + \vec{f}$, where $\tau + \lambda_1 \overset{\nabla}{\tau} + \frac{\lambda_5}{2}(tr\tau)\dot{\gamma} = \mu\left(\dot{\gamma} + \lambda_2 \overset{\nabla}{\dot{\gamma}}\right)$.
Oldroyd 8 – constant fluid (Bird *et al.*, 1987)	Continuity equation $\nabla \cdot \vec{v} = 0$, Linear momentum equation $\rho \frac{\partial \vec{v}}{\partial t} + \rho(\vec{v} \cdot \nabla)\vec{v} = -\nabla p + \nabla \cdot \tau + \vec{f}$, where $\tau + \lambda_1 \overset{\nabla}{\tau} + \frac{1}{2}\lambda_5 (tr\tau)\dot{\gamma} + \frac{1}{2}\lambda_3(\tau \cdot \dot{\gamma} + \dot{\gamma} \cdot \tau) + \frac{1}{2}\lambda_7(\dot{\gamma} : \dot{\gamma})\mathbf{I}$ $+ \frac{1}{2}\lambda_6(\tau : \dot{\gamma})\mathbf{I} = \mu\left\{\dot{\gamma} + \lambda_2 \overset{\nabla}{\dot{\gamma}} + \lambda_4(\dot{\gamma} \cdot \dot{\gamma})\right\}$.

(*Continued*)

TABLE 7.7 (CONTINUED)
Governing Equations of Motion for Viscoelastic Fluids

Model	Equations of Motion
Jeffery (Bird *et al.*, 1987)	Continuity equation $\nabla \cdot \vec{v} = 0$, Linear momentum equation $\rho \frac{\partial \vec{v}}{\partial t} + \rho(\vec{v} \cdot \nabla)\vec{v} = -\nabla p + \nabla \cdot \tau + \vec{f}$, where $\tau + \lambda_1 \frac{D\tau}{Dt} = \mu\left(\dot{\gamma} + \lambda_2 \frac{D\dot{\gamma}}{Dt}\right)$.
Second-grade fluid (Ericksen, 1955)	Continuity equation $\nabla \cdot \vec{v} = 0$, Linear momentum equation $\rho \frac{\partial \vec{v}}{\partial t} + \rho(\vec{v} \cdot \nabla)\vec{v} = -\nabla p + \mu \nabla \cdot \dot{\gamma} + \alpha_1 \nabla \cdot \dot{\gamma}_{(1)} + \alpha_2 \nabla \cdot \dot{\gamma}^2 + \vec{f}$.
Fourth-grade fluid (Truesdell and Noll, 1992)	Continuity equation $\nabla \cdot \vec{v} = 0$, Linear momentum equation $\rho \frac{\partial \vec{v}}{\partial t} + \rho(\vec{v} \cdot \nabla)\vec{v} = -\nabla p + \mu \nabla \cdot \dot{\gamma} + \alpha_1 \nabla \cdot \dot{\gamma}_{(1)} + \alpha_2 \nabla \cdot \dot{\gamma}^2 + \beta_1 \nabla \cdot \dot{\gamma}_{(2)} +$ $\beta_2 \nabla \cdot \left(\dot{\gamma}\dot{\gamma}_{(1)} + \dot{\gamma}_{(1)}\dot{\gamma}\right) + \beta_3 \nabla \cdot \dot{\gamma}^3 + \varepsilon_1 \nabla \cdot \dot{\gamma}_{(3)} + \varepsilon_2 \nabla \cdot \left(\dot{\gamma}\dot{\gamma}_{(2)} + \dot{\gamma}_{(2)}\dot{\gamma}\right) +$ $\varepsilon_3 \nabla \cdot \dot{\gamma}_{(1)}^2 + \varepsilon_4\left(\dot{\gamma}^2\dot{\gamma}_{(1)} + \dot{\gamma}_{(1)}\dot{\gamma}^2\right) + \varepsilon_5 \nabla \cdot \dot{\gamma}^4 + \vec{f}$. The governing equations for fourth grade fluid will be converted into governing equations for **third-grade fluid** (Steinhauer et al., 2006) when $\epsilon_i, i = 1,2,3,4,5$ become zero.

to tackle more realistic situations, researchers rely on numerical methods such as the finite difference method (FDM), finite element method (FEM), finite volume method (FVM), and spectral method. These numerical techniques are capable of dealing with real-world practical fluid flow problems in irregular and complex domains along with multi-physics interactions. Non-Newtonian fluid flows have an extensive list of applications, including:

1. Biomedical engineering – blood flow in arteries (Casson and Herschel-Bulkley models), drug delivery, and synovial fluid dynamics.
2. Polymer processing methods – melt extrusion and injection molding (Carreau and Phan-Thien-Tanner models).
3. Petroleum and chemical industries – drilling mud flow, hydraulic fracturing, improved oil recovery, and sludge transmission.
4. Food and Cosmetic Sciences – Creams, gels, and sauces with shear-thinning, texture control, and spray coating.

TABLE 7.8
Governing Equations of Motion for Microstructured Fluids

Model	Equations of Motion
Couple stress fluid (Stokes, 1984)	Continuity equation $\nabla \cdot \vec{v} = 0$, Linear momentum equation $\rho \frac{\partial \vec{v}}{\partial t} + \rho(\vec{v} \cdot \nabla)\vec{v} = -\nabla p + \mu \nabla^2 \vec{v} - \eta \nabla^4 \vec{v} + \vec{f}$.
Micropolar fluid (Eringen, 1966)	Continuity equation $\nabla \cdot \vec{v} = 0$, Linear momentum equation $\rho\left(\frac{\partial \vec{v}}{\partial t} + \vec{v} \cdot \nabla \vec{v}\right) = -\nabla p + (\mu + \kappa)\nabla^2 \vec{v} + \kappa \nabla \times \vec{\omega} + \vec{f}$, Angular momentum equation $\rho j\left(\frac{\partial \vec{\omega}}{\partial t} + \vec{v} \cdot \nabla \vec{\omega}\right) = \alpha \nabla^2 \vec{\omega} + \beta \nabla(\nabla \cdot \vec{\omega}) - 2\kappa \vec{\omega} + \kappa \nabla \times \vec{v} + \vec{c}$.

5 Microfluidics- chip devices that work with viscoelastic or electrorheological fluids.

Here, we have looked at a number of various classical and widely accepted categories of non-Newtonian fluids, emphasizing their constitutive relationships, governing equations of motion, and uses in biological and engineering systems. Owing to the broad spectrum of rheological behaviors, only selected fluid models have been discussed in this chapter. The continuing growth of non-Newtonian fluid mechanics, particularly in biofluid dynamics and complex multiphase systems, is evidenced by recent advancements, including the generalized Burgers fluid model (Sultan *et al.*, 2014), Saramito fluid model (De Vita *et al.*, 2018), and elastoviscoplastic formulations (Saramito, 2007) etc.. Readers and researchers are suggested to consult specialized monographs and recent research literature for deeper insights into these emerging directions.

REFERENCES

Batchelor, G. K. (2000). *An introduction to fluid dynamics.* Cambridge University Press.

Bautista, F., De Santos, J., Puig, J., and Manero, O. (1999). Understanding thixotropic and antithixotropic behavior of viscoelastic micellar solutions and liquid crystalline dispersions. I. The model. *Journal of Non-Newtonian Fluid Mechanics*, 80(2–3):93–113.

Bingham, E. C. (1917). *An investigation of the laws of plastic flow.* Number 278. US Government Printing Office.

Bingham, E. C. (1922). *Fluidity and plasticity.* McGraw-Hill Book Company, Incorporated.

Bird, R. B., Armstrong, R. C., and Hassager, O. (1987). *Dynamics of polymeric liquids. Vol. 1: Fluid mechanics.* John Wiley and Sons Inc., New York.

Bulıcek, M., Málek, J., and Rajagopal, K. (2012). On kelvin–voigt model and its generalizations. *Evolution Equations and Control Theory*, 1(1):17–42.

Carreau, P. J. (1972). Rheological equations from molecular network theories. *Transactions of the Society of Rheology*, 16(1):99–127.

Casson, N. (1959). Flow equation for pigment-oil suspensions of the printing ink-type. In Mill C. C. (Ed.), *Rheology of disperse systems*, 84–104. Pergamon Press.

Chhabra, R. and Richardson, J. (2008). Non-Newtonian fluid behaviour. In Chhabra R. P. & Richardson J. F. (Eds.), *Non-Newtonian flow and applied rheology*, pages 1–55. Oxford, UK: Butterworth–Heinemann (Elsevier).

Cross, M. M. (1965). Rheology of non-Newtonian fluids: A new flow equation for pseudoplastic systems. *Journal of Colloid Science*, 20(5):417–437.

De Vita, F., Rosti, M. E., Izbassarov, D., Duffo, L., Tammisola, O., Hormozi, S., and Brandt, L. (2018). Elastoviscoplastic flows in porous media. *Journal of Non-Newtonian Fluid Mechanics*, 258:10–21.

Ericksen, J. (1955). Deformations possible in every compressible, isotropic, perfectly elastic material. *Journal of Mathematics and Physics*, 34(1–4):126–128.

Eringen, A. C. (1966). Theory of micropolar fluids. *Journal of Mathematics and Mechanics*, 16(1):1–18.

Ferrás, L., Nóbrega, J., and Pinho, F. (2012). Analytical solutions for Newtonian and inelastic non-Newtonian flows with wall slip. *Journal of Non-Newtonian Fluid Mechanics*, 175:76–88.

Fredrickson, A. (1970). A model for the thixotropy of suspensions. *AIChE Journal*, 16(3):436–441.

Herschel, W. H. and Bulkley, R. (1926). Konsistenzmessungen von gummi-benzollösungen. *Kolloid-Zeitschrift*, 39(4):291–300.

Indiana University, D. O. M. (1952). *Journal of Rational Mechanics and Analysis*, volume 1. University of Indiana. Department of Mathematics.

Khan, A. U., Al-Zubaidi, A., Munir, S., Saleem, S., and Duraihem, F. Z. (2021). Closed form solutions of cross flows of casson fluid over a stretching surface. *Chaos, Solitons & Fractals*, 149:111067.

Khan, M., Munawar, S., and Abbasbandy, S. (2010). Steady flow and heat transfer of a sisko fluid in annular pipe. *International Journal of Heat and Mass Transfer*, 53(7–8):1290–1297.

Kumar, S. and Deo, S. (2025). On exact solution of MHD Jeffery fluid flow equations with heat transfer. *International Journal of Applied and Computational Mathematics*, 11(2):70.

Kumar, S., Deo, S., and Filippov, A. (2022). MHD effects on the flow of second-grade fluid sandwiched between two Newtonian fluid layers through porous medium. *Colloid Journal*, 84(6):794–805.

Kumar, S., Deo, S., Maurya, P. K., and Patel, P. K. (2025). Multiphase flow of three immiscible non-Newtonian fluids through porous cylindrical annulus under magnetic field. *Chinese Journal of Physics*, 97:15–43.

Larson, R. (2015). Constitutive equations for thixotropic fluids. *Journal of Rheology*, 59(3):595–611.

Mallick, B. (2024). Numerical simulation for electrokinetic energy conversion during couple stress fluid flow in microtube. *The European Physical Journal Plus*, 139(12):1075.

Mallick, B., Choudhury, A., and Misra, J. (2023). Irreversibility analysis for ion size-dependent electrothermal transport of micropolar fluid in a microtube. *Journal of Thermal Analysis and Calorimetry*, 148(21):12017–12035.

Mallick, B. and Misra, J. (2019). Peristaltic flow of Eyring-Powell nanofluid under the action of an electromagnetic field. *Engineering Science and Technology, an International Journal*, 22(1):266–281.

Maurya, D. K., Deo, S., and Maurya, P. K. (2024). Hydromagnetic flow of casson fluid over a stretching plane through porous medium. *ZAMM-Journal of Applied Mathematics and Mechanics/Zeitschrift für Angewandte Mathematik und Mechanik*, 104(11):e202300675.

Mederos, G., Bautista, O., Méndez, F., and Arcos, J. (2024). Squeeze force of a Maxwell fluid between circular smooth surfaces with simple harmonic motion. *Physics of Fluids*, 36(9):093130.

Misra, J., Mallick, B., Sinha, A., and Chowdhury, A. R. (2018). Impact of Cattaneo-Christov heat flux on electroosmotic transport of third-order fluids in a magnetic environment. *The European Physical Journal Plus*, 133(5):195.

Nae, H. N. (1993). Introduction to rheology. In Laba D. (Ed.), Rheological properties of cosmetics toiletries, *Cosmetic Science and Technology Series,* 9 13:9–34.

Pal, P. B., Singh, S., and Murthy, P. (2025). Solute dispersion in unsteady and viscous flow regimes of a non-Newtonian fluid flow with periodic body acceleration/deceleration. *Physics of Fluids*, 37(2):023113.

Papanastasiou, T. C. (1987). Flows of materials with yield. *Journal of Rheology*, 31(5):385–404.

Patel, P. K., Deo, S., and Maurya, P. K. (2025). Hydromagnetic effects on couple stress fluid flow through a porous sphere embedded in another porous medium. *Physics of Fluids*, 37(4):043101.

Qi, C. and Ng, C.-O. (2017). Rotating electroosmotic flow of an Eyring fluid. *Acta Mechanica Sinica*, 33(2):295–315.

Rajagopal, K. R. and Srinivasa, A. R. (2000). A thermodynamic frame work for rate type fluid models. *Journal of Non-Newtonian Fluid Mechanics*, 88(3):207–227.

Richardson, J. (2011). *Non-Newtonian flow and applied rheology.* Butterworth-Heinemann.

Rivlin, R. S. and Ericksen, J. L. (1997). Stress-deformation relations for isotropic materials. In Dvorak G. J. & Shield R. T. (Eds.) *Collected Papers of RS Rivlin,* volume I and II, pages 911–1013. New York: Springer–Verlag.

Salahuddin, T., Mahmood, Z., Khan, M., and Awais, M. (2022). A permeable squeezed flow analysis of Maxwell fluid near a sensor surface with radiation and chemical reaction. *Chemical Physics*, 562:111627.

Saramito, P. (2007). A new constitutive equation for elastoviscoplastic fluid flows. *Journal of Non-Newtonian Fluid Mechanics*, 145(1):1–14.

Shit, G., Mondal, A., Sinha, A., and Kundu, P. (2016). Electro-osmotic flow of power-law fluid and heat transfer in a micro-channel with effects of Joule heating and thermal radiation. *Physica A: Statistical Mechanics and its Applications*, 462:1040–1057.

Soomro, F. A., Haq, R.-U., Khan, Z. H., and Zhang, Q. (2017). Passive control of nanoparticle due to convective heat transfer of Prandtl fluid model at the stretching surface. *Chinese Journal of Physics*, 55(4):1561–1568.

Steinhauer, M., Urbano, J. M., and Videman, J. (2006). New global a priori estimates for the third-grade fluid equations. *Mathematical Methods in the Applied Sciences*, 29(11):1339–1348.

Stokes, V. K. (1984). Couple stresses in fluids. *Theories of fluids with microstructure: An introduction*, pages 34–80. Springer.

Sultan, Q., Nazar, M., Imran, M., and Ali, U. (2014). Flow of generalized burgers fluid between parallel walls induced by rectified sine pulses stress. *Boundary Value Problems*, 2014(1):152.

Truesdell, C. and Noll, W. (1992). *The non-linear field theories of mechanics.* Springer.

Ullah, F. and Ashraf, M. B. (2024). Artificial neural network-based computational heat transfer analysis of Carreau fluid over a rotating cone. *Physics of Fluids*, 36(11):113102.

Walait, A., Ashraf, H., Chou, D., and Rehman, H. U. (2024). Stagnant rings and uniform film analysis of Phan-Thien-Tanner fluid film flow on a vertically upward moving tube. *Physics of Fluids*, 36(8):083613.

Williamson, R. V. (1929). The flow of pseudoplastic materials. *Industrial & Engineering Chemistry*, 21(11):1108–1111.

Yasuda, K. (1979). *Investigation of the analogies between viscometric and linear viscoelastic properties of polystyrene fluids*. PhD thesis, Massachusetts Institute of Technology.

Yoon, H. and Ghajar, A. (1987). A note on the Powell-Eyring fluid model. *International Communications in Heat and Mass Transfer*, 14(4):381–390.

8 Magnetohydrodynamics
Study of Magnetized Rotating Ferrofluid Heated from Below

Akanksha Thakur, Sunil, and Reeta Devi

8.1 INTRODUCTION

Thermal convection has been a subject of extensive investigation due to its fundamental role in fluid dynamics, heat transfer, and engineering applications. Convection begins when the Rayleigh number exceeds its critical value, marking the threshold of instability. Linear theory, primarily through normal mode analysis, provides insight into the onset of instability, whereas nonlinear approaches, particularly energy methods, delineate the regions of stability. Consequently, the combined use of linear and nonlinear analyses offers a comprehensive understanding of fluid stability. Chandrasekhar [1] provided a classical treatment of thermal instability under diverse hydrodynamic and hydromagnetic conditions. The origins of the energy method trace back to Reynolds [2] and Orr [3], with significant refinements introduced in later works, including the generalized energy method proposed by Galdi [4] and Galdi and Straughan [5].

Ferrofluids, first systematically studied by Rosensweig [6], are colloidal suspensions of nanoscale ferromagnetic or ferrimagnetic particles in a carrier liquid. Stabilized by surfactants, they exhibit remarkable magnetization in the presence of external magnetic fields, which makes them attractive for diverse technological applications such as targeted drug delivery, biosensing, medical imaging, water treatment, and heat transfer [7]. The study of ferroconvection was pioneered by Finlayson [8], followed by contributions from Shliomis [9], Vaidyanathan [10], Sekar [11], and Venkatasubramanian [12]. More recent works include investigations of rotating ferrofluids, nonlinear oscillatory convection, and chaotic convection [13–15]. Although ferrofluid stability has been studied under a variety of influences [16–22], the role of different boundary conditions remains relatively underexplored, with only a few notable exceptions [23–25].

Rotation introduces additional complexity into ferroconvection. Sunil and Mahajan [26] examined nonlinear stability of rotating magnetized ferrofluids, while Sunil and Mahajan [27] extended their study to porous media. These studies

DOI: 10.1201/9781003664345-8

demonstrated that rotation can exert a stabilizing influence, but the interplay between rotation, magnetization, and boundary conditions is not yet fully understood.

Another factor influencing convection stability is the presence of couple stresses. Couple stresses arise from internal force couples within fluids, altering rheological behavior and stability characteristics. Stokes [28] introduced the concept of couple stress fluids, which has since found relevance in nanofluidics, biomedical flows, and multiphase systems. Several studies have examined convection in couple stress fluids under various conditions, including global stability analysis [29, 30], triple-diffusive systems [31], vertical layers [32], penetrative convection [33], and variable viscosity effects [34]. Despite this progress, the combined influence of couple stresses, rotation, and boundary conditions in ferroconvection has received limited attention.

Boundary conditions play a decisive role in convective stability. While the case of two free boundaries is mathematically convenient and allows analytical insights [35], rigid boundaries often provide more realistic models of confined systems. The study of ferrofluids under different bounding surfaces is of practical significance in materials science, nanotechnology, and biomedical engineering [7].

Motivated by the above considerations, the present study examines the influence of couple stresses on the convective stability of rotating magnetized ferrofluids. A generalized energy functional is employed to evaluate energy dissipation, and both linear and nonlinear stability analyses are conducted, with the corresponding eigenvalue problems solved using the Galerkin method. The investigation emphasizes the role of boundary configurations, rotation, and magnetization in shaping subcritical behavior and thermal stability. The specific objectives of this work are:

- To examine energy dissipation through a generalized energy functional.
- To investigate the existence and variation of subcritical regions in rotating ferrofluids heated from below.
- To analyze the influence of different boundary combinations on subcritical behavior.
- To identify the boundary conditions that enhance thermal stability under rotation.
- To assess the combined effects of rotation, magnetization, and couple stresses on the system.

The structure of this chapter is as follows. Section 2 formulates the governing equations of the system. Section 3 presents the nonlinear analysis and evaluates energy dissipation, while Section 4 develops the associated eigenvalue problems. Section 5 describes the numerical scheme based on the Galerkin method and provides the calculation of linear and nonlinear Rayleigh numbers. Section 6 discusses the results in detail, and Section 7 summarizes the main conclusions.

8.2 FORMULATION OF THE PROBLEM

A horizontal porous layer of thickness d, saturated with a ferrofluid, is considered. The system, initially at rest, is heated from below, establishing a uniform thermal gradient $\beta = \left|\frac{dT}{dz}\right|$, with boundary temperatures T_0 (bottom) and T_d (top), where $T_d < T_0$. The ferrofluid is treated as incompressible and of constant viscosity. A vertical magnetic field $\mathbf{H} = (0,0,H)$ is applied, and the system experiences a gravitational field given by $\mathbf{g} = (0,0,-g)$. The system is rotating with angular velocity $\Omega = (0,0,\Omega)$. A schematic representation of this physical setup is shown in Figure 8.1.

The equations governing the flow of an incompressible rotating magnetized ferrofluid in the presence of couple stresses (using Boussinesq approximation), [6], [8], [36], [37] are as follows:

Continuity Equation: The fluid is assumed to be incompressible, satisfying the mass conservation equation:

$$\nabla \cdot \mathbf{q} = 0 \tag{8.1}$$

where q is the velocity field.

Momentum Equation: The momentum balance equation, incorporating the effects of buoyancy, magnetic fields, rotation, and couple stress forces, is given by

$$\rho_0\left[\frac{\partial \mathbf{q}}{\partial t} + (\mathbf{q}\cdot\nabla)\mathbf{q}\right] = -\nabla p + \rho_0\left[1-\alpha(T-T_a)\right]\mathbf{g} + \mu_0(\mathbf{M}\cdot\nabla)\mathbf{H} + 2\rho_0(\mathbf{q}\times\Omega) + (\mu - \mu'\nabla^2)\nabla^2\mathbf{q}, \tag{8.2}$$

where ρ_0 is the reference density, $\mathbf{q}$ is the velocity, p is the pressure, T is the temperature, Ta is the average temperature, $\mathbf{g}$ is the force of gravity, $\mathbf{M}$ is the magnetization, $\mathbf{H}$ is the magnetic field, μ_0 is magnetic permeability of vacuum, μ is viscosity, and μ' is the couple stress viscosity. The effect of rotation contributes two

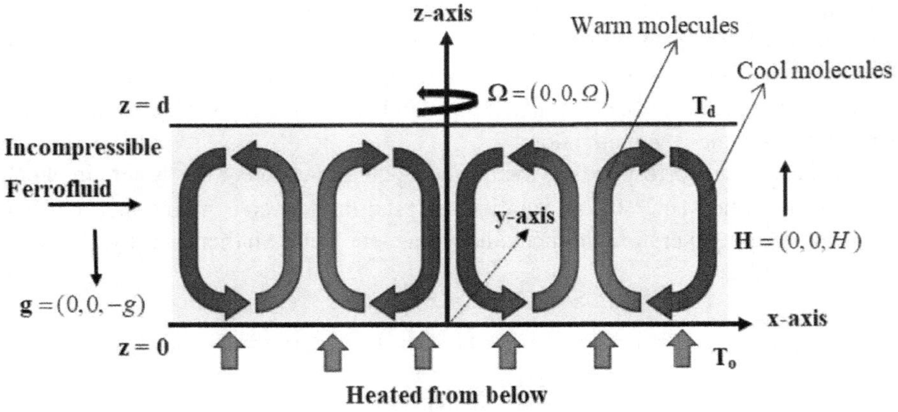

FIGURE 8.1 Schematic representation of the physical problem.

terms: (a) centrifugal force $-\frac{1}{2}grad|\Omega\times \mathrm{r}|^2$ and (b) Coriolis acceleration $2(\Omega\times \mathrm{q})$. In Equation (8.2), $p = p_f - \frac{1}{2}\rho_0|\Omega\times \mathrm{r}|^2$ is the reduced pressure, where p_f stands for fluid pressure.

Energy Equation: The temperature evolution is governed by:

$$(\rho_0 C_0)\frac{\partial T}{\partial t} + (\rho_0 C_0)\mathbf{q}\cdot\nabla T = \nabla\cdot(\kappa_m \nabla T), \tag{8.3}$$

where ρ_0 is reference density, C_0 is specific heat capacity, κ_m is the thermal diffusivity.

Magnetostatic Equations: In the magnetostatic limit, the behavior of the magnetic field $\mathbf{H}$, magnetic induction $\mathbf{B}$, and magnetization $\mathbf{M}$ is governed by Maxwell's equations:

$$\nabla\cdot\mathbf{B} = 0, \nabla\times\mathbf{H} = 0, \mathbf{B} = \mu_0(\mathbf{H}+\mathbf{M}), \tag{8.4}$$

where $\mathbf{B}$ is the magnetic induction (or magnetic flux density), $\mathbf{H}$ is the applied magnetic field, $\mathbf{M}$ is the magnetization, μ_0 is the permeability of free space. The condition $\nabla\cdot\mathbf{B} = 0$ represents the absence of magnetic monopoles, while $\nabla\times\mathbf{H} = 0$ holds in the absence of free currents. The third equation links $\mathbf{B}$ to $\mathbf{H}$ through the material's response, captured by $\mathbf{M}$.

Magnetization Relationship: The magnetization $\mathbf{M}$ is assumed to be anisotropic and dependent on the applied field strength and temperature. It is given by:

$$\mathbf{M} = \frac{\mathbf{H}}{H}\left[M_0 + \chi(H - H_0) - K_1(T - T_a)\right]. \tag{8.5}$$

Here M_0 is the reference magnetization, χ is the magnetic susceptibility, which determines how M varies with H, H_0 is the reference field strength, K_1 is a temperature-dependent coefficient, T is the local temperature, and Ta is the ambient reference temperature. This expression accounts for three key effects: a baseline reference magnetization M_0, a linear response to variations in the magnetic field H with susceptibility χ, and a temperature-dependent term that adjusts magnetization due to thermal fluctuations. The inclusion of $K_1(T - Ta)$ captures the thermomagnetic effect, which is critical for analyzing stability in ferrofluids where temperature variations influence magnetization.

The basic state corresponds to the quiescent configuration of the system in which there is no motion $(\mathbf{q}_b = 0)$, no rotation, and the fluid layer is maintained under a steady vertical temperature gradient and a magnetic field. Mathematically, this state is represented as

$$\begin{gathered} p = p_b(z), \rho_b(z) = \rho_0(1+\alpha\beta z), T_b(z) = -\beta z + T_0, \\ \mathbf{H_b}(z) = \left(\frac{H_0(1+\chi) - K_1\beta z}{1+\chi}\right)\hat{\mathbf{k}}, \mathbf{M_b}(z) = \left(\frac{M_0(1+\chi) + K_1\beta z}{1+\chi}\right)\hat{\mathbf{k}}. \end{gathered} \tag{8.6}$$

Here, $\rho_b(z)$ represents the density distribution at equilibrium, $T_b(z)$ is the linear temperature profile, $H_b(z)$ is the applied magnetic field, and $M_b(z)$ is the corresponding magnetization. The subscript b is used to denote quantities associated with the basic state and $\beta = \dfrac{T_0 - T_d}{d}$.

8.2.1 Perturbation Equations

To investigate stability, small perturbations are introduced around the equilibrium (basic) state. This standard procedure enables examination of how the system responds to infinitesimal disturbances.

$$\mathbf{q} = \mathbf{q}_b + \mathbf{q}',\ p = p_b(z) + p',\ \rho = \rho_b + \rho',\ T = T_b(z) + \theta,\ \mathbf{H}' = \nabla\phi',$$
$$\mathbf{H} = \mathbf{H}_b(z) + \mathbf{H}',\ \mathbf{M} = \mathbf{M}_b(z) + \mathbf{M}',\ \mathbf{M}' = \left(M_1', M_2', M_3'\right). \tag{8.7}$$

Here, $\mathbf{q}'$ denotes the perturbation in the velocity field $\mathbf{q}$, while p' and ρ' represent perturbations in pressure and density, respectively. Similarly, θ corresponds to the perturbation in temperature, and $\mathbf{H}'$ is the perturbation in the magnetic field, expressed in terms of a scalar potential ϕ. The magnetization vector $\mathbf{M}$ is also perturbed, with $\left(M_1', M_2', M_3'\right)$ describing the deviations in its components from the basic state. The set of the perturbed equations is obtained as

$$\nabla \cdot \mathbf{q}' = 0, \tag{8.8}$$

$$\begin{aligned}\rho_0 \frac{\partial \mathbf{q}'}{\partial t} &= -\nabla p' + \rho_0 g\alpha\theta\hat{\mathbf{k}} - \frac{\mu_0 K_1 \beta}{1+\chi}\left[(1+\chi)\phi_z'\hat{\mathbf{k}} - K_1\theta\hat{\mathbf{k}}\right]\\ &+ \frac{\mu_0 M_0}{H_0}\phi_x'\nabla\phi_x' + \frac{\mu_0 M_0}{H_0}\phi_y'\nabla\phi_y' + \mu_0\chi\phi_z'\nabla\phi_z' - \mu_0 K_1\theta\nabla\phi_z'\\ &- \frac{1}{K}\mu\nabla^2\mathbf{q}' + \frac{1}{K}\mu'\nabla^4\mathbf{q}' + \frac{2\rho_0}{\epsilon}(\mathbf{q}'\times\Omega),\end{aligned} \tag{8.9}$$

$$\frac{\partial\theta}{\partial t} + \mathbf{q}'\cdot\nabla\theta = \kappa\nabla^2\theta + \beta\omega, \tag{8.10}$$

$$\left(1 + \frac{M_0}{H_0}\right)\nabla^2\phi' - \left(\frac{M_0}{H_0} - \chi\right)\phi_{zz}' = K_1\theta_z, \tag{8.11}$$

where $\kappa = \dfrac{\kappa_m}{(\rho_0 C_0)_f}$ denotes the thermal diffusivity.

At $z = 0, d$, the boundary conditions are $\mathbf{q}' = 0,\ \theta = 0,\ \dfrac{\partial\phi'}{\partial z} = 0$ together with the plane tiling periodicity for $\mathbf{q}', \theta$ and ϕ'.

8.3 NONLINEAR STABILITY ANALYSIS

The governing equations are nondimensionalized using the same set of characteristic scales and parameters introduced in [24], following the classical approach of Chandrasekhar [1]. The perturbed equations in nondimensional form (dropping asterisks for simplicity) are given by

$$\nabla\cdot\mathbf{q}=0, \tag{8.12}$$

$$\begin{aligned}\frac{1}{\Pr}\left(\frac{\partial\mathbf{q}}{\partial t}+\mathbf{q}\cdot\nabla\mathbf{q}\right)&=-\nabla p+R^{1/2}\left(1+M_1\right)\theta\hat{\mathbf{k}}-R^{1/2}M_1\phi_z\hat{\mathbf{k}}-M_1\theta\nabla\phi_z\\&+M_1\left(M_3-\frac{1}{1+\chi}\right)\phi_x\nabla\phi_x+\left(\frac{\chi}{1+\chi}\right)M_1\phi_z\nabla\phi_z\\&+M_1\left(M_3-\frac{1}{1+\chi}\right)\phi_y\nabla\phi_y+\nabla^2\mathbf{q}+F\nabla^4\mathbf{q}+Ta^{1/2}\left(\mathbf{q}\times\hat{\mathbf{k}}\right),\end{aligned} \tag{8.13}$$

$$\frac{\partial\theta}{\partial t}+\mathbf{q}\cdot\nabla\theta=\nabla^2\theta+R^{1/2}\omega, \tag{8.14}$$

$$M_3\nabla^2\phi-\left(M_3-1\right)\phi_{zz}=\theta_z. \tag{8.15}$$

Here, Ta denotes the Taylor number.

Applying the vertical curl and double curl operations, namely $\hat{\mathbf{k}}\cdot\left(\nabla\times\mathbf{q}\right)$ and $\hat{\mathbf{k}}\cdot\nabla\times\left(\nabla\times\mathbf{q}\right)$ to Equation (8.13), isolates the z-component of vorticity and the corresponding double-curl contribution. Denoting the vertical vorticity by

$$\zeta=\hat{\mathbf{k}}\cdot\left(\nabla\times\mathbf{q}\right)$$

the two resulting evolution equations read

$$\frac{1}{\Pr}\left(\frac{\partial\zeta}{\partial t}\right)=\frac{1}{\Pr}\left(\hat{\mathbf{k}}\ curl\ \mathbf{q}\cdot\nabla\mathbf{q}\right)+\nabla^2\zeta-F\nabla^4\zeta+M_1\left(\theta_y\phi_{zx}-\theta_x\phi_{zy}\right)+T_A^{1/2}\omega_z, \tag{8.16}$$

$$\begin{aligned}\frac{1}{\Pr}\frac{\partial\nabla^2\omega}{\partial t}&=\frac{1}{\Pr}\left(\hat{\mathbf{k}}\cdot curl\ curl\ \mathbf{q}\cdot\nabla\mathbf{q}\right)+R^{1/2}\left(M_1+1\right)\nabla_1^2\theta+\nabla^2\omega-F\nabla^6\omega\\&+M_1R^{1/2}\nabla_1^2\phi_z+M_1\left(-\nabla_1^2\theta\phi_{zz}+\theta_z\nabla_1^2\phi_z\right)\\&+M_1\left(-\theta_x\phi_{zzx}+\theta_{zx}\phi_{zx}-\theta_y\phi_{zzy}+\theta_{zy}\phi_{zy}\right)-T_A^{1/2}\zeta.\end{aligned} \tag{8.17}$$

To complement these, differentiate Equation (8.14) with respect to z to obtain the evolution of θ_z:

$$\frac{\partial\theta_z}{\partial t}+\mathbf{q}_z\cdot\nabla\theta+\mathbf{q}\cdot\nabla\theta_z=\nabla^2\theta_z+R^{1/2}\omega_z. \tag{8.18}$$

Since higher-order equations are involved, the following derivative boundary conditions are imposed at $z=0,1$:

$$\frac{\partial^2\theta}{\partial z^2}=0,\quad \frac{\partial\zeta}{\partial z}=0,\quad \frac{\partial^2\omega}{\partial z^2}=0.$$

To derive integrated energy relations (useful to estimate growth/decay of perturbations) we multiply the governing equations by convenient test functions and integrate over the domain V. Concretely, multiply Equation (8.17) by ω, Equation (8.16) by ζ, Equation (8.18) by θ, and Equation (8.15) by ϕ. The purpose of these multiplications is that by multiplying by ω and integrating yields an evolution law for the kinetic-type quantity $\|\nabla w\|^2$, multiplying by ζ yields an evolution for the enstrophy-type quantity $\|\nabla\zeta\|^2$, multiplying by θ gives an evolution for thermal-gradient norms such as $\|\theta_z\|^2$ and multiplying by ϕ (from Equation 8.15) controls the magnetic-potential energy contribution.

After integration by parts (using the boundary conditions above to eliminate boundary terms), and collecting dissipative, linear coupling and nonlinear advection contributions, one obtains the following energy identities:

$$\frac{1}{2\,\mathrm{Pr}}\frac{d\|\nabla\omega\|^2}{dt}=\left\|\nabla^2\omega\right\|^2-F\left\|\nabla^3\omega\right\|^2+R^{1/2}\left(1+M_1\right)\left\langle\nabla_1\omega\nabla_1\theta\right\rangle$$
$$-M_1R^{1/2}\left\langle\nabla_1\omega\nabla_1\phi_z\right\rangle-M1\left(-\left\langle\omega\phi_{zz}\nabla_1^2\theta\right\rangle+\left\langle\omega\theta_z\nabla_1^2\phi_z\right\rangle-\left\langle\omega\theta_x\phi_{zzx}\right\rangle\right)$$
$$+T_A^{1/2}\left\langle\omega\varsigma_z\right\rangle+M_1\left(\left\langle\omega\theta_x\phi_{zzx}\right\rangle+\left\langle\omega\theta_y\phi_{zzy}\right\rangle+\left\langle\omega\theta_{zy}\phi_{zy}\right\rangle\right)+\frac{1}{Pr}\left\langle\omega\hat{\mathbf{k}}\cdot curl\ curl\ \mathbf{q}\cdot\nabla\mathbf{q}\right\rangle, \tag{8.19}$$

$$\frac{1}{2\mathrm{Pr}}\frac{d\|\nabla\zeta\|^2}{dt}=-\|\nabla^2\zeta\|^2-F\|\nabla^2\zeta\|^2+T_A^{1/2}\left\langle w_z\zeta\right\rangle$$
$$+M_1\left(\zeta\theta_y\phi_{zx}-\zeta\theta_x\phi_{zy}\right)-\frac{1}{Pr}\left\langle\zeta\hat{\mathbf{k}}\cdot curl\mathbf{q}\cdot\nabla\mathbf{q}\right\rangle. \tag{8.20}$$

$$\frac{1}{2}\frac{d\|\theta_z\|^2}{dt}=-\|\nabla\theta_z\|^2+\theta_{zz}\mathbf{q}\cdot\nabla\theta+R^{1/2}\left\langle w_z\theta_z\right\rangle, \tag{8.21}$$

$$M_3\|\nabla\phi\|^2-\left(M_3-1\right)\|\phi_z\|^2+\left\langle\theta_z\phi\right\rangle=0. \tag{8.22}$$

Here, $\cdot$ denotes integration over V, $\|\cdot\|$ denotes the $L^2(V)$ norm and V denotes usual periodicity cell. Then, an L^2 energy $E(t)$ is constructed and change in $E(t)$ is given by

$$\frac{dE}{dt} = I_0 - D_0 + N_0,$$

where E, I_0, D_0 and N_0, are expressed as below

$$E = \frac{1}{2}\|\theta_z\|^2 + \frac{\lambda_1}{2\mathrm{Pr}}\|\nabla\omega\|^2 - \frac{\lambda_3}{2}\|\zeta\|^2, \tag{8.23}$$

$$\begin{aligned} I_0 &= \lambda_1(1+M_1)R^{1/2}\nabla_1\omega\nabla_1\theta - R^{1/2}\lambda_1 M_1\nabla_1\omega\nabla_1\phi_z \\ &\quad + T_A^{1/2}\lambda_1\langle\omega\zeta_z\rangle - T_A^{1/2}\lambda_3\langle\zeta\omega_z\rangle - \lambda_2\langle\phi\theta_z\rangle + R^{1/2}\lambda_1\langle\theta_z\omega_z\rangle, \end{aligned} \tag{8.24}$$

$$\begin{aligned} D_0 &= \|\nabla\theta_z\|^2 + \lambda_1\|\nabla^2\omega\|^2 + \lambda_1 F\|\nabla^3\omega\|^2 - \lambda_3\|\nabla\zeta\|^2 \\ &\quad - \lambda_3 F\|\nabla^2\zeta\|^2 + \lambda_2 M_3\|\nabla\phi\|^2 - \lambda_2(M_3-1)\|\phi_z\|^2 \end{aligned} \tag{8.25}$$

and

$$\begin{aligned} N_0 &= \lambda_1 M_1\left(-\langle\omega\phi_{zz}\nabla_1^2\theta\rangle + \langle\omega\theta_z\nabla_1^2\phi_z\rangle - \langle\omega\theta_x\phi_{zzx}\rangle\right) \\ &\quad + \lambda_1 M_1\left(\langle\omega\theta_{zx}\phi_{zx}\rangle + \langle\omega\theta_y\phi_{zzy}\rangle + \langle\omega\theta_{zy}\phi_{zy}\rangle\right) + \langle\mathbf{q}.\theta_{zz}\nabla\theta\rangle \\ &\quad - \lambda_3 M_1\left(\langle\zeta\theta_y\phi_{zx}\rangle - \langle\zeta\theta_x\phi_{zy}\rangle\right) - \frac{\lambda_3}{Pr}\langle\zeta\,\hat{\mathbf{k}}\cdot curl\mathbf{q}\cdot\nabla\mathbf{q}\rangle + \frac{\lambda_1}{Pr}\langle w\hat{\mathbf{k}}\cdot curl\,curl\,\mathbf{q}\cdot\nabla\mathbf{q}\rangle, \end{aligned} \tag{8.26}$$

where $\lambda_1, \lambda_2, \lambda_3 \geq 0$ are parameters used for coupling.

The negative sign in the third term of Equation (8.23) reflects the energy drain due to rotation, which consumes part of the system's available energy. To ensure the system remains determinate, we impose the condition

$$\lambda_3\|\zeta\|^2 < \frac{\lambda_1}{2}\|\nabla\omega\|^2.$$

Next, following the energy method, we define the stability parameter

$$m = \max_H, \frac{I_0}{D_0}, \tag{8.27}$$

where H is the set of admissible solutions. This leads to the inequality

$$\frac{dE}{dt} \leq -a_0 D_0 + N_0, a_0 = 1 - m,$$

which expresses that the energy functional decays in time provided the dissipation D_0 dominates the nonlinear contributions N_0.

8.3.1 Generalized Energy Functional

The nonlinear term N_0 contains cubic contributions such as $\theta_{zz}, \phi_{zz}, \phi_{zzy}$. These are difficult to control with the basic energy functional because they involve higher derivatives. To handle them, we enrich the energy by introducing an auxiliary term

$$E_1(t) = \frac{1}{2} \| \nabla\theta \|^2 + \frac{1}{2\mathrm{Pr}} \| \nabla \mathbf{q} \|^2 .$$

This term penalizes higher derivatives of the fields, thereby ensuring that nonlinearities remain bounded.

We then define the interaction, dissipation, and nonlinear contributions associated with E_1:

$$I_1 = R^{1/2}(2 + M_1)\langle \nabla w \cdot \nabla\theta \rangle - R^{1/2} M_1 \langle \nabla w \cdot \nabla \phi_z \rangle, \tag{8.28}$$

$$D_1 = \| \nabla^2 \mathbf{q} \|^2 + \| \nabla^2 \theta \|^2 + F \| \nabla^3 \mathbf{q} \|^2 \tag{8.29}$$

and

$$\begin{aligned} N_1 \quad &= \langle \nabla^2\theta\mathbf{q} \cdot \nabla\theta \rangle + \frac{1}{\mathrm{Pr}} \langle \mathbf{q} \cdot \nabla\mathbf{q} \cdot \nabla^2 \mathbf{q} \rangle + M_1 \langle \theta \nabla^2 \mathbf{q} \cdot \nabla \phi_z \rangle \\ -M_1 &\left(M_3 - \frac{1}{1+\chi} \right) \langle \phi_x \nabla^2 \mathbf{q} \cdot \nabla \phi_x \rangle - M_1 \left(M_3 - \frac{1}{1+\chi} \right) \langle \phi_y \nabla^2 \mathbf{q} \cdot \nabla \phi_y \rangle \\ &-M_1 \left(\frac{\chi}{1+\chi} \right) \langle \phi_z \nabla^2 \mathbf{q} \cdot \nabla \phi_z \rangle, \end{aligned} \tag{8.30}$$

where the terms in N_1 capture magnetization effects. The evolution of E_1 is then governed by

$$\frac{dE_1}{dt} \le I_1 - D_1 + N_1.$$

Generalized Energy Functional: To combine the controls offered by E and E_1, we follow Galdi and Straughan [5] and introduce the generalized energy functional

$$V_g(t) = E(t) + b_0 E_1(t). \tag{8.31}$$

where $b_0 > 0$ is a coupling parameter chosen to balance dissipation and interaction terms. The evolution inequality becomes

$$\frac{dV_g}{dt} \leq -a_0 D_0 + N_0 + b_0 I_1 - b_0 D_1 + b_0 N_1. \quad (8.32)$$

8.3.2 Estimates via Embedding Theorems

To bound the nonlinear terms, we apply embedding results from Sobolev spaces (Adams [38]) and inequalities such as Cauchy–Schwarz and Young. These give relations like:

$$\| \nabla\phi \| \leq \| \theta \|, \| \nabla^2 \phi_z \| \leq \| \nabla^2 \theta \|, \| \nabla\omega \| \leq \| \nabla q \|,$$
$$\sup |G| \leq C^* \| \nabla^2 G \|.$$

Here $C^* > 0$ is a constant, which depends entirely on V, and it was determined by Galdi and Straughan [5]. Such bounds ensure that higher-order derivatives of magnetic and velocity perturbations can be controlled by thermal perturbations.

Using these estimates, we deduce from Equation (8.28) that

$$b_0 I_1 \leq b_0 \varepsilon_0^2 D_1 + \frac{b_0 R}{2\varepsilon_0^2 \pi^2} \left\{ (2 + M_1)^2 + M_1^2 \right\} D_0. \quad (8.33)$$

By choosing $\varepsilon_0^2 = \frac{1}{2}, b_0 = \frac{a_0 \pi^2}{2R\left\{ (2 + M_1)^2 + M_1^2 \right\}}$, we define a combined dissipation term

$$D_2 = \frac{a_0}{2} D_0 + \frac{b_0}{2} D_1. \quad (8.34)$$

This leads to the compact inequality (using Equation (8.34) in Inequation (8.33))

$$b_0 I_1 \leq D_2. \quad (8.35)$$

8.3.3 Control of Nonlinearities

The nonlinear terms N_0 and N_1, are estimated in a similar way, giving

$$N_1 \leq C^* \left(\frac{2}{b_0} \right)^{3/2} \left[\frac{1}{\mathrm{Pr}^{1/2}} + \left\{ 1 + M_1 \left(2M_3 + \frac{2\chi - 1}{1 + \chi} \right) \right\} \right] D_2 V_g^{1/2} \quad (8.36)$$

and

$$N_0 \leq C^* \left(\frac{2}{b_0} \right)^{3/2} \left[1 + 6\lambda_1 M_1 + \frac{2\lambda_1}{\mathrm{Pr}^{1/2}} + \left(\frac{2b_0 \lambda_3}{a_0} \right) \left(M_1 + \frac{1}{\mathrm{Pr}^{1/2}} \right) \right] D_2 V_g^{1/2}. \quad (8.37)$$

Substituting these bounds into Equation (8.32) gives the compact energy inequality:

$$\dot{V}_g(t) \le -D_2\left(1 - H'V_g^{1/2}\right), \tag{8.38}$$

here,

$$\begin{aligned} H' &= 2C^*\left(\frac{2}{b_0}\right)^{3/2}\left[\frac{1}{\mathrm{Pr}^{1/2}} + \left\{1 + M_1\left(2M_3 + \frac{2\chi - 1}{1+\chi}\right)\right\} + 1 \right. \\ &\left. + 6\lambda_1 M_1 + \frac{2\lambda_1}{\mathrm{Pr}^{1/2}} + \left(\frac{2b_0\lambda_3}{a_0}\right)\left(M_1 + \frac{1}{\mathrm{Pr}^{1/2}}\right)\right]. \end{aligned} \tag{8.39}$$

This inequality establishes that the generalized energy functional V_g decays monotonically provided the initial energy is sufficiently small, i.e. $V_g^{1/2} < 1/H$. Physically:

- The linear dissipation terms dominate, ensuring stability.
- The rotation contributes as an energy sink (negative term).
- The couple stresses and magnetization modify both dissipation and nonlinear growth, shifting the critical threshold.

Thus, the generalized functional method provides a rigorous framework to account for higher-order nonlinearities and ensures boundedness of perturbations, yielding conditions for nonlinear stability.

Theorem 1. *Let us assume* $0 < m < 1$ *and* $V_g(0) < H'^{-2}$, *with* H' *expressed by Equation (8.39). Then,* $\exists\, K^* > 0$, *such that*

$$V_g(t) \le V_g(0)\exp\left(-K^*\left(1 - AH'V_g^{1/2}(0)\right)t\right), \quad t \ge 0. \tag{8.40}$$

Proof. From Inequation (8.38) it can be obtained that

$$\dot{V}_g(0) < 0. \tag{8.41}$$

Therefore, by following the procedure of [24] we can get

$$V_g(t) \le V_g(0)\exp\left(-K^*\left(1 - AH'V_g^{1/2}(0)\right)t\right), \quad t \ge 0, \tag{8.42}$$

where K^* is given by

$$K^* = \frac{\pi^2}{\left(A + \frac{1}{\mathrm{Pr}}\right)\left(1 + k_0\right)}. \tag{8.43}$$

□

As the formulation (8.31) of $V_g(t)$ doesn't encompass $\|\nabla\phi\|^2$, the term representing the kinetic energy, we use embedding theorems to observe the behavior of $\|\nabla\phi\|^2$, and get

$$\|\nabla\phi\|^2 \le \|\theta\|^2 . \tag{8.44}$$

Thus, Equation (8.31) and Inequation (8.42) guarantee the decay of $\|\nabla\phi\|^2$.

8.4 VARIATIONAL PRINCIPLE

By solving Equation (8.27) at its critical value, and applying the transformations

$$\hat{\mathbf{w}} = \sqrt{\lambda_1}\mathbf{w}, \hat{\phi} = \sqrt{\lambda_2}\phi, \hat{\zeta} = \sqrt{\lambda_3}\zeta,$$

the Euler–Lagrange equations are obtained. These transformations help rescale the variables, simplifying the energy functional and allowing us to treat the problem in terms of normalized perturbation amplitudes. Dropping the hats for convenience, we write the system as:

$$R^{1/2}\lambda_1^{1/2}\left(1+M_1\right)\nabla_1^2\theta - \frac{\lambda_1^{1/2}}{\lambda_2^{1/2}}R^{1/2}M_1\nabla_1^2\phi_z - \frac{\lambda_1^{1/2}}{\lambda_3^{1/2}}T_A^{1/2}\zeta_z - \frac{\lambda_3^{1/2}}{\lambda_1^{1/2}}T_A^{1/2}\zeta_z$$

$$+\frac{R^{1/2}}{\lambda_1^{1/2}}\theta_{zz} + 2F\nabla^4\omega - 2F\nabla^6\omega = p, \tag{8.45}$$

$$2\nabla^2\theta_{zz} + R^{1/2}\lambda_1^{1/2}\left(1+M_1\right)\nabla_1^2\omega - \lambda_2^{1/2}\phi_z + \frac{R^{1/2}}{\lambda_1^{1/2}}\omega_{zz} = 0, \tag{8.46}$$

$$R^{1/2}M_1\frac{\lambda_1^{1/2}}{\lambda_2^{1/2}}\nabla_1^2\omega_z - 2M_3\nabla^2\phi + \lambda_2^{1/2}\theta_z + 2\left(M_3-1\right)\phi_{zz} = 0, \tag{8.47}$$

$$\frac{\lambda_1^{1/2}}{\lambda_3^{1/2}}T_A^{1/2}\omega_z + \frac{\lambda_3^{1/2}}{\lambda_1^{1/2}}T_A^{1/2}\omega_z + 2F\nabla^2\zeta - 2F\nabla^4\zeta = 0. \tag{8.48}$$

Here, $p$represents a Lagrange multiplier that enforces the incompressibility condition, $\nabla\cdot\mathbf{q} = 0$.

To solve this system, we assume perturbations of the form:

$$(\omega, \theta, \phi) = \left[W(z), \Theta(z), \Phi(z)\right] f(x, y), \tag{8.49}$$

where the horizontal dependence is captured by

$$\nabla_1^2 f + a^2 f = 0,$$

with a denoting the horizontal wave number. This separation into vertical and horizontal modes is standard in linear stability analysis, as it isolates the governing dynamics into tractable one-dimensional problems.

Substituting Equation (8.49) into Equations (8.45)–(8.48), we arrive at:

$$\begin{aligned}&2\left(D^2-a^2\right)^2 W - 2F\left(D^2-a^2\right)^3 W - R^{1/2}\lambda_1^{1/2}\left(1+M_1\right)a^2\Theta \\ &+\frac{\lambda_1^{1/2}}{\lambda_2^{1/2}}R^{1/2}M_1 a^2 D\Phi - \frac{\lambda_1^{1/2}}{\lambda_3^{1/2}}T_A^{1/2}D\Psi - \frac{\lambda_3^{1/2}}{\lambda_1^{1/2}}T_A^{1/2}D\Psi + \frac{R^{1/2}}{\lambda_1^{1/2}}D^2\Theta = 0,\end{aligned} \tag{8.50}$$

$$2\left(D^2-a^2\right)D^2\Theta - R^{1/2}\lambda_1^{1/2}\left(1+M_1\right)a^2 W - \lambda_2^{1/2}D\Phi + \frac{R^{1/2}}{\lambda_1^{1/2}}D^2 W = 0, \tag{8.51}$$

$$R^{1/2}M_1\frac{\lambda_1^{1/2}}{\lambda_2^{1/2}}a^2 DW + 2\left(D^2 - M_3 a^2\right)\Phi - \lambda_2^{1/2}D\Theta = 0 \tag{8.52}$$

$$2\left(D^2-a^2\right)\Psi - 2F\left(D^2-a^2\right)^2\Psi + \frac{\lambda_1^{1/2}}{\lambda_3^{1/2}}T_A^{1/2}DW + \frac{\lambda_3^{1/2}}{\lambda_1^{1/2}}T_A^{1/2}DW = 0. \tag{8.53}$$

The corresponding boundary conditions depend on the type of bounding surfaces

1. Bounding surfaces: Free–free

$$W = D^2W = D^4W = \Theta = D^2\Theta = D\Psi = D^3\Psi = D\Phi = 0, \text{ at } z = 0,1. \tag{8.54}$$

2. Bounding surfaces: Rigid–free

$$\begin{aligned}W = DW = D^3W = \Theta = D^2\Theta = \Psi = D^2\Psi = D\Phi = 0, \quad \text{at } z = 0,\\ W = D^2W = D^4W = \Theta = D^2\Theta = D\Psi = D^3\Psi = D\Phi = 0, \quad \text{at } z = 1.\end{aligned} \tag{8.55}$$

3. Bounding surfaces: Rigid–free

$$W = DW = D^3W = \Theta = D^2\Theta = \Psi = D^2\Psi = D\Phi = 0, \quad \text{at } z = 0,1. \tag{8.56}$$

Finally, for the linear stability analysis, we use the above normal-mode decomposition on Equations (8.8)–(8.11) while discarding the nonlinear terms. This leads directly to an eigenvalue problem, whose solution provides the critical Rayleigh number and corresponding eigenfunctions. The eigenvalue problem, supplemented with the appropriate boundary conditions, forms the foundation for determining the stability characteristics of the system.

8.5 METHOD OF SOLUTION

8.5.1 Free–Free Boundaries

The Galerkin method is used on Equations (8.50)–(8.53) to obtain critical Rayleigh numbers. The Galerkin method is a weighted residual technique that transforms a system of partial differential equations into a set of algebraic equations by expanding the dependent variables in terms of suitable trial functions. These trial functions are chosen to automatically satisfy the imposed boundary conditions, thereby reducing the problem to determining a finite number of unknown coefficients.

In this approach, the residuals obtained after substituting the assumed trial functions into the governing equations are forced to be orthogonal to the space spanned by the basis functions. This requirement ensures that the error is minimized in an average sense across the entire domain, rather than at individual points. Mathematically, this is achieved by taking inner products of the residuals with each basis function and setting them equal to zero.

For the present problem, sine and cosine functions (for the free–free case) and suitably constructed polynomials (for rigid–free and rigid–rigid cases) are employed as basis functions, since these satisfy the respective boundary conditions. Each dependent variable—vertical velocity W, temperature perturbation Θ, microrotation Ψ, and magnetic potential perturbation Φ – is expanded in terms of these functions with unknown coefficients. The Galerkin process then reduces the system of partial differential equations into a matrix eigenvalue problem involving these coefficients.

A non-trivial solution to this system exists only if the determinant of the coefficient matrix vanishes, leading to the characteristic equation for the critical parameter N_e. This procedure is carried out separately for each boundary configuration, yielding explicit expressions for N_e in terms of the system parameters. We expand the dependent variables in terms of basis functions that satisfy the prescribed boundary conditions (Equation 8.54). For the free–free case, these expansions are taken as

$$W = \sum_{i=1}^{n} A_i W_i, W_i(z) = \sin i\pi z; \quad \Theta = \sum_{i=1}^{n} B_i \Theta_i, \Theta_i(z) = \sin i\pi z;$$

$$\Psi = \sum_{i=1}^{n} C_i \Psi_i, \Psi_i(z) = \cos i\pi z; \quad \Phi = \sum_{i=1}^{n} D_i \Phi_i, \Phi_i(z) = \cos i\pi z. \tag{8.57}$$

Here, A_i, B_i, C_i and D_i represent unknown constants. These trial functions are substituted into Equations (8.50)–(8.53), and enforcing the Galerkin condition (i.e., the residuals are orthogonal to the space spanned by the basis functions), leads to a system of algebraic equations:

$$
\begin{aligned}
&L_{ji}A_i + M_{ji}B_i + N_{ji}C_i + J_{ji}D_i = 0,\\
&O_{ji}A_i + P_{ji}B_i + Q_{ji}C_i + K_{ji}D_i = 0,\\
&G_{ji}A_i + F_{ji}B_i + H_{ji}C_i + I_{ji}D_i = 0,\\
&R_{ji}A_i + S_{ji}B_i + T_{ji}C_i + E_{ji}D_i = 0.
\end{aligned} \tag{8.58}
$$

where the coefficients coefficients E_{ji} to T_{ji} are obtained from inner products of the form

$$
\cdots = \int_0^1 (\cdots)\, dz,
$$

which represent weighted contributions of the basis functions and their derivatives. For a non-trivial solution to exist, the determinant of the coefficient matrix must vanish:

$$
\begin{vmatrix}
L_{ji} & M_{ji} & N_{ji} & J_{ji}\\
O_{ji} & P_{ji} & Q_{ji} & K_{ji}\\
G_{ji} & F_{ji} & H_{ji} & I_{ji}\\
R_{ji} & S_{ji} & T_{ji} & E_{ji}
\end{vmatrix} = 0.
$$

This condition forms the characteristic equation of the system, from which the critical parameter $N_e$is determined. We use $\lambda_1, \lambda_2^{'}$ and λ_3 as

$$
\lambda_1 = \lambda_3 = \frac{1}{x(1+M_1)} \quad \text{and} \quad \lambda_2' = \frac{(1+x)M_1}{1+M_1}.
$$

and get

$$
N_e = M_1 R_e = \frac{(3+4xM_3)\left((1+x)^3 + 2(1+x)^4 F_1 + (1+x)^5 F_1^2 + T_1\right)}{2x(\ 1+(1+x)F_1)(1+2xM_3)}. \tag{8.59}
$$

for large value of M_1. Here,

$$
R_e = \frac{R}{\pi^4}, x = \frac{a^2}{\pi^2}, F_1 = \pi^2 F, J_1 = \pi^2 J, T_1 = \pi^2 Ta \text{ and } \lambda_2' = \frac{\lambda_2}{\pi^4}.
$$

When $F = 0$, this result (Equation 8.59) aligns entirely with the findings presented by Sunil and Mahajan [27]. Additionally, Devi et al. [39] validate this result for $M_3 = 0$. For linear analysis,

$$N_e = M_1 R_e = \frac{(1+x)(1+xM_3)\left\{\left(T_1(1+x)\left(1+(F_1+D_aV_r)(1+x)\right)^2 D_a^2\right)\right\}}{D_a x^2 (1+x)\left(1+(F_1+D_aV_r+F_1x+D_aV_rx)\right)M_3}. \quad (8.60)$$

8.5.2 Rigid–Free Boundaries

For rigid–free bounding surfaces (Equation 8.55), appropriate trial functions must satisfy a rigid condition at one boundary and a free condition at the other. The following set of functions is chosen accordingly:

$$\begin{aligned} W_i(z) &= 0.8z^{2i+4} - 3.05z^{2i+3} + 3.25z^{2i+2} - z^{2i}, \\ \Theta_i(z) &= z^{4i} - 2z^{2i+1} + z, \\ \Phi_i(z) &= 3z^{2i} - 2z^{2i+1}, \\ \Psi_i(z) &= z^{2i+4} - 3z^{2i+3} + 10z^{2i+1} - 21z^{i}. \end{aligned} \quad (8.61)$$

These basis functions ensure that the boundary conditions are satisfied at $z = 0,1$ while providing sufficient flexibility for the Galerkin expansion. Substituting (8.61) into the governing equations (8.50)–(8.53), and following the same Galerkin procedure described in Section 8.5.1, we obtain the following expression for the critical parameter:

$$N_e = \frac{(84+85x)\left\{\begin{array}{l} 59822070757T_1 + 381600x + 164990x^2 + 25250x^3 \\ \quad + 20F_1(2038608 + 2046330x + 907830x^2 \\ \quad + 231235x^3 + 25250x^4) + 200F_1^2(4077216 \\ \quad + 5096520x + 255856x^2 + 779130x^3 \\ \quad + 148740x^4 + 126250x^5) \end{array}\right\}(211+832xM_3)}{153x\left(99+25x+10F_1\left(396+198x+25x^2\right)\right)(7407+45500xM_3)}. \quad (8.62)$$

8.5.3 Rigid–Rigid Boundaries

For rigid–rigid bounding surfaces (Equation 8.56), both velocity and microrotation must vanish at the boundaries. To satisfy these conditions, the following trial functions are chosen:

$$\begin{aligned} W_i(z) &= z^{2i+4} - 3z^{2i+3} + 2.5z^{2i+2} - 0.5z^{2i}, \\ \Theta_i(z) &= z^{4i} - 2z^{2i+1} + z, \\ \Phi_i(z) &= 3z^{2i} - 2z^{2i+1}, \\ \Psi_i(z) &= 6z^{2i+3} + 15z^{2i+2} + 10z^{2i+1} - z^{i}. \end{aligned} \quad (8.63)$$

Substituting (8.63) into Equations (8.50)–(8.53) and applying the Galerkin process yields the following expression for N_e:

$$N_e = \frac{(84+85x)\left\{\begin{array}{c} 509652+325T_1+381600x+164990x^2 \\ +25250x^3+20F_1(2038608+2046330x \\ +907830x^2+231235x^3+25250x^4) \\ +200F_1^2(4077216+5096520x+255856x^2 \\ +779130x^3+148740x^4+126250x^5) \end{array}\right\}(211+832xM_3)}{153x\left(99+25x+10F_1\left(396+198x+25x^2\right)\right)\left(7407+45500xM_3\right)}. \tag{8.64}$$

8.6 RESULTS AND DISCUSSION

As obvious from the expressions obtained in the previous section, the critical parameters N_{cl} (linear theory) and N_{ce} (nonlinear theory) depend on the magnetization, couple stresses, and rotation. The effects of these parameters are then studied by varying these parameters and observing the corresponding values of Rayleigh numbers. The variation with respect to these parameters is presented in Figures 8.2–8.9. Since nonlinear effects are central to the present study, comparatively larger values of the magnetic nonlinearity parameter M_3 are considered, following the observation of Finlayson [8]. We now study the effects of these parameters one by one:

- **Effect of Magnetization:** Figures 8.2–8.4 illustrate the variation of N_{cl} and N_{ce} with respect to M_3 for free–free, rigid–free, and rigid–rigid boundaries. In all cases, the two thresholds differ, signifying the existence of a subcritical region. At lower values of M_3, this region is relatively wide; however, as M_3 increases, N_{cl} and N_{ce} approach each other, thereby reducing the extent of subcriticality.
  - Physically, M_3 represents the nonlinearity in the magnetization law. As its value increases, the magnetic body force becomes stronger and more nonlinear in nature. This enhanced force amplifies buoyancy-driven perturbations in the ferrofluid, reducing the threshold Rayleigh numbers and promoting the onset of convection. The effect is most pronounced for free–free boundaries, where the fluid is less constrained, allowing perturbations to grow more rapidly. By contrast, rigid boundaries suppress vertical motion and vorticity at the surfaces, which narrows the subcritical region and delays instability compared to free boundaries.
  - The overall outcome is that magnetization destabilizes the system by effectively lowering the energy barrier required for convective motion. These findings agree with Sunil and Mahajan [40], and confirm that even in the presence of couple stresses and rotation, nonlinear magnetization continues to promote earlier onset of convection and diminish the subcritical regime.

Figures 8.5–8.7 depict the variation of N_{cl} and N_{ce} with respect to the couple stress parameter F_1, while keeping M_3, J_1, and T_1 fixed at small values so as not to significantly alter the stability region. The parameter F_1 is varied between 0 and 1, where $F_1 = 0$ corresponds to the absence of couple stresses [36].

It is observed that increasing F_1 raises both N_{cl} and N_{ce}, thereby enlarging the subcritical region and enhancing system stability. This stabilizing influence is evident across all boundary configurations, though its manifestation differs with boundary type. Free–free surfaces show a wider subcritical region, reflecting the ease with which perturbations grow in unconstrained systems. In contrast, rigid–rigid surfaces exhibit the largest increase in Rayleigh numbers, signifying stronger thermal stability under rigid confinement. Rigid–free surfaces display intermediate behavior.

Physically, couple stresses introduce an additional resistance to microrotation in the fluid, which acts as a damping mechanism against small-scale disturbances. This suppresses the amplification of perturbations and delays the onset of convection. As a result, the system requires higher Rayleigh numbers to become unstable. Importantly, this stabilizing role of F_1 persists even in the presence of magnetization and rotation, highlighting the robustness of couple stresses in counteracting destabilizing influences.

Effect of Rotation: Figures 8.6–8.9 show the influence of the rotational parameter T_1 on the critical thresholds N_{cl} and N_{ce} for free–free, rigid–free, and rigid–rigid boundaries. The parameter T_1 is varied from zero to 100,000 [1], covering a wide range of rotational effects. For all boundary configurations, the linear and nonlinear thresholds differ, confirming the existence of a subcritical region.

At low rotation rates, rigid–rigid boundaries exhibit the highest critical values, indicating greater stability due to the strong constraint on vertical motion. However, as T_1 increases, this trend reverses: free–free boundaries become comparatively more stable, whereas rigid–rigid cases show reduced resistance to convection.

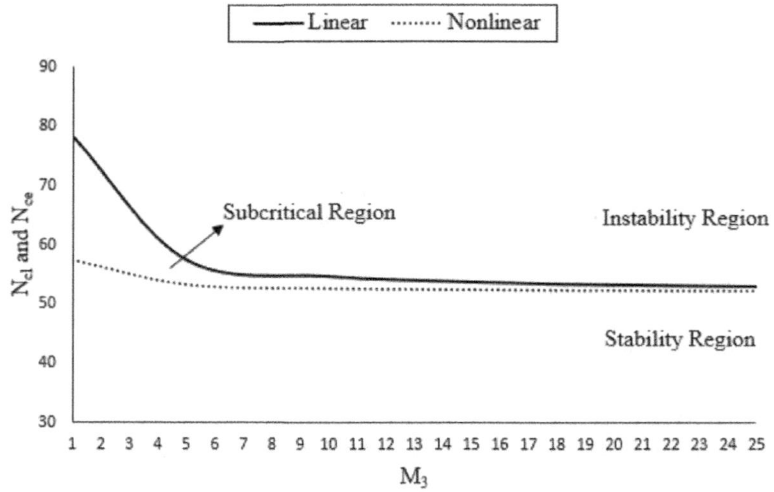

FIGURE 8.2 Effect of magnetic parameter M_3 on N_{cl} and N_{ce} for free–free boundaries with $F_1 = 0.4$, $J_1 = 0.03$ and $T_1 = 100$.

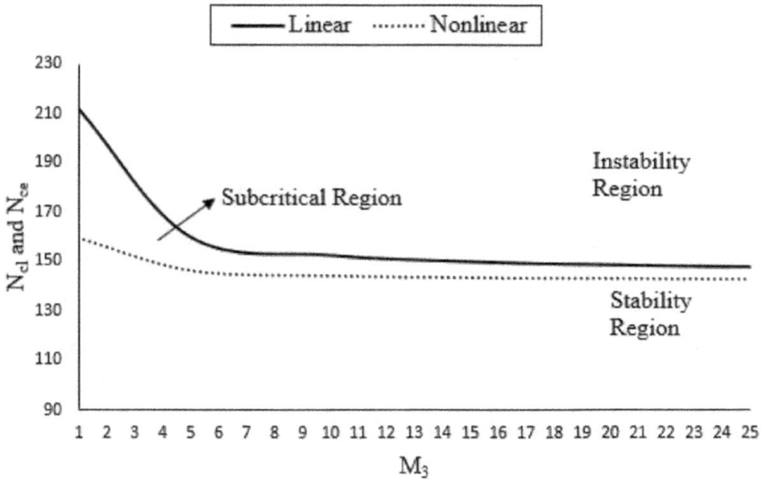

FIGURE 8.3 Effect of magnetic parameter M_3 on N_{cl} and N_{ce} for rigid-free boundaries with $F_1 = 0.4$, $J_1 = 0.03$ and $T_1 = 100$

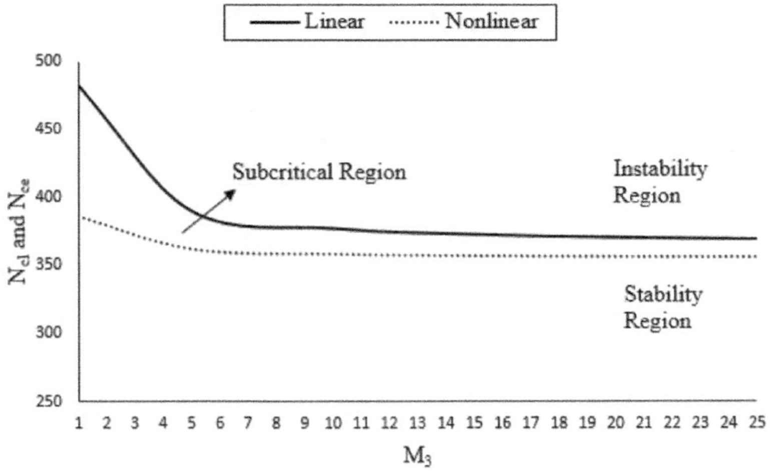

FIGURE 8.4 Effect of magnetic parameter M_3 on N_{cl} and N_{ce} for for rigid-rigid boundaries with $F_1 = 0.4$, $J_1 = 0.03$ and $T_1 = 100$.

Additionally, the size of the subcritical region expands with increasing T_1, with the most pronounced growth observed under rigid–free boundary conditions.

Physically, rotation introduces Coriolis forces, which suppress vertical fluid motions and inhibit the growth of convective disturbances. This explains the stabilizing effect observed at low T_1. At higher rotation rates, however, rotational effects interact strongly with boundary constraints. Free boundaries, being less restrictive, allow the Coriolis force to redistribute energy more effectively, leading to enhanced stability compared to rigid surfaces. The rigid–free configuration, on the other hand,

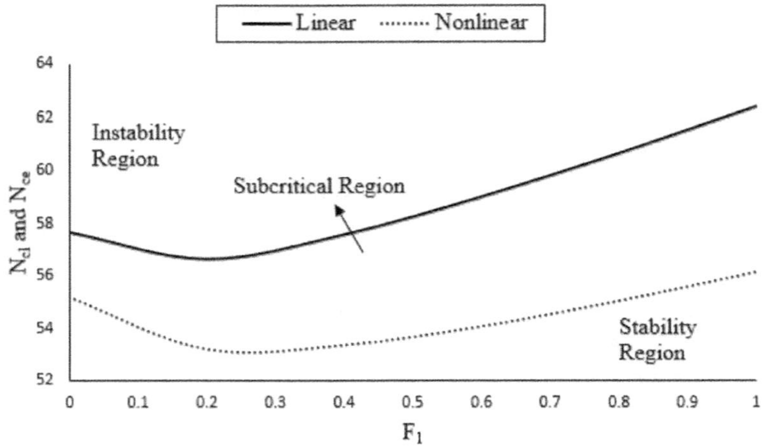

FIGURE 8.5 Effect of couple stress parameter F_1 on N_{cl} and N_{ce} for free-free boundaries with $M_3 = 5$, $J_1 = 0.03$ and $T_1 = 100$.

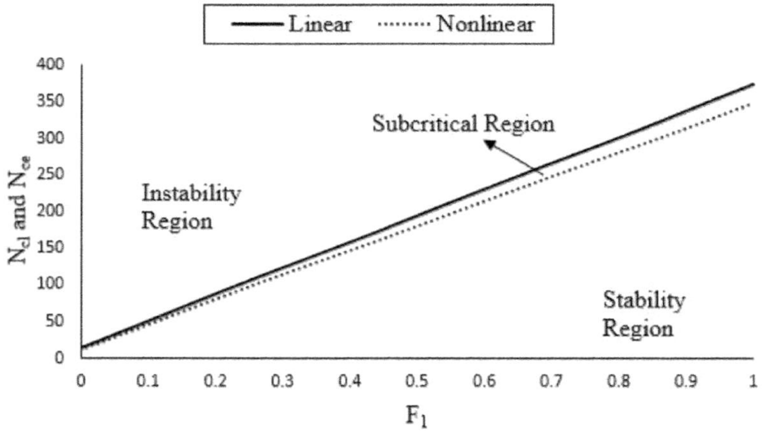

FIGURE 8.6 Effect of couple stress parameter F_1 on N_{cl} and N_{ce} for rigid-free boundaries with $M_3 = 5$, $J_1 = 0.03$ and $T_1 = 100$.

experiences competing influences from one constrained and one unconstrained boundary, resulting in the largest expansion of the subcritical region.

In the absence of rotation, the present results reduce to those of Thakur et al. [24], thereby validating the analysis.

8.7 CONCLUSIONS

The present work investigates the ferroconvection in the presence of rotation and couple stress forces. The analysis is carried out for three different types of boundary

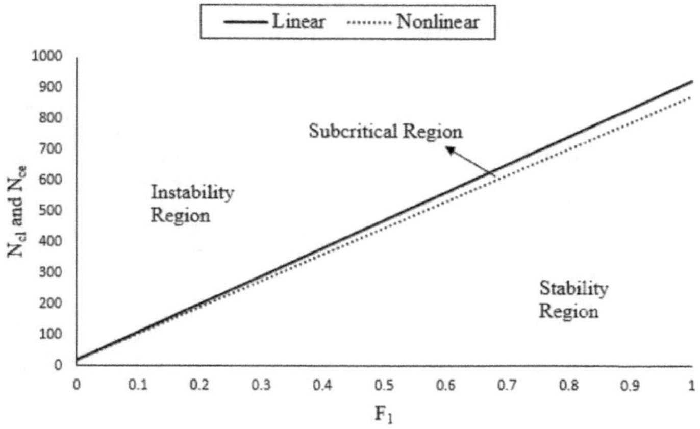

FIGURE 8.7 Effect of couple stress parameter F_1 on N_{cl} and N_{ce} for rigid-rigid boundaries with $M_3 = 5$, $J_1 = 0.03$ and $T_1 = 100$.

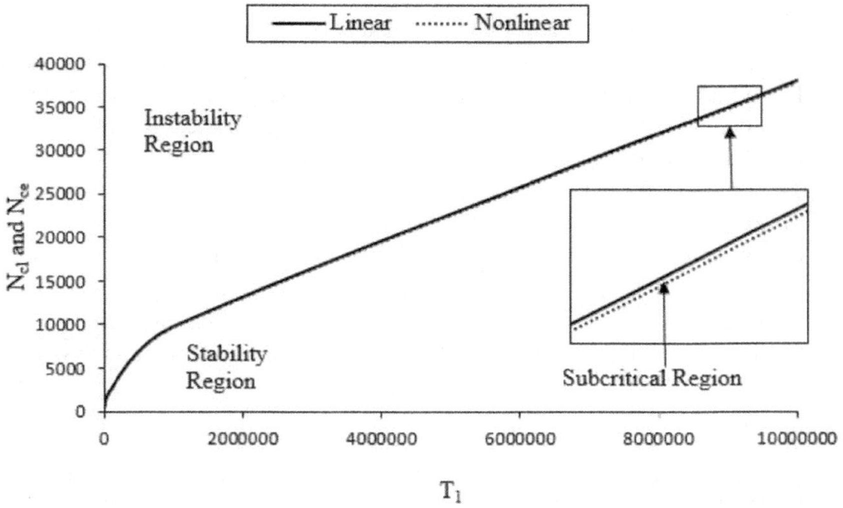

FIGURE 8.8 Effect of rotational parameter T_1 on N_{cl} and N_{ce} for free-free boundaries with $M_3 = 5$, $J_1 = 0.03$ and $T_1 = 0.4$.

conditions. Linear stability is examined using the normal mode method, while nonlinear stability is assessed through the energy method. The resulting eigenvalue problems are solved via a single-term Galerkin approximation. Alongside the findings, which were visually presented, there was a thorough discussion. The following is a summary of the main conclusions:

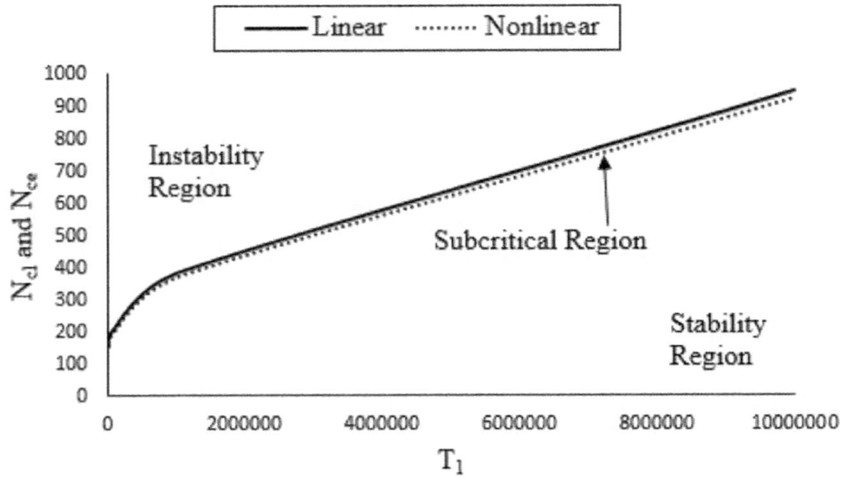

FIGURE 8.9 Effect of rotational parameter T_1 on N_{cl} and N_{ce} for rigid-free boundaries with $M_3 = 5$, $J_1 = 0.03$ and $F_1 = 0.4$.

- The results confirm the existence of a subcritical region, since the linear Rayleigh number is consistently higher than the nonlinear Rayleigh number.
- Ferrofluids confined between free–free boundaries exhibit greater thermal stability at higher rotation rates, which is favorable for the onset of convection. This finding addresses the second objective of the study.
- The extent of the subcritical region decreases when the boundary configuration is changed from free to rigid surfaces.
- Couple stresses and rotation were the parameters that delayed the commencement of convection, while magnetization aided in the process.
- In the absence of couple stresses, the results are in good agreement with those reported in earlier studies, thereby validating the present analysis.

REFERENCES

1. S. Chandrasekhar, *Hydrodynamic and Hydromagnetic Stability* (Dover Publications, New York, 1981)
2. O. Reynolds, IV. On the dynamical theory of incompressible viscous fluids and the determination of the criterion. Philos. Trans. R. Soc. London, Ser. A 186:123–164 (1895). https://doi.org/10.1098/rsta.1895.0004
3. W.M. Orr, The stability or instability of the steady motions of a perfect liquid and of a viscous liquid. in *Proceedings of the Royal Irish Academy. Section A: Mathematical and Physical Sciences* (1907), pp. 69–138. https://www.jstor.org/stable/20490591
4. G.P. Galdi, Nonlinear stability of the magnetic B´enard problem via a generalized energy method. *Arch. Ration. Mech. Anal.* **87**, 167–186 (1985). https://doi.org/10.1007/BF00280699

5. G.P. Galdi, B. Straughan, A nonlinear analysis of the stabilizing effect of rotation in the B´enard problem. *Proc. R. Soc. London, Ser. A* **402**, 257–283 (1985). https://doi.org/10.1098/rspa.1985.0118
6. R.E. Rosensweig, *Ferrohydrodynamics* (Cambridge University Press, Cambridge, 1985)
7. M. Kole, S. Khandekar, Engineering applications of ferrofluids: A review. *J. Magn. Magn. Mater.* **537**, 168222 (2021). https://doi.org/10.1016/j.jmmm.2021.168222
8. B.A. Finlayson, Convective instability of ferromagnetic fluids. *J. Fluid Mech.* **40**(4), 753–767 (1970). https://doi.org/10.1017/S0022112070000423
9. M. Shliomis, Convective instability of a ferrofluid. *Fluid Dyn.* **8**(6), 957–961 (1973). https://doi.org/10.1007/BF01014273
10. G. Vaidyanathan, R. Sekar, R. Balasubramanian, Ferroconvective instability of fluids saturating a porous medium. *Int. J. Eng. Sci.* **29**(10), 1259–1267 (1991). https://doi.org/10.1016/0020-7225(91)90029-3
11. R. Sekar, G. Vaidyanathan, Convective instability of a magnetized ferrofluid in a rotating porous medium. *Int. J. Eng. Sci.* **31**(8), 1139–1150 (1993). https://doi.org/10.1016/0020-7225(93)90087-B
12. S. Venkatasubramanian, P. Kaloni, Effects of rotation on the thermoconvective instability of a horizontal layer of ferrofluids. *Int. J. Eng. Sci.* **32**(2), 237–256 (1994). https://doi.org/10.1016/0020-7225(94)90004-3
13. A. Mahajan, H. Parashar, Linear and weakly nonlinear stability analysis on a rotating anisotropic ferrofluid layer. *Phys. Fluids* **32**(2), 024101 (2020). https://doi.org/10.1063/1.5133102
14. J.S. Dhiman, S. Sood, Linear and weakly non-linear stability analysis of oscillatory convection in rotating ferrofluid layer. *Appl. Math. Comput.* **430**, 127239 (2022). https://doi.org/10.1016/j.amc.2022.127239
15. D. Laroze, P. Siddheshwar, H. Pleiner, Chaotic convection in a ferrofluid. *Commun. Nonlinear Sci. Numer. Simul.* **18**(9), 2436–2447 (2013). https://doi.org/10.1016/j.cnsns.2013.01.016
16. Sunil, A. Mahajan, A nonlinear stability analysis for thermoconvective magnetized ferrofluid saturating a porous medium. *Transp. Porous Media* **76**, 327–343 (2009). https://doi.org/10.1007/s11242-008-9249-1
17. Sunil, P. Sharma, A. Mahajan, A nonlinear stability analysis of a rotating double-diffusive magnetized ferrofluid. *Appl. Math. Comput.* **218**(6), 2785–2799 (2011). https://doi.org/10.1016/j.amc.2011.08.021
18. I. Shivakumara, J. Lee, C. Nanjundappa, M. Ravisha, Ferromagnetic convection in a rotating ferrofluid saturated porous layer. *Transp. Porous Media* **87**, 251–273 (2011). https://doi.org/10.1007/s11242-010-9678-5
19. C. Nanjundappa, I. Shivakumara, H. Prakash, Effect of coriolis force on thermomagnetic convection in a ferrofluid saturating porous medium: A weakly nonlinear stability analysis. *J. Magn. Magn. Mate*r. **370**, 140–149 (2014). https://doi.org/10.1016/j.jmmm.2014.06.035
20. Sunil, Divya, V. Sharma, Effect of dust particles on a rotating ferromagnetic fluid heated from below saturating a porous medium. *J. Colloid and Interface Sci.* **291**(1), 152–161 (2005). https://doi.org/10.1016/j.jcis.2005.04.106.
21. T. Javed, Z. Mehmood, Z. Abbas, Natural convection in square cavity filled with ferrofluid saturated porous medium in the presence of uniform magnetic field. *Physica B* **506**, 122–132 (2017). https://doi.org/10.1016/j.physb.2016.11.008
22. R. Canu, M.C. Renoult, Linear stability analysis of a newtonian ferrofluid cylinder under a magnetic field. *J. Fluid Mech.* **915**, A137 (2021). https://doi.org/10.1017/jfm.2021.171

23. K. Dev, O.P. Suthar, Nonlinear stability analysis of penetrative convection in ferrofluids via internal heating. *J. Magn. Magn. Mater.* **564**(2), 170064 (2022). https://doi.org/10.1016/j.jmmm.2022.170064
24. A. Thakur, S. Kumar, R. Devi, A nonlinear stability analysis for magnetized ferrofluid heated from below in the presence of couple stresses for combination of different bounding surfaces. *Numer. Heat Transf. B: Fundam.* (2023). https://doi.org/10.1080/10407790.2023.2275728
25. A. Thakur, Sunil, R. Devi, The effect of couple stresses on stability analysis of magnetized ferrofluid saturating a porous medium heated from below. *J. Porous Media* (in press). https://doi.org/10.1615/JPorMedia.2023050054
26. Sunil, A. Mahajan, A nonlinear stability analysis for rotating magnetized ferrofluid heated from below. *Appl. Math. Comput.* **204**(1), 299–310 (2008). https://doi.org/10.1016/j.amc.2008.06.043
27. Sunil, A. Mahajan, A nonlinear stability analysis for rotating magnetized ferrofluid heated from below saturating a porous medium. Z. Angew. *Math. Phys.* **60**(2), 344–362 (2009). https://doi.org/10.1007/s00033-008-7133-z
28. V. Stokes, *Theories of Fluids with Microstructure: An Introduction* (Springer Science & Business Media, 2012). https://doi.org/10.1007/978-3-642-82351-0
29. Sunil, R. Devi, A. Mahajan, Global stability for thermal convection in a couple-stress fluid. *Int. Commun. Heat Mass Transfer* **38**(7), 938–942 (2011). https://doi.org/10.1016/j.icheatmasstransfer.2011.03.030
30. Sunil, R. Devi, Global stability for thermal convection in a couple stress fluid: Rigid boundaries. *Heat Transfer Res.* **45**(1), 43–57 (2014). https://doi.org/10.1615/HeatTransRes.2013003569
31. I. Shivakumara, S.B.N. Kumar, Linear and weakly nonlinear triple diffusive convection in a couple stress fluid layer. *Int. J. Heat Mass Transfer* **68**, 542–553 (2014). https://doi.org/10.1016/j.ijheatmasstransfer.2013.09.051
32. B.M. Shankar, J. Kumar, I.S. Shivakumara, Stability of natural convection in a vertical couple stress fluid layer. *Int. J. Heat Mass Transfer* **78**, 447–459 (2014). https://doi.org/10.1016/j.ijheatmasstransfer.2014.06.087
33. A. Mahajan, R. Nandal, On the stability of penetrative convection in a couple-stress fluid. *Int. J. Appl. Comput. Math.* **3**, 3745–3758 (2017). https://doi.org/10.1007/s40819-017-0324-6
34. S. Choudhary, R. Devi, A. Mahajan, Sunil, Stability analysis in a couple-stress fluid layer with variable viscosity heated from below: Different conducting boundaries. *Chin. J. Phys.* **83**, 94–102 (2023). https://doi.org/10.1016/j.cjph.2023.02.007
35. E.A. Spiegel, Convective instability in a compressible atmosphere. I. *Astrophys. J.* **141**, 1068–1090 (1965). https://doi.org/10.1086/148197
36. V. Stokes, Couple stress in fluids. *Phys. Fluids* **9**, 1709–1715 (1966). https://doi.org/10.1063/1.1761925
37. D.D. Joseph, *Stability of Fluid Motions I*, vol. 27 (Springer Science & Business Media, 2013). https://doi.org/10.1007/978-3-642-80991-0
38. R. Adams, *Sobolev Spaces* (Academic Press, New York, 1975)
39. R. Devi, Sunil, R. Paul, The effect of rotation on thermal convection in a couple-stress fluid saturating a porous medium using galerkin method: A nonlinear stability analysis. *Res. J. Sci. Tech* **5**(1), 10–19 (2013). https://doi.org/https://rjstonline.com/AbstractView.aspx?PID=2013-5-1-2
40. Sunil, A. Mahajan, A nonlinear stability analysis for magnetized ferrofluid heated from below. *Proc. R. Soc. A* **464**(2089), 83–98 (2008). https://doi.org/10.1098/rspa.2007.1906

9 Magnetothermal Instabilities in Fluid Systems

Deepak Kumar, Sweta Sharma, and Sunil

9.1 INTRODUCTION

Thermal convection refers to heat transfer through fluid motion driven by temperature differences, leading to circulation of warm and cool regions. Chandrasekhar [14] offered a detailed account using linear stability theory, which identifies conditions for definite instability but not for stability. In contrast, the nonlinear energy method provides conditions ensuring stability but cannot fully establish instability [58]. Originating with Reynolds [50] and Orr [46], the method was later extended through the generalized approach by Galdi [18] and Galdi and Straughan [20].

The study of viscoelastic fluids has gained importance across science and engineering because of their combined viscous and elastic properties. Joseph [27] offered a comprehensive account of fluid rheology covering viscosity and elasticity, along with models for different viscoelastic fluids. Contributions by Khayat [29], Li and Khayat [37], Malashetty et al. [38], Hayat et al. [24], and Demir [15] further advanced the field. Foundational theories are given by the Maxwell, Oldroyd, and Kelvin–Voigt models. Straughan [60] provided constitutive equations for these models and showed that solutions of boundary–initial value problems for first-order Kelvin–Voigt fluids depend continuously on viscoelastic coefficients, viscosity, and Kelvin–Voigt parameters. Models for the Kelvin–Voigt fluid, presented by Oskolkov [47], describe linear viscoelastic behavior under small deformations, while his later work [48] examined nonlocal problems of the equations of motion. The zeroth-order Kelvin–Voigt fluid corresponds to the Navier–Stokes–Voigt fluid [61], which is used to study materials with subtle viscoelastic features.

Research on the Navier–Stokes–Voigt (NSV) fluid has expanded considerably. Studies by Levant et al. [36], Sukacheva and Matveeva [63], Berselli and Bisconti [12], and Baranovskii [8] examined its fundamental properties. More recently, stability analysis of the Navier–Stokes–Voigt fluid heated from below under various influences has been investigated by Straughan [59, 62], Basavarajappa and Bhatta [10], Kavitha et al. [28], Afluk and Harfash [3], Sharma et al. [55, 56], and Kumar et al. [32, 33].

DOI: 10.1201/9781003664345-9

The influence of magnetic fields on fluid flow and heat transfer has been a longstanding research interest, spanning diverse fluid systems and stability analyses. Early contributions include Sarpkaya [53] on non-Newtonian fluids in magnetic fields, Iroshnikov [25] on turbulence in conducting fluids, and Moffatt [41] on flows induced by rotating magnetic fields. Neuringer [43] studied viscous ferrofluid flows under combined thermal and magnetic gradients, while Sawada et al. [54] examined natural convection of magnetic fluids in horizontal annuli with nonuniform fields. Wilson [65, 65] investigated uniform magnetic field effects on the onset of Marangoni and Bénard–Marangoni convection, and Takashima [64] analyzed stability of modified Poiseuille flow in transverse fields. Odenbach [44] summarized advances in magnetic fluid research, complemented by experimental and applied studies such as Ohira et al. [45] on tuned liquid dampers, Mekheimer [39, 40] and Ebaid [16] on peristaltic transport under magnetic fields, and Hayat et al. [23] on Carreau fluids. Heat transfer effects were further explored by Motozawa et al. [42], Lajvardi et al. [35], Ghasemi et al. [21], Azizian et al. [7], and Kraszewska et al. [30], while Aliabadi and Taklifi [5] investigated bubble dynamics in viscoelastic fluids and Akbar et al. [4] studied Eyring–Powell fluids.

The nonlinear and stability aspects of magnetized fluids have been addressed by Harfash and Straughan [22], Jahanbakhshi et al. [26], Abro et al. [1], Adıgüzel and Atalık [2], Benhacine et al. [11], and Aminian et al. [6]. More recent works highlight advanced investigations, including Skumiel et al. [57] on rotating magnetic fields, Lai et al. [34] on stabilizing effects in magnetic Bénard fluids, Barman et al. [9] on micropolar fluids in porous layers, and Raghunatha et al. [52] on thermosolutal convection in Kelvin–Voigt fluids. Together, these studies demonstrate the wide-ranging impact of magnetic fields on convection, stability, and transport phenomena across Newtonian, non-Newtonian, ferrofluid, and viscoelastic systems.

This chapter focuses on the effect of a magnetic field on the thermal convection of a Navier–Stokes–Voigt fluid, an aspect not extensively explored in existing literature. The analysis is carried out using both linear and nonlinear stability approaches under three types of bounding surface combinations: free–free, rigid–free, and rigid–rigid. A free boundary allows both velocity and temperature perturbations to vanish at the surface without constraining fluid motion tangentially, while a rigid boundary imposes no-slip conditions, requiring the fluid velocity to be zero at the surface. The free–free case represents stress-free surfaces at both boundaries, the rigid–free case represents one no-slip and one stress-free boundary, and the rigid–rigid case imposes no-slip conditions at both boundaries. These boundary conditions provide a comprehensive framework to examine how the interplay of viscoelasticity and magnetic fields affects the onset and stability of convection.

9.2 MATHEMATICAL FORMULATION

An infinite horizontal layer of incompressible NSV fluid with thickness d is considered, confined between two parallel boundaries. The layer is subjected to uniform heating from below, resulting in a constant vertical temperature gradient $\beta = \left|\frac{dT}{dz}\right|$.

The system is further influenced by a uniform vertical magnetic field. Gravity $\mathbf{g}$ acts downward along the negative z-axis. At the lower surface ($z = 0$), the temperature is maintained at T_0, while at the upper surface ($z = d$), it is fixed at T_d. A schematic representation of the physical configuration is shown in Figure 9.1.

To examine the influence of the imposed magnetic field, we employ the quasi-static magnetohydrodynamic (MHD) approximation introduced by Galdi and Straughan [19]. Under this approximation, the electric field $\mathbf{E}$ is assumed to be derivable from a scalar potential ϕ, such that $\mathbf{E} = -\nabla\phi$ [22]. The magnetic and electric fields must further satisfy Maxwell's equations:

$$\nabla \times \mathbf{H} = \mathbf{j},\ \nabla \times \mathbf{E} = -\frac{\partial \mathbf{B}}{\partial t},\ \nabla \cdot \mathbf{j} = 0,\ \nabla \cdot \mathbf{B} = 0,$$

where $\mathbf{j}$ represents the current density, while the magnetic induction field is given by $\mathbf{B} = \mu_m \mathbf{H}$, with μ_m denoting the magnetic permeability.

Following Galdi and Straughan [19], the Lorentz force term $\mathbf{j} \times \mathbf{B}$ takes the form

$$\mathbf{j} \times \mathbf{B} = \sigma_1 \left(\mathbf{q} \times \mathbf{B}_0\right) \times \mathbf{B}_0,$$

where $\mathbf{B}_0 = \left(0, 0, B_0\right)$ denotes the imposed vertical magnetic field and σ_1 is the electrical conductivity of the fluid.

The governing equations of the present system, formulated under the Boussinesq approximation, are given as ([51], [31]):

Continuity equation

$$\nabla \cdot \mathbf{q} = 0. \tag{9.1}$$

Momentum equations

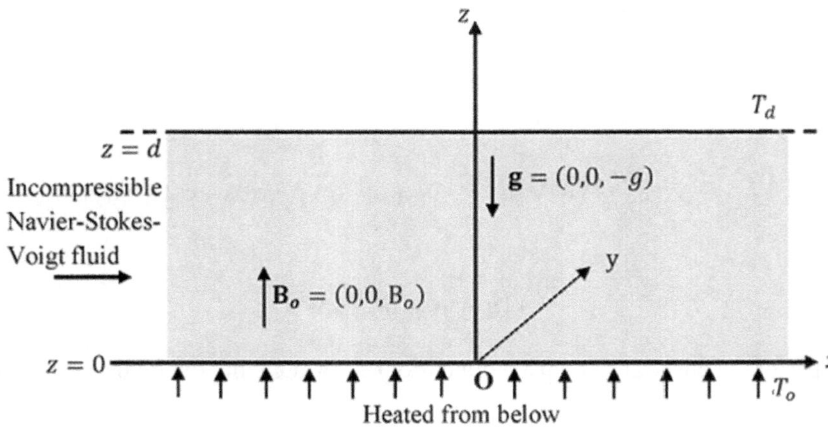

FIGURE 9.1 Geometrical configuration of the problem.

$$\rho_o\left(\frac{\partial \mathbf{q}}{\partial t}-\left(\hat{\lambda}\nabla^2\right)\frac{\partial \mathbf{q}}{\partial t}+(\mathbf{q}\cdot\nabla)\mathbf{q}\right)=\rho_o g\alpha T\hat{\mathbf{k}}+\mu\nabla^2\mathbf{q}-\nabla p+\sigma_1(\mathbf{q}\times\mathbf{B_o})\times\mathbf{B_o}. \tag{9.2}$$

Energy equation

$$\frac{\partial T}{\partial t}+(\mathbf{q}\cdot\nabla)T=\kappa\nabla^2 T. \tag{9.3}$$

Equation of state

$$\rho=\rho_o\left(1-\alpha\left(T-T_o\right)\right). \tag{9.4}$$

In these expressions, $\mathbf{q}$ denotes the velocity vector, ρ the density, p the pressure, ρ_o the reference density, t the time, μ the dynamic viscosity, κ the thermal diffusivity, $\hat{\lambda}$ the Kelvin–Voigt coefficient, and α the thermal expansion coefficient.

The steady basic state (denoted by the subscript 'b') is described by the following expressions:

$$\begin{aligned}\mathbf{q}= \quad &(0,0,0)=\mathbf{q}_b,\ \rho=\rho_o(1+\alpha\beta z)=\rho_b(z),\\ &T=-\beta z+T_o=T_b(z),\ p=p_b(z).\end{aligned} \tag{9.5}$$

For examining how the NSV fluid responds to disturbances, perturbations are applied to the base state in the following manner:

$$p=p_b(z)+p',\ \mathbf{q}=\mathbf{q}_b+\mathbf{q}',\ T=T_b(z)+\theta,\ \rho=\rho_b(z)+\rho',$$

In this context, the perturbation components are defined as follows: p' for pressure, $\mathbf{q}'=(u',v',w')$ for velocity, and θ for temperature.

The system of Equations (9.1)–(9.4), incorporating the perturbations, is presented as follows (prime $'$ is neglected in the following for simplicity):

$$\nabla\cdot\mathbf{q}=0, \tag{9.6}$$

$$\rho_o\left(\frac{\partial \mathbf{q}}{\partial t}-\left(\hat{\lambda}\nabla^2\right)\frac{\partial \mathbf{q}}{\partial t}+(\mathbf{q}\cdot\nabla)\mathbf{q}\right)=-\nabla p+\mu\nabla^2\mathbf{q}+\rho_o g\alpha\theta\hat{\mathbf{k}}+\sigma_1(\mathbf{q}\times\mathbf{B_o})\times\mathbf{B_o}, \tag{9.7}$$

$$\frac{\partial\theta}{\partial t}+(\mathbf{q}\cdot\nabla)\theta=\beta w+\kappa\nabla^2\theta. \tag{9.8}$$

The perturbed equations are made dimensionless by applying the following scales

$$z = dz^*, \quad p = \frac{\mu^2}{\rho_o d^2} p^*, \mathbf{q} = \frac{\mu}{\rho_o d} \mathbf{q}^*, t = \frac{\rho_o d^2}{\mu} t^*, \\ \theta = \frac{\mu}{\rho_o d} \sqrt{\frac{\beta\mu}{\kappa g \alpha \rho_o}} \theta^*, \hat{\lambda} = t\nu\lambda^*, \tag{9.9}$$

and the following dimensionless parameters are introduced

$$Pr = \frac{\nu}{\kappa}, M^2 = \frac{B_o^2 d^2 \sigma_1}{\rho_o \nu}, Ra = \frac{\alpha g \beta d^4}{\nu\kappa}, \tag{9.10}$$

namely Prandtl number, Hartmann number, and Rayleigh number, respectively.

By applying the scaling from Equation (9.9) along with the dimensionless parameters defined in Equation (9.10) to Equations (9.6)–(9.8), we obtain the dimensionless form of the governing equations as follows:

$$\nabla \cdot \mathbf{q} = 0, \tag{9.11}$$

$$\left(\frac{\partial \mathbf{q}}{\partial t} - \left(\lambda \nabla^2\right) \frac{\partial \mathbf{q}}{\partial t} + (\mathbf{q} \cdot \nabla)\mathbf{q} \right) = -\nabla p + \nabla^2 \mathbf{q} + Ra^{1/2} \theta \hat{\mathbf{k}} + M^2 \left(\mathbf{q} \times \hat{\mathbf{k}} \right) \times \hat{\mathbf{k}}, \tag{9.12}$$

$$Pr \left(\frac{\partial \theta}{\partial t} + (\mathbf{q} \cdot \nabla)\theta \right) = Ra^{1/2} w + \nabla^2 \theta. \tag{9.13}$$

The boundary conditions (BCs) for the system are expressed as:

$$w = 0, \theta = 0, \text{ at } z = 0, 1. \tag{9.14}$$

9.3 NONLINEAR ANALYSIS

In this approach, we apply the classical nonlinear energy method ([58]) by multiplying each equation with suitable variables: Equation (9.12) by $\mathbf{q}$ and Equation (9.13) by θ. The resulting expressions are then integrated over the periodic cell V. By incorporating the boundary conditions specified in (9.14), we arrive at the following set of equations:

$$\frac{1}{2} \frac{d}{dt} \| \mathbf{q} \|^2 + \frac{\lambda}{2} \frac{d}{dt} \| \nabla \mathbf{q} \|^2 = Ra^{1/2} \langle w\theta \rangle - M^2 \left(\| u \|^2 + \| v \|^2 \right) - \| \nabla \mathbf{q} \|^2, \tag{9.15}$$

$$\frac{Pr}{2} \frac{d}{dt} \| \theta \|^2 = Ra^{1/2} \langle w\theta \rangle - \| \nabla \theta \|^2 . \tag{9.16}$$

Here, $\|\cdot\|$ denotes the L^2-norm over the domain V, while $\langle\cdot\rangle$ represents integration over V.

Using Eqs. (9.15)–(9.16), we construct the energy functional $E(t)$. By introducing the positive coupling parameters λ_1 and λ_2, the time evolution of $E(t)$ can be expressed as

$$\frac{dE}{dt} = I_o - D_o. \tag{9.17}$$

Here,

$$E = \frac{Pr}{2} \quad \| \theta \|^2 + \frac{\lambda_1}{2} \| \mathbf{q} \|^2 + \frac{\lambda\lambda_1}{2} \| \nabla \mathbf{q} \|^2, \tag{9.18}$$

$$D_o = \lambda_1 M^2 \left(\| u \|^2 + \| v \|^2 \right) + \| \nabla\theta \|^2 + \lambda_1 \| \nabla \mathbf{q} \|^2, \tag{9.19}$$

$$I_o = \left(1 + \lambda_1\right) Ra^{1/2} \left\langle w\theta \right\rangle. \tag{9.20}$$

In this context, D_o and I_o denote the dissipation and production terms, respectively.

Let us define the parameter m as

$$m = \max_{\mathrm{H}} \frac{I_o}{D_o}.$$

Here H denotes the space of admissible (feasible) solutions.

For the energy estimate to be valid, the following condition must be satisfied:

$$\frac{dE}{dt} \leq -D_o a_o. \tag{9.21}$$

This inequality is ensured only if the parameter $m < 1$. Here, a_o is defined as $a_o = 1 - m$, which remains positive under this constraint. Therefore, the requirement $m < 1$ ensures the validity of inequality (9.21).

By applying the Poincaré inequality to Equation (9.19), we obtain

$$D_o \geq \mathrm{K} E(t), \tag{9.22}$$

where $\mathrm{K} = 2\pi^2 \min\left(\frac{1}{Pr}, \frac{1}{\frac{1}{\lambda_1} + \lambda\pi^2} \right)$.

Integrating equation (9.21) w.r.t. time after inserting inequality (9.22) yields

$$E(t) \le E(0)e^{-a_0 \mathrm{K} t}. \tag{9.23}$$

This result indicates that the energy E decays exponentially over time.

9.3.1 Variational Problem

We employ the calculus of variations to analyze the maximum problem at the critical case $m = 1$. Using the transformations $w = \frac{\hat{w}}{\sqrt{\lambda_1}}$ and $\mathbf{q} = \frac{\hat{\mathbf{q}}}{\sqrt{\lambda_1}}$ in Equations (9.19) and (9.20), the corresponding Euler–Lagrange relation $\delta I_o - \delta D_o = 0$ yields the results

$$\left(\frac{1+\lambda_1}{\sqrt{\lambda_1}}\right) Ra^{1/2}\theta\,\hat{\mathbf{k}} + 2\nabla^2\mathbf{q} - 2M^2\left(\mathbf{q} - w\,\hat{\mathbf{k}}\right) = 0, \tag{9.24}$$

$$\left(\frac{1+\lambda_1}{\sqrt{\lambda_1}}\right) Ra^{1/2} w + 2\nabla^2\theta = 0. \tag{9.25}$$

By applying curl on Equation (9.24) and subsequently extracting the third component, we obtain the following result

$$\left(\frac{1+\lambda_1}{\sqrt{\lambda_1}}\right) Ra^{1/2}\nabla_1^{\,2}\theta + 2\nabla^4 w - 2M^2 w_{zz} = 0, \tag{9.26}$$

where $\nabla_1^2 \equiv \frac{\partial^2}{\partial x^2} + \frac{\partial^2}{\partial y^2}$ represents the horizontal Laplacian.

The plane tiling form is now assumed as

$$\{w, \theta\} = \{W(z), \Theta(z)\}\,\Psi(x, y), \tag{9.27}$$

where 'Ψ' denotes the shape of the stability cell (see Chandrasekhar [14], pp. 43–52) and $\Psi(x, y)$ satisfies

$$\nabla_1^{\,2}\Psi = -a^2\Psi. \tag{9.28}$$

Here, 'a' is the wave number.

Using (9.27) in Equations (9.26) and (9.25), the resulting expressions lead to the following eigenvalue problem

$$2\left(D^2 - a^2\right)^2 W \quad - \left(\frac{1+\lambda_1}{\sqrt{\lambda_1}}\right) Ra^{1/2} a^2 \Theta - 2M^2 D^2 W = 0, \tag{9.29}$$

$$2\left(D^2-a^2\right)\Theta+\left(\frac{1+\lambda_1}{\sqrt{\lambda_1}}\right)Ra^{1/2}W=0, \tag{9.30}$$

with the boundary conditions

$$W=0,\ D^2W=0,\ \Theta=0,\ \text{at } z=0,1(\text{for free-free}), \tag{9.31}$$

$$W=0,\ DW=0,\ \Theta=0,\ \text{at } z=0,1\ (\text{for rigid-rigid}), \tag{9.32}$$

$$\left.\begin{aligned} W=0,\ DW=0,\ \Theta=0,\ \text{at } z=0,\\ W=0,\ D^2W=0,\ \Theta=0,\ \text{at } z=1\end{aligned}\right\}(\text{for rigid-free}). \tag{9.33}$$

9.4 LINEAR ANALYSIS

We consider Equations (9.12)–(9.13), which are simplified by discarding the nonlinear terms. The solutions are taken in the form ([55]): $\mathbf{q}(\mathbf{x},t)=\mathbf{q}(\mathbf{x})e^{\sigma t}$, $\theta(\mathbf{x},t)=\theta(\mathbf{x})e^{\sigma t}$, $\phi(\mathbf{x},t)=\phi(\mathbf{x})e^{\sigma t}$, $p(\mathbf{x},t)=p(\mathbf{x})e^{\sigma t}$, where σ represents the growth rate of the disturbances, and $\mathbf{x}=(x,y,z)$ denotes the Cartesian coordinates. The corresponding linearized equations are then expressed as:

$$\sigma\left(1-\lambda\nabla^2\right)\mathbf{q}=\nabla^2\mathbf{q}+Ra^{1/2}\theta\hat{\mathbf{k}}-\nabla p+M^2\left(\mathbf{q}\times\hat{\mathbf{k}}\right)\times\hat{\mathbf{k}}, \tag{9.34}$$

$$Pr\sigma\theta=\nabla^2\theta+Ra^{1/2}w. \tag{9.35}$$

Applying $\hat{\mathbf{k}}\cdot$curlcurl to Equation (9.38) gives

$$\sigma\left(1-\lambda\nabla^2\right)\nabla^2 w=\nabla^4 w+Ra^{1/2}\nabla_1{}^2\theta-M^2\frac{\partial^2 w}{\partial z^2}. \tag{9.36}$$

Using (9.27) in Equations (9.36) and (9.35), we obtain the following result

$$\sigma\left(D^2-a^2\right)W-\sigma\lambda\left(D^2-a^2\right)^2W=-a^2Ra^{1/2}\Theta+\left(D^2-a^2\right)^2W-M^2D^2W, \tag{9.37}$$

$$Pr\sigma\Theta=\left(D^2-a^2\right)\Theta+Ra^{1/2}W, \tag{9.38}$$

under the boundary conditions described by (9.31)–(9.33).

9.4.1 Principle of Exchange of Stabilities

The principle of the exchange of stabilities (PES) is valid if a stationary pattern of motions prevails at the onset of instability [14]. To verify this, we have to show that σ is real.

Multiply Equation (9.37) by $W^{\star}$ and Equation (9.38) by $\Theta^{\star}$ ($\star$ represents complex conjugate), and subsequently integrating the resulting expression over the z-domain while applying boundary conditions (9.31)–(9.33), we obtain the following results

$$-\sigma\mathcal{I}_1 - \sigma\lambda\mathcal{I}_2 = \mathcal{I}_2 + M^2\mathcal{I}_3 - a^2 Ra^{1/2}\int_0^1 W^{\star}\Theta\, dz, \tag{9.39}$$

$$Pr\sigma\mathcal{I}_4 = -\mathcal{I}_5 + Ra^{1/2}\int_0^1 W\Theta^{\star}\, dz, \tag{9.40}$$

where

$$\mathcal{I}_1 = \int_0^1\left(a^2|W|^2 + |DW|^2\right)dz, \mathcal{I}_2 = \int_0^1\left(a^4|W|^2 + 2a^2|DW|^2 + |D^2W|^2\right)dz,$$
$$\mathcal{I}_3 = \int_0^1\left(|DW|^2\right)dz, \mathcal{I}_4 = \int_0^1\left(|\Theta|^2\right)dz, \mathcal{I}_5 = \int_0^1\left(a^2|\Theta|^2 + |D\Theta|^2\right)dz. \tag{9.41}$$

Here, the integrals $\mathcal{I}_1$, $\mathcal{I}_2$, $\mathcal{I}_3$, $\mathcal{I}_4$, $\mathcal{I}_5$ are positive.

Adding Equation (9.39) with the product of a^2 and the complex conjugate of Equation (9.40), we get

$$-\sigma\mathcal{I}_1 - (\sigma\lambda + 1)\mathcal{I}_2 - M^2\mathcal{I}_3 + a^2 Pr\sigma^{\star}\mathcal{I}_4 + a^2\mathcal{I}_5 = 0, \tag{9.42}$$

where $\sigma^{\star}$ is the complex conjugate of σ.

Substituting $\sigma = \sigma_r + \iota\sigma_i$ in Equation (9.42), yields the imaginary part of the resulting equation as follows

$$\sigma_i\left(\mathcal{I}_1 + \lambda\mathcal{I}_2 + a^2 Pr\mathcal{I}_4\right) = 0, \tag{9.43}$$

which implies that $\sigma_i = 0$, since $\left(\mathcal{I}_1 + \lambda\mathcal{I}_2 + a^2 Pr\mathcal{I}_4\right) > 0$. This validates the PES, indicating the lack of oscillatory mode of convection. Thus, the shift from stability to instability must proceed through a stationary convection state.

Consequently, the equations for the marginal state can be derived by setting $\sigma=0$ in Equation (9.37) and Equation (9.38). Therefore, the eigenvalue problem can be formulated as

$$\left(D^2 - a^2\right)^2 W - a^2 Ra^{1/2}\Theta - M^2 D^2 W = 0, \tag{9.44}$$

$$\left(D^2 - a^2\right)\Theta + Ra^{1/2} W = 0, \tag{9.45}$$

under the boundary conditions described by (9.31)–(9.33).

9.5 METHOD OF SOLUTION

The eigenvalue problems were solved using the single-term Galerkin method, implemented in Mathematica, to compute the Ra for both linear and nonlinear analyses. As illustrated by Postelnicu [49], Finlayson [17], and more recently Chandel et al. [13], by choosing appropriate trial functions, the governing equations are reduced to a matrix eigenvalue problem, facilitating the stability analysis. Applied to multiple boundary configurations—free–free, rigid–rigid, and rigid–free—this method ensures that the solutions adhere to the relevant boundary conditions. The trial functions chosen for these cases are shown below:

For Free–Free Bounding Surfaces

As per boundary conditions (9.31), the set of exact solutions is as follows ([14])

$$W_i = \sin\left(i\pi z\right),\ \Theta_i = \sin\left(i\pi z\right). \tag{9.46}$$

For Rigid–Rigid Bounding Surfaces

As per boundary conditions (9.32), the basis functions are as follows ([31])

$$W_i = -0.5z^{2i} + 2.5z^{2i+2} - 3z^{2i+3} + z^{2i+4},\ \Theta_i = z^i - 2z^{i+2} + z^{i+3}. \tag{9.47}$$

For Rigid–Free Bounding Surfaces

As per boundary conditions (9.33), the basis functions are as follows ([31])

$$W_i = -z^{2i} + 3.25z^{2i+2} - 3.05z^{2i+3} + 0.8z^{2i+4},\ \Theta_i = z^i - 2z^{i+2} + z^{i+3}. \tag{9.48}$$

9.6 RESULTS AND DISCUSSION

The single-term Galerkin method is employed to solve the eigenvalue problems in both the nonlinear and linear frameworks. Ra obtained from the nonlinear stability analysis, denoted by Ra_N, expressed as a function of M^2, λ_1 and a. The optimal value of $\lambda_1 = 1$ is computed by applying the condition for extremizing the Rayleigh number with respect to λ_1. Upon substituting this optimal value into the expression for Ra_N, the resulting Rayleigh number matches exactly with the value Ra_L obtained independently from the linear stability analysis. This agreement confirms the consistency between the nonlinear and linear analyses under the applied approximations.

The resulting expressions for $Ra_N = Ra_L$ are provided below for different thermal boundary conditions:

For free–free bounding surfaces

$$Ra_N = \frac{\left(a^4 + 2a^2\pi^2 + \pi^2\left(M^2 + \pi^2\right)\right)\left(a^2 + \pi^2\right)}{a^2} = Ra_L. \tag{9.49}$$

For rigid–rigid bounding surfaces

$$Ra_N = \frac{\left(\left(1.0381a^6 + 36.9703a^4 + 792.905a^2 + 5222.93\right) + \left(13.3616a^2 + 131.892\right)M^2\right)}{a^2} = Ra_L. \tag{9.50}$$

For rigid–free bounding surfaces

$$Ra_N = \frac{\left(\left(1.07057a^6 + 36.3051a^4 + 514.057a^2 + 2566.47\right) + \left(12.8688a^2 + 127.027\right)M^2\right)}{a^2} = Ra_L. \tag{9.51}$$

The identical Ra values in the nonlinear and linear analyses demonstrate that subcritical instabilities are not present. This alignment is a strong indicator of global stability, confirming that instability arises only when Ra exceeds the critical threshold. Furthermore, the analysis reveals that Ra is influenced by the wave number and the magnetic field, reflecting their direct roles in modulating the convective threshold. Ra_c is determined by substituting the critical wave number, which is obtained by applying the condition $\frac{\partial Ra}{\partial a} = 0$. In contrast, λ does not appear explicitly in the critical Rayleigh number, indicating that the onset of convection remains unaffected by it. Nonetheless, λ contributes significantly to the energy dissipation in the system, as observed through its role in the damping mechanisms, thereby enhancing the fluid's long-term stability without altering the linear or nonlinear thresholds.

Figure 9.2 illustrates the variation of the Ra_c with increasing values of M^2 for three different thermal boundary conditions: free–free, rigid–free, and rigid–rigid. It is observed that, for all three cases, the Ra_c increases monotonically with M^2. This trend signifies a delay in the onset of convection, thereby highlighting the stabilizing influence of the magnetic field on the system. Lorentz forces opposing the fluid motion are induced by the presence of a magnetic field, suppressing perturbations and enhancing the fluid's resistance to buoyancy-driven instability.

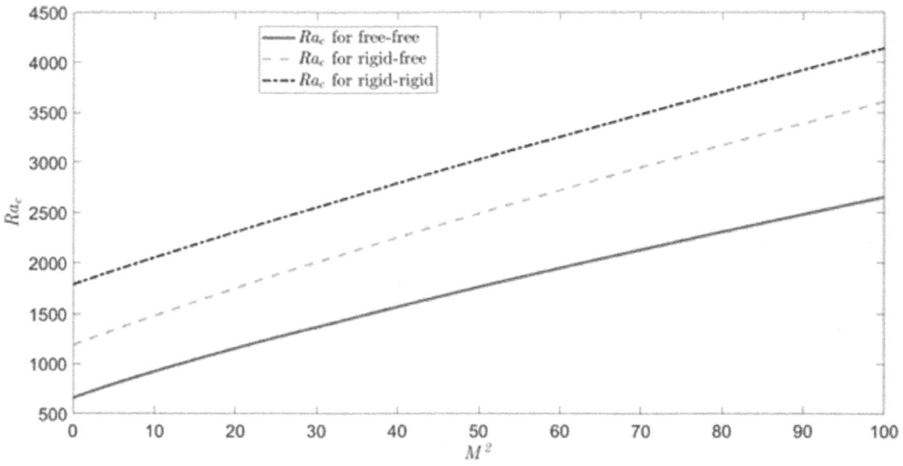

FIGURE 9.2 Critical Rayleigh number (Ra_c) vs Hartmann number (M_2) for different combinations of bounding surfaces.

Among the three configurations, the lowest Ra_c is obtained for the free–free bounding surfaces, indicating that the system becomes unstable more easily under this configuration. However, the rigid–rigid bounding surfaces yield the highest critical Rayleigh number, signifying the greatest resistance to the onset of convection. The rigid–free configuration lies between the other two, with a stability threshold higher than free–free but lower than rigid–rigid.

9.7 CONCLUSIONS

The nonlinear and linear analyses have been performed on the thermal convection of NSV fluid with a magnetic field for various combinations of boundary conditions. The findings of this study have been summarized as follows:

- Across various parameter variations and bounding surfaces, the thresholds for linear instability and nonlinear stability are found to coincide. This alignment suggests the absence of a subcritical instability region in the present analysis.
- The validation of PES of stabilities confirms the lack of oscillatory mode of convection.
- The thermal stability of the fluid is notably enhanced when confined within rigid–rigid bounding surfaces, suggesting that such configurations are particularly favorable for convection in NSV fluids.
- Magnetic fields are found to delay the onset of thermal convection, indicating their role in stabilizing the fluid system against convective disturbances.
- Interestingly, the Kelvin–Voigt parameter does not affect the Rayleigh numbers, implying that its impact is primarily on the energy decay rather than on the convection onset.

9.8 APPLICATION AND FUTURE SCOPE

The findings of this study have significant applications in engineering and scientific processes where thermal management and fluid stability are critical. The stabilizing effect of magnetic fields on viscoelastic fluids, such as those described by the Navier–Stokes–Voigt model, can be utilized in magnetohydrodynamic (MHD) systems to control heat transfer and suppress unwanted convective instabilities. This has practical relevance in polymer processing, crystal growth, and cooling of electronic devices, where precise thermal regulation is necessary. Additionally, the insights on the influence of bounding surfaces can guide the design of industrial reactors and heat exchangers to enhance stability and efficiency, particularly in systems involving viscoelastic fluids. The study also provides a framework for optimizing energy dissipation in viscoelastic flows, which is valuable in microfluidic devices and biomedical engineering applications, such as lab-on-a-chip systems and drug delivery technologies.

The present analysis, while offering meaningful insights into the stability of magnetoconvection in Navier–Stokes–Voigt fluids, has several limitations. The study is confined to linear normal mode analysis and the nonlinear energy method, which, although effective for determining stability thresholds, do not capture finite-amplitude convection, complex nonlinear dynamics, or secondary instabilities. Only idealized boundary conditions are considered, and the assumption of stationary convection based on the principle of exchange of stabilities limits the exploration of more intricate oscillatory or chaotic modes. The analysis also neglects additional physical effects such as rotation, chemical reactions, variable viscosity, and anisotropic magnetic fields, all of which may substantially influence stability in practical systems. The formulation is two-dimensional, without examining fully three-dimensional convection structures that could yield richer dynamical behavior. Furthermore, the study lacks numerical simulations and experimental validation, which are essential for verifying analytical predictions and assessing their applicability to real engineering and scientific systems. Finally, the role of the Kelvin–Voigt parameter is restricted to energy decay in the present model, and more comprehensive viscoelastic rheological models may reveal additional effects not captured here. Future Scope: Building upon the limitations of the present work, several avenues exist for extending and deepening the analysis of magnetoconvection in Navier–Stokes–Voigt fluids. Future studies may incorporate finite-amplitude and fully nonlinear convection analyses to capture complex post-critical behavior, including secondary instabilities, pattern transitions, and long-time transient dynamics that cannot be assessed through linear or energy methods alone. Extending the study to include oscillatory, chaotic, or time-dependent convection modes—enabled by relaxing the assumption of the principle of exchange of stabilities—could offer a more comprehensive understanding of realistic MHD environments. More advanced physical effects such as rotation, chemical reactions, thermal radiation, variable viscosity, and spatially varying magnetic fields can be incorporated to better represent industrial, geophysical, and astrophysical conditions. Future models may also investigate three-dimensional convection patterns, as 3D instabilities often dominate practical systems and can exhibit far richer dynamical structures. Numerical simulations using high-resolution computational fluid dynamics (CFD) and experimental validation in laboratory-scale MHD setups would strengthen the applicability of the analytical findings and uncover behaviors that theory alone cannot predict. Additionally, exploring more general viscoelastic models beyond the NSV framework—such as Oldroyd-B, Maxwell, or Giesekus fluids—may reveal deeper insights into how complex rheology interacts with magnetic fields to influence convective stability. Further extensions to multicomponent fluids, porous media, microfluidic systems, and bioengineering flows could broaden the relevance of this study to next-generation MHD devices, geothermal energy systems, cooling technologies, and biomedical applications.

REFERENCES

1. K A Abro, M M Baig, and M Hussain. Influences of magnetic field in viscoelastic fluid. *Int. J. Numer. Anal. Appl.*, 9:99–109, 2018.
2. Ö B Adıgüzel and K Atalık. Magnetic field effects on newtonian and non-newtonian ferrofluid flow past a circular cylinder. *Appl. Math. Modell.*, 42:161–174, 2017.
3. Z A Afluk and A J Harfash. Instability of thermosolutal convection in a Brinkman-Darcy-Kelvin-Voigt fluid. *J. Porous Media*, 28:1–19, 2025.
4. N S Akbar, A Ebaid, and Z H Khan. Numerical analysis of magnetic field effects on eyring-powell fluid flow towards a stretching sheet. *J. Magn. Magn. Mater.*, 382:355–358, 2015.
5. A Aliabadi and A Taklifi. The effect of magnetic field on dynamics of gas bubbles in visco-elastic fluids. *Appl. Math. Modell.*, 36:2567–2577, 2012.
6. E Aminian, H Moghadasi, and H Saffari. Magnetic field effects on forced convection flow of a hybrid nanofluid in a cylinder filled with porous media: A numerical study. *J. Therm. Anal. Calorim.*, 141:2019–2031, 2020.
7. R Azizian, E Doroodchi, T McKrell, J Buongiorno, L W Hu, and B Moghtaderi. Effect of magnetic field on laminar convective heat transfer of magnetite nanofluids. *Int. J. Heat Mass Transfer*, 68:94–109, 2014.
8. E S Baranovskii. Strong solutions of the incompressible Navier–Stokes–Voigt model. *Mathematics*, 8:181, 2020.
9. P Barman and D Srinivasacharya. Linear and nonlinear stability of double diffusive convection in a micropolar fluid saturated porous layer with magnetic field and throughflow effects. *Phys. Fluids*, 36:104124, 2024.
10. M Basavarajappa and D Bhatta. Nonlinear stability analysis of Rayleigh-Bénard problem for a Navier-Stokes-Voigt fluid. *Int. J. Non-Linear Mech.*, 162:104712, 2024.
11. H Benhacine, B Mahfoud, and M Salmi. Stability effect of an axial magnetic field on fluid flow bifurcation between coaxial cylinders. *Int. J. Comput. Mater. Sci. Eng.*, 10: 2150023, 2021.
12. L C Berselli and L Bisconti. On the structural stability of the Euler–Voigt and Navier–Stokes–Voigt models. *Nonlinear Anal. Theory Methods Appl.*, 75:117–130, 2012.
13. V Chandel and Sunil. Stability analysis of partially ionized plasma in a porous medium with local thermal non-equilibrium effects. *J. Plasma Phys.*, 91:E1, 2025.
14. S Chandrasekhar. *Hydrodynamic and Hydromagnetic Stability.* Dover, New York, 1981.
15. H Demir. Thermal convection of viscoelastic fluid with biot boundary conduction. *Math. Comput. Simul.*, 56:277–296, 2014.

16. A Ebaid. Effects of magnetic field and wall slip conditions on the peristaltic transport of a Newtonian fluid in an asymmetric channel. *Phys. Lett. A*, 372:4493–4499, 2008.
17. B A Finlayson. *The Method of Weighted Residuals and Variational Principles*. Society for Industrial and Applied Mathematics, Philadelphia, 2013.
18. G P Galdi. Non-linear stability of the magnetic Bénard problem via a generalized energy method. *Arch. Ration. Mech. Anal.*, 87:167–186, 1985.
19. G P Galdi and B Straughan. Exchange of stabilities, symmetry, and nonlinear stability. *Arch. Rational Mech. Anal.*, 89:211–228, 1985.
20. G P Galdi and B Straughan. A nonlinear analysis of the stabilizing effect of rotation in the Bénard problem. *Proc. R. Soc. London A*, 402:257–283, 1985.
21. B Ghasemi, S M Aminossadati, and A Raisi. Magnetic field effect on natural convection in a nanofluid-filled square enclosure. *Int. J. Therm. Sci.*, 50:1748–1756, 2011.
22. A J Harfash and B Straughan. Magnetic effect on instability and nonlinear stability in a reacting fluid. *Meccanica*, 47:1849–1857, 2012.
23. T Hayat, N Saleem, and N Ali. Effect of induced magnetic field on peristaltic transport of a carreau fluid. *Commun. Nonlinear Sci. Numer. Simul.*, 15(9):2407–2423, 2010.
24. T Hayat, M B Ashraf, H H Alsulami, and M S Alhuthali. Three-dimensional mixed convection flow of viscoelastic fluid with thermal radiation and convective conditions. *PLoS One*, 9:e90038, 2014.
25. P S Iroshnikov. Turbulence of a conducting fluid in a strong magnetic field. *Soviet Astronomy*, 7:566, 1964.
26. A Jahanbakhshi, A A Nadooshan, and M Bayareh. Magnetic field effects on natural convection flow of a non-newtonian fluid in an L-shaped enclosure. *J. Therm. Anal. Calorim.*, 133:1407–1416, 2018.
27. D D Joseph. *Fluid Dynamics of Viscoelastic Liquids*. Springer Science & Business Media, New York, 2013.
28. G N Kavitha, B M Shankar, and I S Shivakumara. On the magnetohydrodynamic stability of channel flow of Navier–Stokes–Voigt fluid. *Phys. Fluids*, 36:043105, 2024.
29. R E Khayat. Non-linear overstability in the thermal convection of viscoelastic fluids. *J. Non-Newtonian Fluid Mech.*, 58:331–356, 1995.
30. A Kraszewska, L Pyrda, and J Donizak. High magnetic field impact on the natural convection behaviour of a magnetic fluid. *Heat and Mass Transfer*, 54:2383–2394, 2018.
31. D Kumar, Sunil, and R Devi. Couple stress effects on thermosolutal convection in Navier–Stokes–Voigt fluid. *Chin. J. Phys.*, 96:1020–1034, 2025.
32. D Kumar, Sunil, and R Devi. Effect of couple stresses on convective stability of rotating Navier–Stokes–Voigt fluid saturating a porous medium. *Z. Naturforsch., A: Phys. Sci.*, 80:587–602, 2025.
33. D Kumar, Sunil, and R Devi. Influence of rotation and couple stresses on convective stability in Navier–Stokes–Voigt fluid. *Arch. Mech.*, 77:305–332, 2025.
34. S Lai, L Shen, X Ye, and X Zhao. The stabilizing effect of temperature and magnetic field on a 2d magnetic bénard fluids. *J. Differ. Equations*, 409:851–880, 2024.
35. M Lajvardi, J Moghimi-Rad, I Hadi, A Gavili, T D Isfahani, F Zabihi, and J Sabbaghzadeh. Experimental investigation for enhanced ferrofluid heat transfer under magnetic field effect. *J. Magn. Magn. Mater.*, 322:3508–3513, 2010.
36. B Levant, F Ramos, and E S Titi. On the statistical properties of the 3D incompressible Navier-Stokes-Voigt model. *Commun. Math. Sci.*, 8:277–293, 2010.
37. Z Li and R E Khayat. Three-dimensional thermal convection of viscoelastic fluids. *Phys. Rev. E*, 71:066305, 2005.
38. M S Malashetty, M S Swamy, and W Sidram. Thermal convection in a rotating viscoelastic fluid saturated porous layer. *Int. J. Heat Mass Transfer*, 53:5747–5756, 2010.

39. K S Mekheimer. Peristaltic flow of blood under effect of a magnetic field in a non-uniform channels. *Appl. Math. Comput.*, 153:763–777, 2004.
40. K S Mekheimer and Y A Elmaboud. The influence of heat transfer and magnetic field on peristaltic transport of a Newtonian fluid in a vertical annulus: application of an endoscope. *Phys. Lett. A*, 372:1657–1665, 2008.
41. H K Moffatt. On fluid flow induced by a rotating magnetic field. *J. Fluid Mech.*, 22: 521–528, 1965.
42. M Motozawa, J Chang, T Sawada, and Y Kawaguchi. Effect of magnetic field on heat transfer in rectangular duct flow of a magnetic fluid. *Physics Procedia*, 9:190–193, 2010.
43. J L Neuringer. Some viscous flows of a saturated ferro-fluid under the combined influence of thermal and magnetic field gradients. *Int. J. Non-Linear Mech.*, 1:123–137, 1966.
44. S Odenbach. Recent progress in magnetic fluid research. *J. Phys.: Condens. Matter*, 16:R1135, 2004.
45. Y Ohira, H Houda, and T Sawada. Effect of magnetic field on a tuned liquid damper using a magnetic fluid. *Int. J. Appl. Electromagn. Mech.*, 13:71–78, 2001.
46. W M Orr. Stability and instability of steady motions of a perfect liquid. *Proc. Royal Irish Acad. A*, 27:69–138, 1907.
47. A P Oskolkov. Initial-boundary value problems for equations of motion of Kelvin–Voight fluids and Oldroyd fluids. *Trudy Math. Inst. Imeni Steklov*, 179:126–164, 1988.
48. A P Oskolkov. Nonlocal problems for the equations of motion of Kelvin-Voight fluids. *J. Math. Sci.*, 75:2058–2078, 1995.
49. A. Postelnicu. The onset of a Darcy–Brinkman convection using a thermal nonequilibrium model. *Int. J. Therm. Sci.*, 47:1587–1594, 2008.
50. O Reynolds. On the dynamical theory of incompressible viscous fluids and the determination of the criterion. *Philos. Trans. Royal Soc. London A*, 186:123–164, 1895.
51. Sangamesh, K R Raghunatha, and A J Chamkha. Linear stability analysis of thermohaline and magneto-convection in a viscoelastic fluid layer. *Chin. J. Phys.*, 94: 720–735, 2025.
52. Sangamesh, K R Raghunatha, and A J Chamkha. Nonlinear stability of thermosolutal convection in a Kelvin-Voigt fluid subject to a vertical magnetic field. *Chin. J. Phys.*, 96:377–398, 2025.
53. T Sarpkaya. Flow of non-newtonian fluids in a magnetic field. *AIChE Journal*, 7:324–328, 1961.
54. T Sawada, H Kikura, A Saito, and T Tanahashi. Natural convection of a magnetic fluid in concentric horizontal annuli under nonuniform magnetic fields. *Exp. Therm Fluid Sci.*, 7:212–220, 1993.
55. S Sharma, Sunil, and P Sharma. Stability analysis of thermosolutal convection in a rotating Navier–Stokes–Voigt fluid. *Z. Naturforsch. A*, 79, 2024.
56. S Sharma, Sunil, and P Sharma. Thermosolutal convection in a rotating Navier–Stokes–Voigt fluid saturating a porous medium under thermal non-equilibrium conditions. *J. Korean Phys. Soc.*, 87:1–15, 2025.
57. A Skumiel, P Kopcansky, M Timko, M Molcan, K Paulovicova, and R Wojciechowski. The influence of a rotating magnetic field on the thermal effect in magnetic fluid. *Int. J. Therm. Sci.*, 171:107258, 2022.
58. B Straughan. *The Energy Method, Stability, and Nonlinear Convection*. Springer Verlag, New York, 2004.
59. B Straughan. Thermosolutal convection with a Navier–Stokes–Voigt fluid. *Appl. Math. Optim.*, 84:2587–2599, 2021.

60. B Straughan. Continuous dependence and convergence for a Kelvin–Voigt fluid of order one. *Ann. dell'Universita di Ferrara*, 68:49–61, 2022.
61. B Straughan. Instability thresholds for thermal convection in a Kelvin–Voigt fluid of variable order. *Rend. Circ. Mat. Palermo Ser II*, 71:187–206, 2022.
62. B Straughan. Nonlinear stability for convection with temperature dependent viscosity in a Navier–Stokes–Voigt fluid. *Eur. Phys. J. Plus*, 138:438, 2023.
63. T G Sukacheva and O P Matveeva. On a homogenous thermoconvection model of the non-compressible viscoelastic Kelvin-Voight fluid of the non-zero order. *J. Samara State Tech. Univ. Ser. Phys. Math. Sci.*, 14:33–41, 2010.
64. M Takashima. The stability of the modified plane poiseuille flow in the presence of a transverse magnetic field. *Fluid Dyn. Res.*, 17:293, 1996.
65. S K Wilson. The effect of a uniform magnetic field on the onset of marangoni convection in a layer of conducting fluid. *Q. J. Mech. Appl. Math.*, 46:211–248, 1993.

10 Instability Modeling of Viscoelastic Fluids in Porous Medium

Suma N. Manjunath, Jayalatha G., and Vanishree R. K.

10.1 INTRODUCTION

Thermal convection in fluid-saturated porous media has been extensively investigated owing to its relevance in a wide range of natural and engineering applications such as geothermal energy extraction, enhanced oil recovery, polymer processing, ceramic manufacturing, and nuclear waste disposal. The fundamental aspects of buoyancy-driven convection were first demonstrated experimentally in a horizontal fluid layer heated from below (Bénard, 1900) and later explained theoretically through linear stability analysis (Rayleigh, 1916). Since then, the Rayleigh–Bénard problem has served as a classical paradigm in the study of hydrodynamic stability, and comprehensive treatments of thermal convection and instability theory have been documented in standard monographs (Chandrasekhar, 1961).

In many industrial and geophysical processes, the working fluids deviate significantly from Newtonian behavior. In particular, viscoelastic fluids are frequently encountered in polymeric solutions, biological fluids, and geological materials. Such fluids exhibit both viscous and elastic responses, which cannot be adequately described using Newtonian constitutive relations (Joseph, 1990). It has been recognized that thermal instability in viscoelastic fluids differs fundamentally from that in Newtonian fluids, especially with respect to the occurrence of oscillatory convection at the onset of instability. Early investigations demonstrated that oscillatory modes may arise in Oldroyd-type viscoelastic fluids under infinitesimal disturbances (Herbert, 1963; Green, 1968). Subsequently, conditions for the onset of convection and overstability were derived, and it was shown that fluid elasticity tends to destabilize the system, while rigid boundaries exert a stabilizing influence (Vest and Arpaci, 1969). Further studies confirmed that the principle of exchange of stabilities is generally violated in viscoelastic fluids, allowing oscillatory convection to precede stationary convection (Sokolov and Tanner, 1972; Rosenblat, 1986; Khayat, 1994; Demir, 2003; Mahesha et al., 2013; Jayalatha and Suma, 2021 (references therein)).

In most early investigations of thermal convection, the viscosity of the fluid was assumed to be constant in order to simplify the mathematical formulation. However,

DOI: 10.1201/9781003664345-10

experimental evidence suggests that viscosity is often strongly dependent on temperature, particularly in highly viscous and polymeric fluids. It has been pointed out that viscosity is more sensitive to temperature variations than most other physical properties of fluids (Straughan, 2002). The influence of temperature-dependent viscosity on thermal convection has therefore received increasing attention, and it has been demonstrated that viscosity variations can significantly alter the stability characteristics of the system (Torrance and Turcotte, 1971; Trompert and Hansen, 1998). For viscoelastic fluids, only a limited number of studies have addressed thermal convection in the presence of temperature-dependent viscosity and additional mechanisms controlling convection (Sekhar and Jayalatha, 2009, 2010; Jayalatha et al., 2024 and references therein).

The study of thermal convection in porous media was initiated through the pioneering works on fluid flow in permeable materials (Horton and Rogers, 1945; Lapwood, 1948). Since then, extensive research has been carried out to understand the influence of porous structures on convective heat transfer. Detailed reviews and monographs on convection in porous media are available in the literature (Ingham and Pop, 2005; Nield and Bejan, 2017). When porous media are saturated with viscoelastic fluids, the resulting flow dynamics become more complex due to the combined effects of elasticity, viscous dissipation, and porous resistance. It has been demonstrated that oscillatory convection may occur in viscoelastic-fluid-saturated porous layers under suitable conditions (Rudraiah et al., 1989; Kim et al., 2003; Yoon et al., 2004; Shivakumara et al., 2006). The effects of rotation and anisotropy on thermal convection in porous media with temperature-dependent viscosity have also been reported by Vanishree and Siddheshwar (2010).

In many investigations, the basic state of the system has been assumed to be quiescent. However, in several practical situations, a background flow exists across the porous layer, commonly referred to as throughflow. The presence of throughflow modifies the basic temperature distribution from linear to nonlinear and can significantly influence the onset of convection. The effect of throughflow on convection in porous layers saturated with Newtonian fluids has been examined in detail and it has been shown that throughflow may either stabilize or destabilize the system depending on the flow direction and boundary conditions (Wooding, 1960; Nield, 1987; Shivakumara, 1999; Khalili and Shivakumara, 2003). The combined effects of throughflow and internal heat generation in anisotropic porous media have also been investigated (see Vanishree, 2014 and references therein).

Several studies have addressed thermal convection in viscoelastic fluid saturated porous media under different physical mechanisms such as rotation, internal heating, and variable properties, and abnormal behavior of the oscillation frequency at the critical state has been reported in connection with polymeric liquids (Tan and Masuoka, 2007; Malashetty and Swamy, 2007; Sheu et al., 2008; Wang and Tan, 2008; Kumar and Bhadauria, 2011; Bhadauria and Kiran, 2014).

Despite these efforts, the influence of throughflow on convection in viscoelastic-fluid-saturated porous media has received comparatively less attention. A notable contribution in this direction demonstrated that oscillatory convection in an Oldroyd-B fluid-saturated porous layer can be either enhanced or suppressed depending on viscoelastic parameters and boundary conditions, although the analysis was restricted to constant-viscosity fluids (Shivakumara and Sureshkumar, 2007). More recently, the effect of temperature-dependent viscosity on thermal instability in viscoelastic fluids saturating porous media has been analyzed in the absence of throughflow (Dhiman et al., 2024).

In many realistic applications, such as geothermal reservoirs, polymer-assisted oil recovery and alloy solidification processes, both temperature-dependent viscosity and throughflow are inherently present. However, the combined influence of these two effects on thermal convection in viscoelastic-fluid-saturated porous media has not been examined so far.

Motivated by this gap in the literature, the present chapter investigates the onset of thermal convection in a horizontal porous layer saturated with a Jeffreys-type viscoelastic fluid by incorporating both temperature-dependent viscosity and uniform vertical throughflow. The Darcy–Brinkman–Jeffreys model is employed to account for viscoelastic effects, viscous diffusion, and porous resistance. Linear stability theory is applied and the resulting eigenvalue problem is solved using the Galerkin method for stress-free isothermal boundary conditions. The objective of the present study is to elucidate the coupled roles of viscosity variation, throughflow and viscoelasticity on the nature and threshold of convective instability.

10.2 MATHEMATICAL FORMULATION

A horizontal porous layer of infinite lateral extent saturated with a Jeffreys-type viscoelastic fluid is considered. The layer is heated uniformly from below and subjected to a uniform vertical through flow (see Figures 10.1a and 10.1b). The fluid motion is assumed to obey the Boussinesq approximation, and the viscosity is taken to be a function of temperature.

The governing equations for the system consist of the conservation of mass, momentum, and energy, together with the equation of state and the rheological relation for a Jeffreys fluid, are as follows:

The conservation of mass is expressed as

$$\nabla.\vec{q} = 0 \tag{10.1}$$

The conservation of momentum for flow through a porous medium, incorporating viscoelastic effects, is given by

$$\rho_0\left[1+\lambda_1\frac{\partial}{\partial t}\right]\frac{\partial\vec{q}}{\partial t} = \left[1+\lambda_1\frac{\partial}{\partial t}\right]\left[-\nabla p+\rho\vec{g}\right]+\left[1+\lambda_2\frac{\partial}{\partial t}\right]\left[\nabla.\mu_e(T)\left\{\nabla\vec{q}.\nabla\vec{q}^T\right\}-\frac{\mu_f(T)}{k}\vec{q}\right] \tag{10.2}$$

The thermal energy equation is

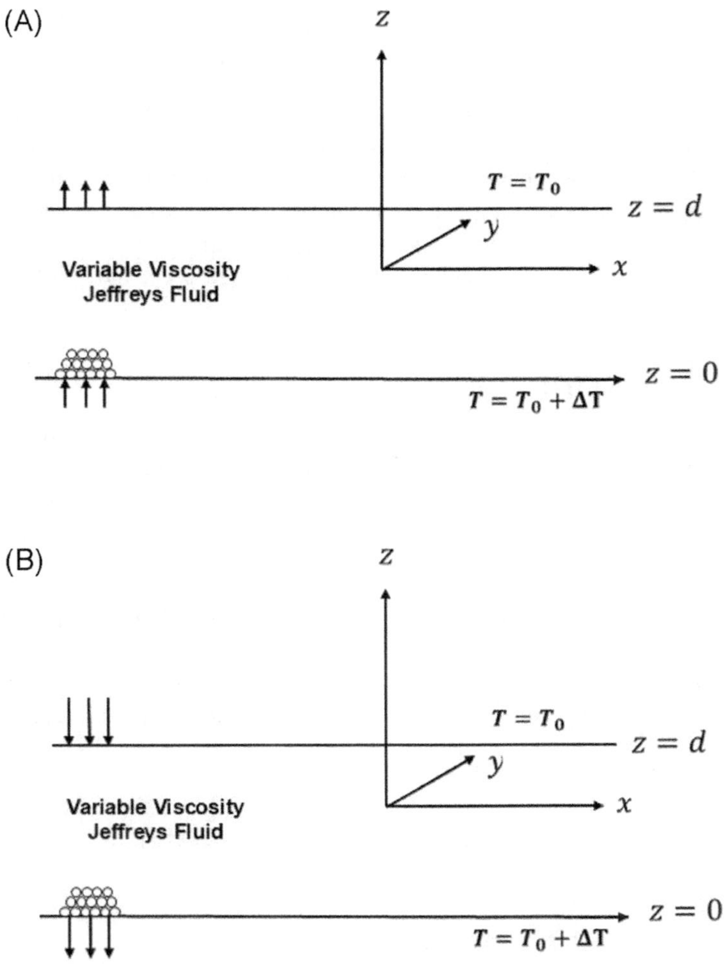

FIGURE 10.1A,B (A) Upward Through Flow; (B) Downward Through Flow.

$$\frac{\partial T}{\partial t} = \left(\vec{q}.\nabla\right)T + \kappa\nabla^2 T \tag{10.3}$$

The density variation responsible for buoyancy is represented through the linearized equation of state:

$$\rho = \rho_0\left[1 - \alpha\left(T - T_0\right)\right] \tag{10.4}$$

The thermorheological behavior of the viscoelastic fluid is described following Nield (1996):

$$\mu_f(T) = \frac{\mu_1}{1-\Gamma(T-T_0)}, \mu_e(T) = \frac{\mu_2}{1-\Gamma(T-T_0)} \tag{10.5}$$

where $\vec{q} = (u, v, w)$ are the components of the velocity of the liquid, ρ is the density, ρ_0 is the density at the reference temperature T_0, p is the pressure, $\mu_f(T)$and $\mu_e(T)$ are the temperature-dependent viscosity of the liquid and the effective viscosity, respectively, $\vec{g} = (0, 0, -g)$, are the components of the gravitational acceleration, $\lambda_1$is the stress relaxation coefficient, λ_2 is the strain retardation coefficient, k is the permeability, T is the temperature and κ is the thermal diffusivity, and α is the constant coefficient of thermal expansion.

10.3 BASIC STATE

The basic (undisturbed) state corresponds to a steady solution in which the fluid motion is solely due to the imposed vertical throughflow and all physical quantities depend only on the vertical coordinate. Accordingly, the basic state is assumed in the form

$$\frac{\partial()}{\partial t} = 0, \vec{q}_b = w_0 \hat{k}, p = p_b(z), \rho = \rho_b(z), T = T_b(z), \mu_f = \mu_{f_b}(z), \mu_e = \mu_{e_b}(z) \tag{10.6}$$

Substitution of the basic state variables from Equation (10.6) into Equations (10.1)–(10.5) yields the basic state solution for velocity, pressure, temperature, and density fields.

$$\frac{dp_b}{dz} + \rho_b g = 0, \bar{\theta}_b(z) = \frac{\Delta T}{1-e^{Pe}}\left[e^{\frac{w_0}{\kappa}z} - e^{Pe}\right], \rho_b = \rho_0\left[1-\alpha\bar{\theta}_b(z)\right], \mu_{e_b} = \frac{\mu_1}{1-\Gamma\bar{\theta}_b(z)},$$

$$\mu_{f_b} = \frac{\mu_2}{1-\Gamma\bar{\theta}_b(z)} \tag{10.7}$$

10.4 PERTURBED STATE AND LINEARIZATION

To investigate the stability of the basic state, infinitesimal perturbations are superimposed on it. Each dependent variable is decomposed into a basic state component and a small perturbation component.

The governing equations are then linearized by neglecting nonlinear terms involving products of perturbation quantities. This procedure yields the linearized equations governing the evolution of the perturbations.

$$\vec{q} = w_0\hat{k} + \vec{q}', \rho = \rho_b + \rho', p = p_b + p', T = T_b + \theta, \mu = \mu_b + \mu' \tag{10.8}$$

10.5 NON-DIMENSIONALIZATION

The linearized equations are rendered dimensionless using appropriate characteristic scales for length, velocity, time, temperature, and pressure.

$$t^* = \frac{t}{d^2/\kappa}, \theta^* = \frac{\theta}{\Delta T}, \left(x^*, y^*, z^*\right) = \left(\frac{x}{d}, \frac{y}{d}, \frac{z}{d}\right), w^* = \frac{w}{\kappa/d} \tag{10.9}$$

Upon introducing these non-dimensional variables and dropping the asterisks for simplicity, the governing perturbation equations take the form

$$\frac{1}{Pr}\left[1+\Lambda_1\frac{\partial}{\partial t}\right]\frac{\partial}{\partial t}\left(\nabla^2 w\right) = \left[1+\Lambda_1\frac{\partial}{\partial t}\right]R\nabla_1^2\theta + \left[1+\Lambda_2\frac{\partial}{\partial t}\right]$$

$$\left[\begin{array}{l} -\dfrac{2\Lambda_{vf}V}{(1+Vz)^2}\dfrac{\partial}{\partial z}\left(\nabla^2 w\right) + \dfrac{\Lambda_{vf}}{1+Vz}\nabla^4 w + \dfrac{2\Lambda_{vf}V^2}{(1+Vz)^3}\left\{\dfrac{\partial^3 w}{\partial z^3} - \nabla_1^2 w\right\} \\ -\dfrac{Da^{-1}}{1+Vz}\nabla^2 w + \dfrac{Da^{-1}V}{(1+Vz)^2}\dfrac{\partial w}{\partial z} \end{array}\right] \tag{10.10}$$

$$\frac{\partial\theta}{\partial t} + Pe\frac{d\theta}{dz} + \left[\frac{Pe}{1-e^{Pe}}e^{Pe\,z}\right]w = \nabla^2\theta \tag{10.11}$$

The dimensionless parameter appearing in Equations (10.10) and (10.11) are Prandtl number, $Pr(=\frac{\mu_0}{\rho_0\,\kappa})$, Rayleigh number $R, \left(=\frac{\alpha\rho_0 g d^3\Delta T}{\mu_0\kappa}\right)$, Peclet number $Pe, \left(=\frac{w_0 d}{\kappa}\right)$, stress relaxation parameter or Deborah number $\Lambda_1, \left(=\frac{\lambda_1\kappa}{d^2}\right)$, strain retardation parameter $\Lambda_2, \left(=\frac{\lambda_2\kappa}{d^2}\right)$, viscosity factor, $\Lambda_{vf}\left(=\frac{\mu_e}{\mu_f}\right)$ and variable viscosity parameter, $V\left(=\Gamma\Delta T\right)$.

10.6 BOUNDARY CONDITIONS

The boundary conditions appropriate to the problem are imposed at the upper and lower boundaries of the porous layer and are expressed as

$$w = \nabla^2 w = \theta = 0 \text{ at } z = 0,1 \tag{10.12}$$

10.7 NORMAL MODE ANALYSIS

The infinitesimal perturbations w and θ are assumed to be periodic waves and hence these permit normal mode solutions in the form (see Chandrasekhar, 1961)

$$\left.\begin{aligned} w &= e^{i\omega t} W(z) e^{i(lx+my)} \\ \theta &= e^{i\omega t} \Theta(z) e^{i(lx+my)} \end{aligned}\right\} \tag{10.13}$$

where $a^2 = l^2 + m^2$ is the non-dimensional wave number.

Substitution of the normal mode representation (10.13) into Equations (10.10) and (10.11) reduces the governing equations to a coupled system of ordinary differential equations, given by,

$$R\left[1+i\Lambda_1\omega\right]a^2\Theta = -\frac{1}{\Pr}\left[1+i\Lambda_1\omega\right]i\omega\left(D^2-a^2\right)W$$
$$+\left[1+i\Lambda_2\omega\right]\left[\begin{aligned} &-\frac{2\Lambda_{vf}V}{(1+Vz)^2}\left(D^2-a^2\right)DW + \frac{\Lambda_{vf}}{1+Vz}\left(D^2-a^2\right)^2 W + \\ &\frac{2\Lambda_{vf}V^2}{(1+Vz)^3}\left(D^2+a^2\right)W - \frac{Da^{-1}}{1+Vz}\left(D^2-a^2\right)W + \frac{Da^{-1}V}{(1+Vz)^2}DW \end{aligned}\right] \tag{10.14}$$

$$i\omega\Theta + Pe\, D\Theta - \left(D^2-a^2\right)\Theta = -\left[\frac{Pe}{1-e^{Pe}}e^{Pe\,z}\right]W \tag{10.15}$$

10.8 GALERKIN VARIANT OF WEIGHTED-RESIDUALS TECHNIQUE

The eigenvalue problem represented by Equations (10.14) and (10.15) is solved by employing the Galerkin approach of the weighted residual method, as outlined by Finlayson (1972). This technique is well suited for obtaining approximate solutions to eigenvalue problems and is particularly effective in determining the lowest eigenvalue, which is of primary interest in stability studies. In this approach, suitable trial functions are selected for the velocity and temperature perturbations. These functions are chosen such that they satisfy the imposed boundary conditions, although they may not satisfy the governing differential equations exactly. When these trial functions are substituted into the differential equations, non-zero residuals arise due to this approximation.

The essential idea of the weighted residual method is to minimize these residuals in an average sense. This is achieved by requiring the residuals to be orthogonal to each of the chosen trial functions over the domain of interest. In the Galerkin formulation, the weighting functions are taken to be the same as the trial functions themselves. Accordingly, the velocity and temperature perturbations are expanded in terms of these trial functions as follows

$$\left.\begin{aligned} W(z,t) &= \Sigma A_i(t) W_i(z) \\ \Theta(z,t) &= \Sigma B_i(t) \Theta_i(z) \end{aligned}\right\} \tag{10.16}$$

where $W_i(z)$ and $\Theta_i(z)$ denote the chosen trial functions, which are selected to satisfy the boundary conditions specified in Equation (10.12). We assume the trial function in the form $w = \sin \pi z, T = \sin \pi z$ which is equivalent to higher order technique results. To demonstrate the procedure clearly, the formulation is restricted to a single-term approximation in the present analysis.

Multiplying Equations (10.14) and (10.15) by W and Θ respectively and integrating the resulting equations by parts with respect to z between 0 and 1, and taking $W = AW_1$, and $\Theta = B\Theta_1$ in which A and B are constants, and W_1 and Θ_1 are trial functions satisfying Equation (10.12), following Equations (10.17) and (10.18) are obtained.

$$BRa^2\left[1+i\Lambda_1\omega\right]C_3 = \frac{C_2}{\Pr}\left[i\omega - \Lambda_1\omega^2\right]A + C_1\left[1+i\Lambda_2\omega\right]A \tag{10.17}$$

$$B\left[i\omega C_4 + Pe\,C_5 + C_6\right] = A\,K\,C_7 \tag{10.18}$$

where

$$C_1 = \Lambda_{vf}\left\langle G(z)^{-1} W_1 D^4 W_1\right\rangle - 2\Lambda_{vf}\left\langle VG(z)^{-2} W_1 D^3 W_1\right\rangle + \left(2\Lambda_{vf} Va^2 Da^{-1}\right)\left\langle G(z)^{-2} W_1 DW_1\right\rangle$$

$$-\left(2\Lambda_{vf}a^2 + Da^{-1}\right)\left\langle G(z)^{-2} W_1 D^2 W_1\right\rangle + a^2\left(\Lambda_{vf}a^2 + Da^{-1}\right)\left\langle G(z)^{-1} W_1^2\right\rangle$$

$$+2\Lambda_{vf}V^2\left\langle G(z)^{-3} W_1 D^2 W_1\right\rangle + 2\Lambda_{vf}V^2 a^2\left\langle G(z)^{-3} W_1^2\right\rangle, C_2$$

$$= -\left\langle W_1 D^2 W_1\right\rangle + a^2\left\langle W_1^2\right\rangle, \quad C_3 = \left\langle W_1\theta_1\right\rangle, \quad C_4 = \left\langle \Theta_1^2\right\rangle, \quad C_5 = \left\langle \Theta_1 D\Theta_1\right\rangle,$$

$$C_6 = -\left\langle \Theta_1 D^2\Theta_1\right\rangle + a^2\left\langle \Theta_1^2\right\rangle, C_7 = -\left\langle e^{Pe\,z} W_1\Theta_1\right\rangle,$$

$$\langle\ \rangle = \int_0^1, \quad K = \frac{Pe}{1-e^{Pe}}.$$

Solving the resulting system of algebraic equations provides the condition for the onset of convection and yields the critical Rayleigh number, R, corresponding to oscillatory instability.

$$R = \left[\frac{C_1C_4\omega^2N_2Pr + C_1\left(Pe\,C_5 + C_6\right)N_1Pr - \omega^2C_2C_4}{a^2KC_3C_7\,Pr}\right]$$

$$+i\omega\left[\frac{\left(C_1C_4N_1 - C_1C_4N_2\right)Pr + C_2\left(Pe\,C_5 + C_6\right)}{a^2KC_3C_7\,Pr}\right] \tag{10.19}$$

where $N_2 = \dfrac{1+\Lambda_1^2\Lambda\omega^2}{1+\Lambda_1^2\omega^2}$ and $N_1 = \dfrac{\Lambda_1(1-\Lambda)}{1+\Lambda_1^2\omega^2}, \Lambda = \dfrac{\Lambda_2}{\Lambda_1}$ is the elastic ratio.

Since R is real, the imaginary part is equated to zero, and an expression for ω^2 is obtained.

10.9 RESULTS AND DISCUSSION

The onset of thermal convection in a horizontal porous layer saturated with a Jeffreys-type viscoelastic fluid has been investigated by incorporating the combined effects of temperature-dependent viscosity and uniform vertical throughflow. A linear stability analysis is carried out using the Darcy–Brinkman–Jeffreys model, and the resulting eigenvalue problem was solved by employing the Galerkin variant of the weighted-residuals technique under free–free isothermal boundary conditions. Particular attention was devoted to the oscillatory mode of instability, which arises due to the viscoelastic nature of the fluid.

The frequency ω_c and critical wave number a_c, which govern the critical oscillatory Darcy–Rayleigh number R_{oc} marking the onset of convection in a Jeffreys fluid, have been calculated for a range of parameter values, and the obtained results are presented graphically.

From Figures 10.2–10.4, it is clear that with an increase in the stress relaxation parameter Λ_1, R_{oc} decreases, which indicates the destabilizing effect of Λ_1 on the onset of convection and also a_c and ω_c decrease with an increase in Λ_1. Increasing Λ,

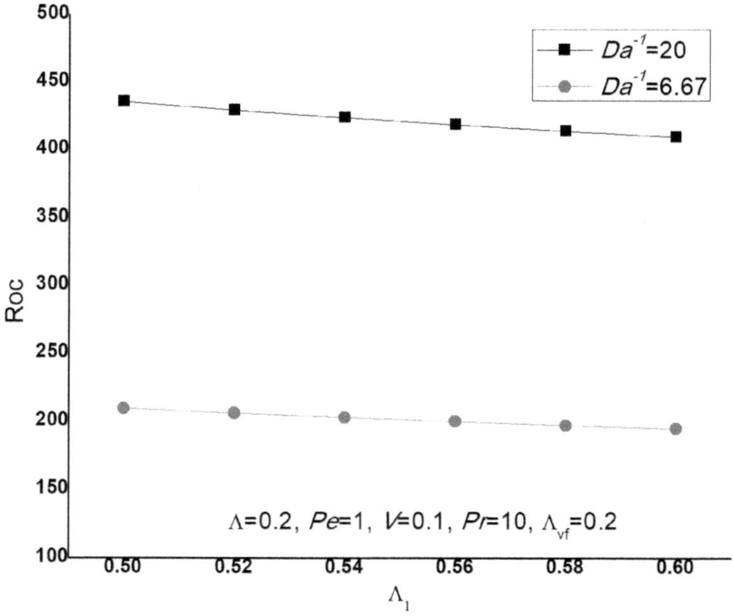

FIGURE 10.2 Plot of R_{oc} versus Λ_1 for different values of Da^{-1}.

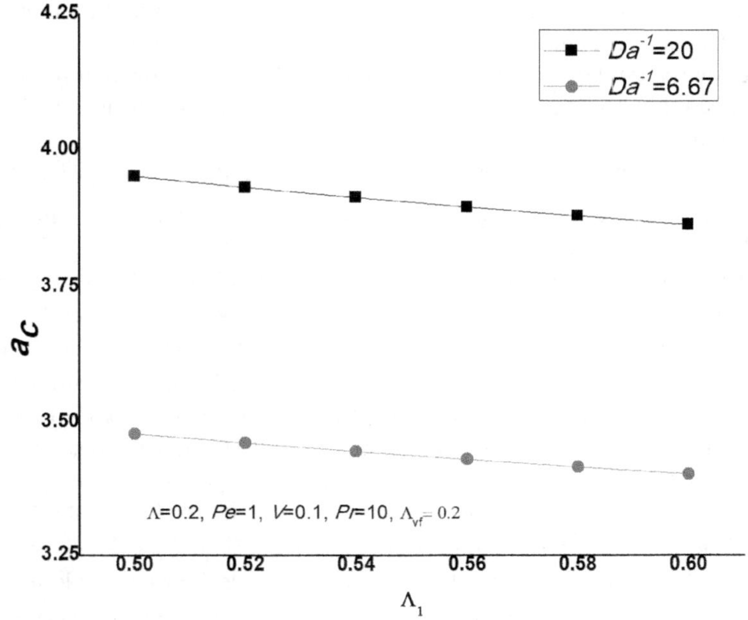

FIGURE 10.3 Plot of a_c versus Λ_1 for different values of Da^{-1}.

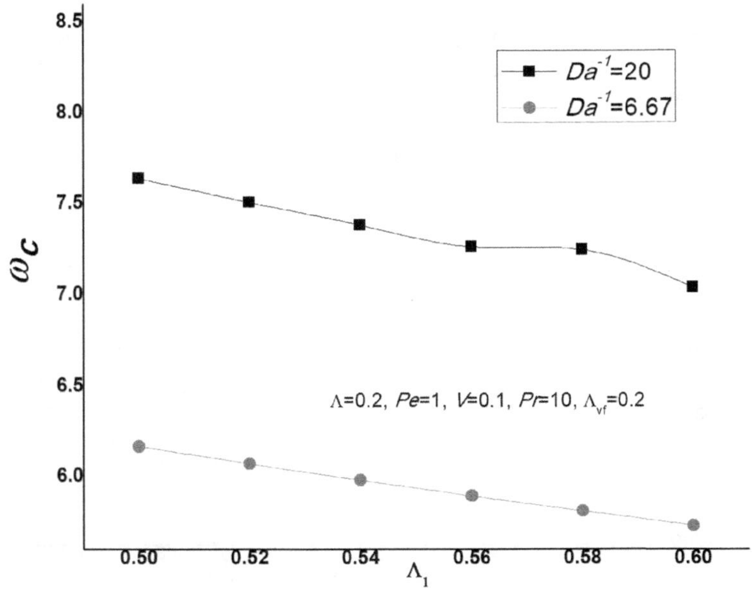

FIGURE 10.4 Plot of ω_c versus Λ_1 for different values of Da^{-1}.

elastic ratio stabilizes strongly by increasing R_{oc}, delays the onset of oscillatory convection, decreases a_c, and ω_c as observed from Figures 10.5–10.7.

Increasing variable viscosity parameter V yields a destabilization (reduction in R_{oc}), with negligible change in a_c and a slight decrease in ω_c which is evident from Figures 10.8–10.10.

From Figures 10.11–10.13, it is observed that the increase in viscosity factor $\Lambda_{vf} = \frac{\mu_e}{\mu_f}$ increases R_{oc}, delaying the onset, which is due to the fact that the increase in effective viscosity (μ_e) increases the thickness of fluid, thereby resisting the flow. Also, the increase in Λ_{vf} decreases, a_c thereby increasing the cell size. It is observed that the increase in Λ_{vf} decreases ω_c slightly for certain Da^{-1} and shows a reverse effect for smaller values of Da^{-1} (For the choice of Λ_{vf} see Givler and Altobelli (1994)).

The variation of the oscillatory Darcy–Rayleigh number R_{oc} with the horizontal wavenumber a_c is presented through the neutral stability curves. The neutral curves exhibit the expected single-minimum ("U-shaped") structure, indicating the existence of a preferred critical mode characterized by R_{oc_min} and the corresponding a_c. For the fixed parameter set Λ=0.2, Λ_1=0.5, $\Lambda_{vf} = 0.2$, Pe=1, Pr=10 and V=0.1, the curves shift markedly with changes in Da^{-1}. From the plotted neutral curves (Figure 10.14), it is observed that increasing Da^{-1} (i.e., reducing permeability / increasing resistance offered by the porous matrix) raises the entire neutral curve and increases the minimum value of R_{oc}. In other words, a more resistive porous

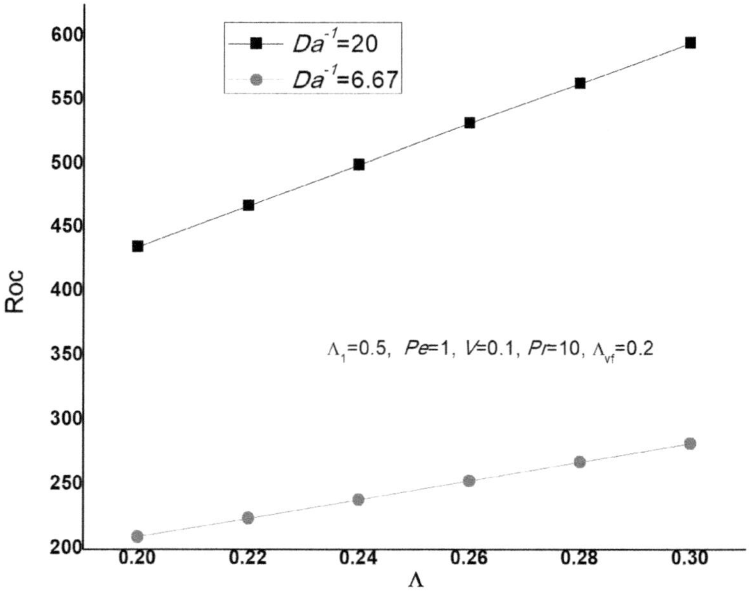

FIGURE 10.5 Plot of R_{ac} versus Λ_1 for different values of Da^{-1}.

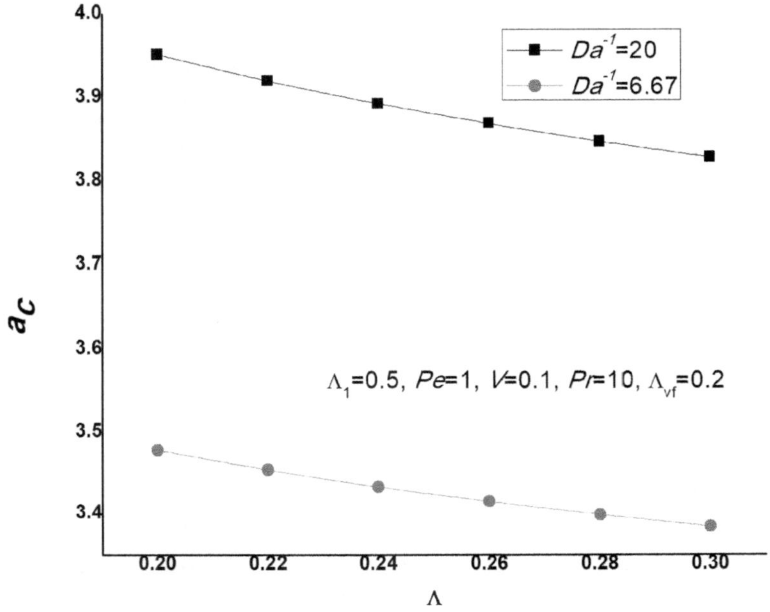

FIGURE 10.6 Plot of a_c versus Λ_1 for different values of Da^{-1}.

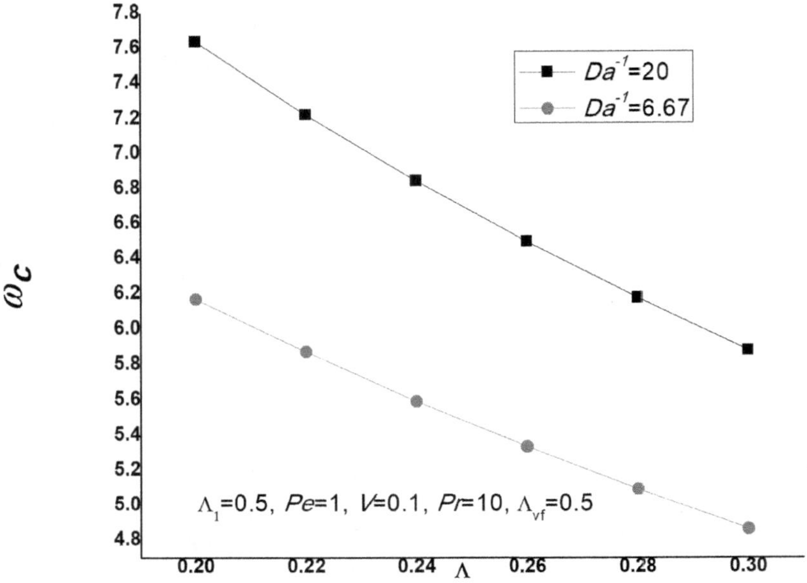

FIGURE 10.7 Plot of ω_c versus Λ_1 for different values of Da^{-1}.

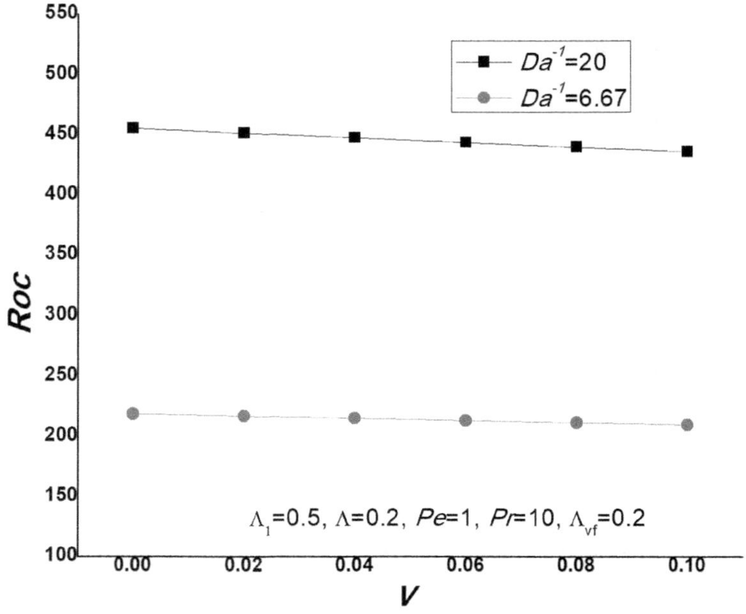

FIGURE 10.8 Plot of R_{ac} versus V for different values of Da^{-1}.

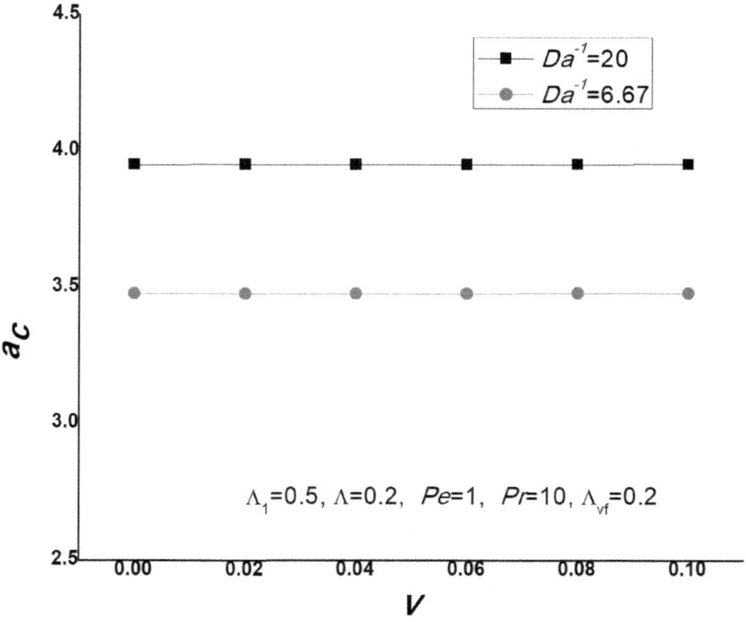

FIGURE 10.9 Plot of a_c versus V for different values of Da^{-1}.

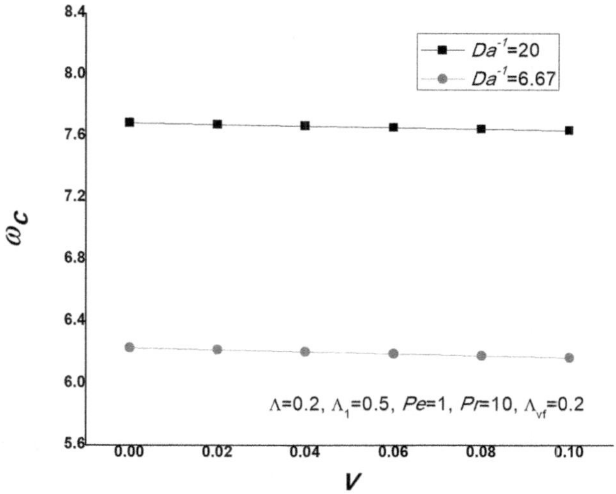

FIGURE 10.10 Plot of ω_c versus V for different values of Da^{-1}.

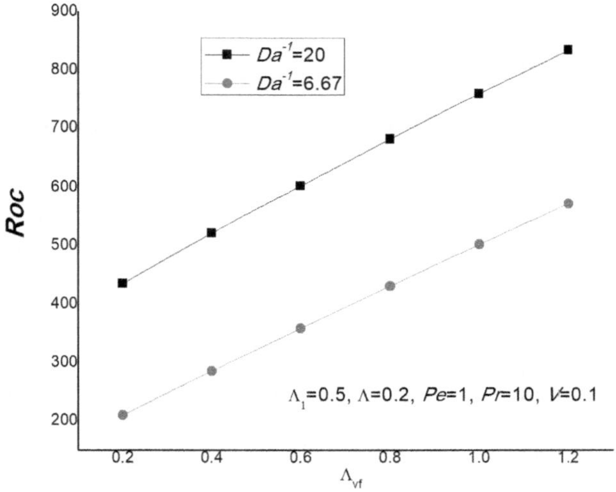

FIGURE 10.11 Plot of R_{ac} versus Λ_{vf} for different values of Da^{-1}.

medium stabilizes the system by demanding a larger buoyancy driving to overcome porous drag. Conversely, as permeability increases (smaller Da^{-1}, convection sets in at lower R_{oc}, implying earlier onset of oscillatory convection. This behavior is consistent with the physical role of Darcy resistance: higher permeability permits easier fluid motion and hence reduces the thermal forcing needed for instability.

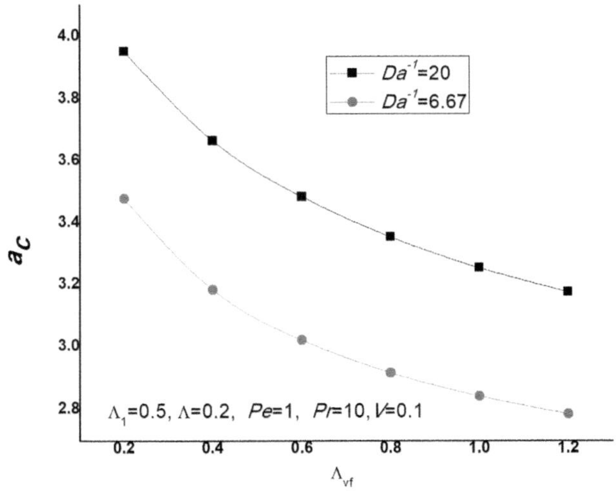

FIGURE 10.12 Plot of a_c versus Λ_{vf} for different values of Da^{-1}.

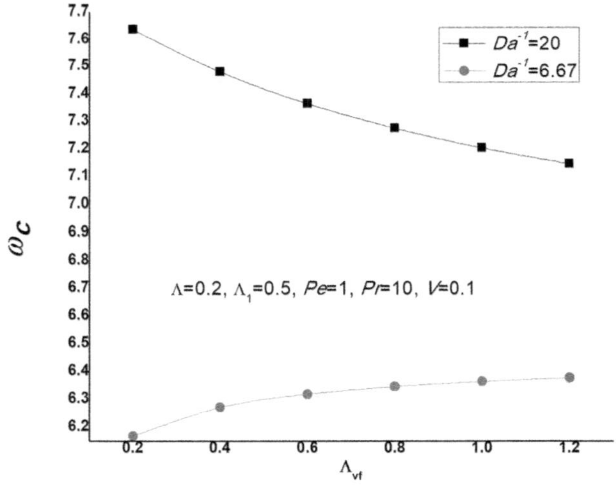

FIGURE 10.13 Plot of ω_c versus Λ_{vf} for different values of Da^{-1}.

The stabilization is also visible in Figure 10.15, where increasing Λ shifts the entire neutral curve upward, increasing R_{oc_min}.

The increasing Λ_1 makes the system more prone to oscillatory convection (destabilizing influence), as reinforced by the neutral curves in Figure 10.16, larger Λ_1

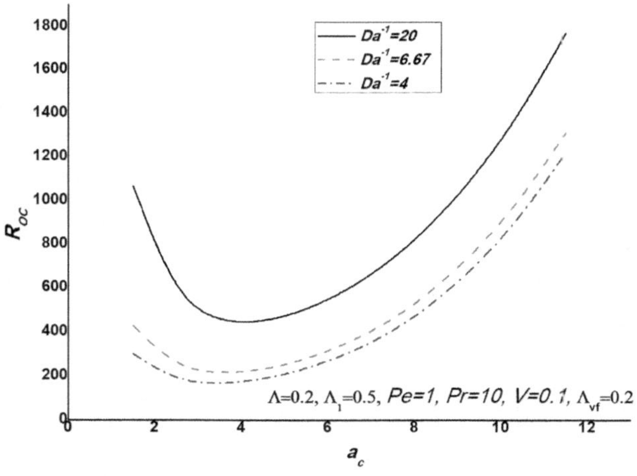

FIGURE 10.14 Plot of R_{ac} versus a_c varying Da for different values of Da^{-1}.

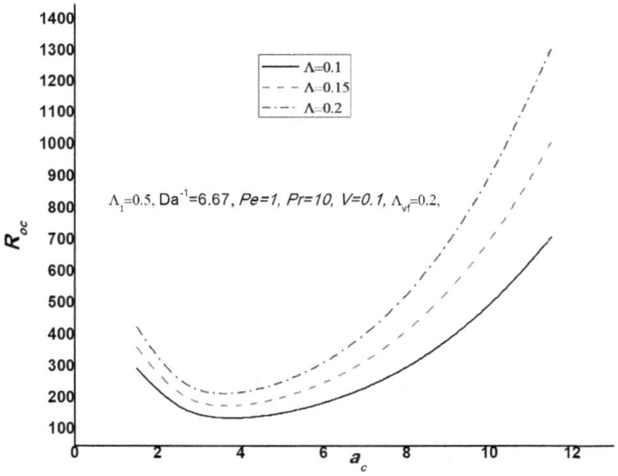

FIGURE 10.15 Plot of R_{ac} versus a_c for different values of Λ.

shifts the neutral curve downward, lowering R_{oc_min}. From the neutral curve plot (Figure 10.17), it is clear that increasing viscosity factor shifts the curve upward, increasing R_{oc_min} and has a stabilizing effect.

The combined influence of permeability and throughflow on oscillatory instability is analyzed by comparing results obtained for two inverse Darcy numbers, Da^{-1}=20 and Da^{-1}=6.67, with the viscosity-variation parameter fixed at V=0.5 as seen in Figure 10.18. For the lower Darcy number case $(Da^{-1} = 20)$, corresponding

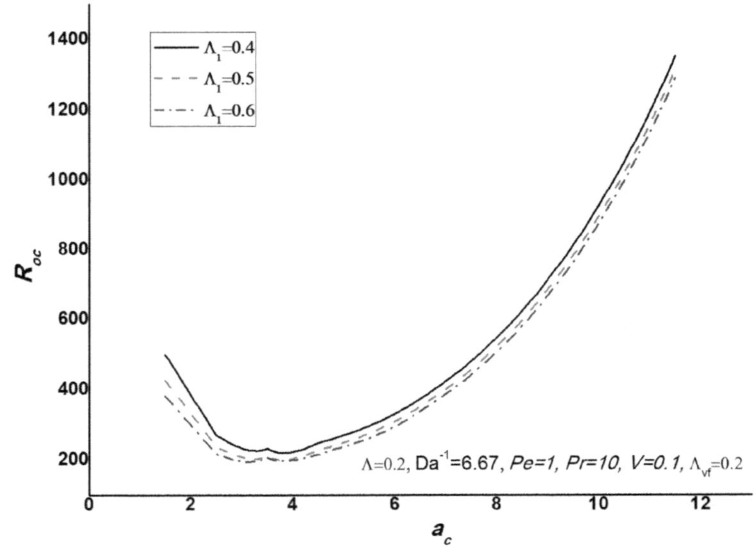

FIGURE 10.16 Plot of R_{ac} versus a_c for different values of Λ_1.

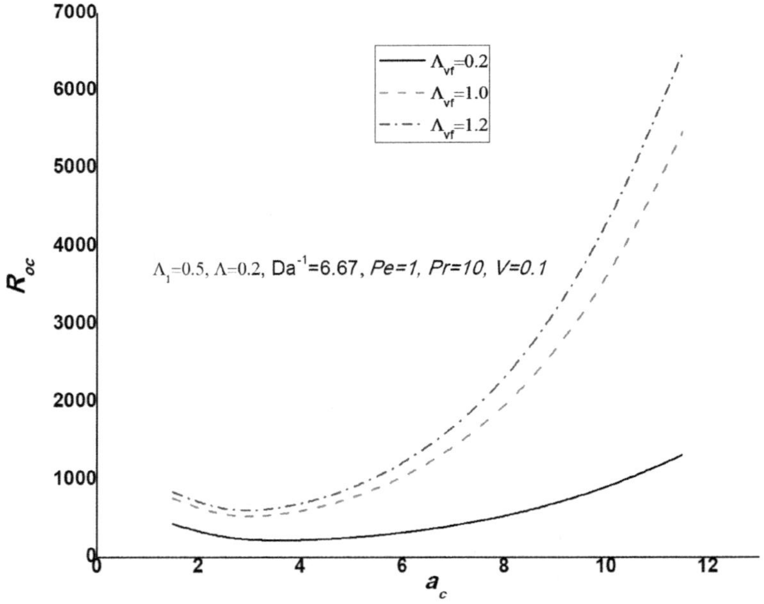

FIGURE 10.17 Plot of R_{ac} versus a_c for different values of Λ_{vf}.

to reduced permeability, the minimum critical Rayleigh number obtained is substantially higher than the corresponding value obtained for the higher permeability case (Da^{-1} =6.67). This marked increase in the critical Rayleigh number demonstrates that **reduced permeability strongly stabilizes oscillatory convection,** as the porous matrix offers greater resistance to fluid motion. As the magnitude of the Péclet number increases, the stabilizing influence of reduced permeability becomes even more pronounced. Thus, for all values of Pe, a lower inverse Darcy number consistently demands significantly higher buoyancy forcing for oscillatory instability to occur. The resulting curves corresponding to both inverse Darcy numbers remain symmetric about $Pe = 0$, indicating that the **magnitude of throughflow,** rather than its direction, governs the qualitative nature of instability. However, the substantial vertical separation between the curves clearly highlights the **dominant stabilizing role of inverse Darcy resistance** relative to the destabilizing influence of viscosity variation.

From Figure 10.19 variation of the critical oscillatory Darcy Rayleigh number R_{oc} with the throughflow Péclet number Pe is examined for the **extreme porosity regime**, corresponding to the inverse Darcy number Da^{-1}= 6.67. The results are presented for two cases: absence of viscosity variation ($V = 0$) and presence of viscosity variation ($V = 0.5$) showing the destabilizing influence of V. A reverse effect is seen with Λ_{vf} (Figure 10.20). For both cases, the curves are observed to be **symmetric**

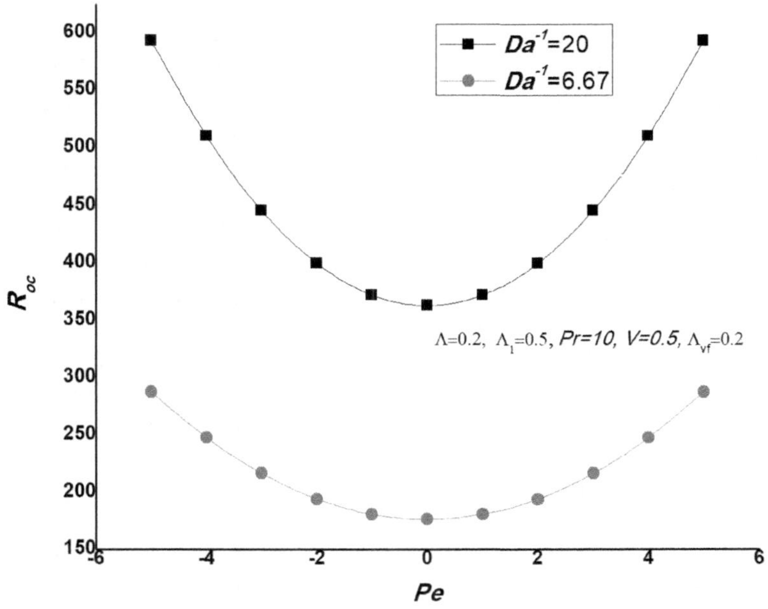

FIGURE 10.18 Plot of R_{ac} versus Pe for different values of Da^{-1}.

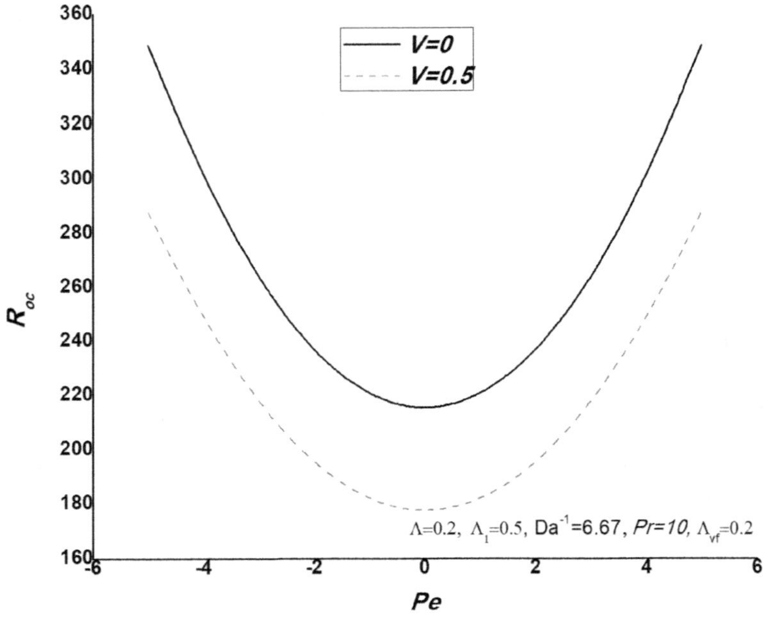

FIGURE 10.19 Plot of R_{ac} versus P_e for different values of V.

about $Pe = 0$ and attain a minimum at $Pe = 0$, indicating the existence of a preferred oscillatory mode at zero throughflow. In the absence of viscosity variation ($V = 0$), as the magnitude of the Péclet number increases, the critical Rayleigh number increases monotonically. This behavior reflects the stabilizing influence of strong throughflow in either direction, as large values of |Pe| enhance advective transport and suppress the growth of oscillatory disturbances. When viscosity variation is introduced ($V = 0.5$), the entire neutral stability curve shifts downward, indicating that **temperature-dependent viscosity has a destabilizing influence** on oscillatory convection in the extreme porosity regime. Physically, viscosity variation reduces effective viscous resistance in regions of higher temperature, thereby facilitating oscillatory motion at lower thermal driving. However, the persistence of symmetric, U-shaped neutral curves confirms that sufficiently strong throughflow, irrespective of its direction, ultimately stabilizes the system by dominating buoyancy-driven oscillations.

The above results demonstrate that the onset of oscillatory convection is governed by a competition between buoyancy forces, porous resistance, and advective transport induced by throughflow. While temperature-dependent viscosity promotes instability by reducing effective viscous damping, increased porous resistance suppresses oscillatory motion by enhancing momentum dissipation.

The symmetry of the neutral curves about $Pe = 0$ confirms that stability is primarily controlled by the strength of throughflow rather than its direction. The

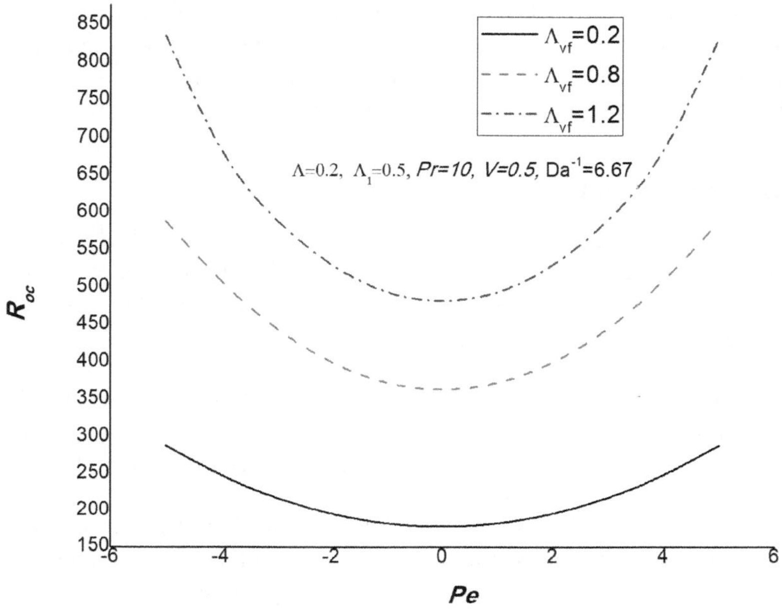

FIGURE 10.20 Plot of R_{ac} versus Pe for different values of Λ_{vf}.

occurrence of minimum critical Rayleigh numbers at $Pe = 0$ further indicates that weak throughflow favors oscillatory instability, whereas strong throughflow in either direction stabilizes the system.

These findings are consistent with earlier observations that viscoelastic effects are responsible for the emergence of oscillatory modes, while permeability and viscosity variation govern the threshold and structure of instability. The present results extend this understanding by quantitatively demonstrating how **throughflow, inverse Darcy number, and viscosity variation interact** to control oscillatory convection in viscoelastic fluid-saturated porous media.

Before concluding the discussion, it is appropriate to verify the consistency of the present results by examining relevant limiting cases. When the throughflow parameter tends to zero ($Pe = 0$), the stability characteristics reduce to those corresponding to thermal convection in a viscoelastic fluid-saturated porous layer with temperature-dependent viscosity alone(Dhiman et al., 2024). In this limit, the trends obtained for the critical oscillatory Rayleigh number, critical wavenumber, and oscillatory frequency are found to be qualitatively consistent with previously reported results for variable-viscosity viscoelastic convection in porous media. Similarly, when the viscosity-variation parameter is set to zero ($V = 0$), the model recovers the constant-viscosity Jeffreys fluid case and the computed stability thresholds and oscillatory features agree with those established for viscoelastic convection without viscosity stratification(Shivakumara and Sureshkumar, 2007). Furthermore, in the absence of viscoelastic effects, the oscillatory mode of instability is no longer observed and the system reduces to the classical Newtonian porous-medium convection problem (Horton and Rogers, 1945). These limiting-case considerations confirm that the present formulation correctly reproduces known results and provide confidence in the validity of the numerical findings discussed above.

10.10 CONCLUSIONS

In conclusion, the present investigation has clarified the individual and combined roles of viscosity variation, permeability, and throughflow on the onset of oscillatory convection in a viscoelastic fluid-saturated porous layer. The main findings of the study may be summarized as follows:

- Temperature-dependent viscosity exerts a **weak but consistent destabilizing influence**, leading to a slight reduction in the critical Rayleigh number, with negligible change in the critical wavenumber and a small decrease in the oscillatory frequency.
- A decrease in the Darcy number (reduced permeability) **strongly stabilizes** the system by increasing the critical Rayleigh number, whereas higher permeability facilitates earlier onset of oscillatory convection.
- The neutral stability curves exhibit a **symmetric U-shaped structure**, confirming the existence of a preferred critical mode characterized by a minimum Rayleigh number and an associated wavenumber.
- Throughflow, represented by the Péclet number, modifies the stability threshold without altering the qualitative nature of instability; **weak throughflow promotes oscillatory convection**, while strong throughflow in either direction stabilizes the system.
- The symmetry of the neutral curves about zero Péclet number indicates that the **magnitude of throughflow**, rather than its direction, governs the stability behaviour.
- Limiting-case analyses demonstrate that the present results correctly reduce to known configurations, thereby **confirming the validity and robustness** of the formulation.

Overall, the study highlights the delicate interplay between viscoelastic effects, temperature-dependent viscosity, throughflow, and porous resistance in controlling oscillatory convection. The results provide useful physical insight relevant to geophysical and industrial processes involving heat transfer in porous media saturated with complex fluids.

REFERENCES

Bénard, H. (1900). Les tourbillons cellulaires dans une nappe liquide. *Revue Générale des Sciences Pures et Appliquées*, 11, 1261–1271.

Bhadauria, B. S., & Kiran, P. (2014). Weakly nonlinear oscillatory convection in a viscoelastic fluid saturating porous medium under temperature modulation. *International Journal of Heat and Mass Transfer*, 77, 843–851.

Chandrasekhar, S. (1961). *Hydrodynamic and Hydromagnetic Stability*. Oxford University Press, London.

Demir, H. (2003). Rayleigh–Bénard convection of viscoelastic fluid. *Applied Mathematics and Computation*, 136, 251–267.

Dhiman, J. S., Sharma, P. K., & Gupta, K. (2024). Effect of temperature-dependent viscosity on thermal instability of Oldroydian viscoelastic fluid layer saturating porous medium. *Journal of Theoretical and Applied Mechanics, Sofia*, 54, 255–277.

Finlayson, B. A. (1972). *Method of Weighted-residuals and Variational Principles*. Academic Press, Cambridge.

Givler, R. C., & Altobelli, A. (1994). A determination of the effective viscosity for Brinkman-Forchhemier model. *Journal of Fluid Mechanics*, 258, 355–367.

Green, T. (1968). Oscillating convection in an elasticoviscous liquid. *Physics of Fluids*, 11(7), 1410–1412.

Herbert, D. M. (1963). On the stability of visco-elastic liquids in heated plane Couette flow. *Journal of Fluid Mechanics*, 17(3), 353–359.

Horton, C. W., & Rogers, F. T. (1945). Convection currents in a porous medium. *Journal of Applied Physics*, 16, 367–370.

Ingham, D. B., & Pop, I. (2005). *Transport Phenomena in Porous Media*. Elsevier, Oxford.

Jayalatha, G., Sakshath, T. N., Siddheshwar, P. G., & Sekhar, G. N. (2024). Impact of variable viscosity and gravity variations on Rayleigh–Bénard instabilities of viscoelastic liquids in energy sustainable systems. Proceedings of the 8th International Conference on Computational System and Information Technology for Sustainable Solutions (CSITSS 2024), Bengaluru, India, 1–6. https://doi.org/10.1109/CSITSS64042.2024.10816883

Jayalatha, G., & Suma, N. (2021). Effect of heat source on Rayleigh–Bénard convection in rotating viscoelastic liquids. *Heat Transfer*, 50(8), 7672–7690.

Joseph, D. D. (1990). *Dynamics of Viscoelastic Liquids*. Springer, New York.

Khalili, A. & Shivakumara, I. S. (2003). Non-Darcian effects on the onset of convection in a porous layer with throughflow. *Transport in Porous Media*, 53, 245–63.

Khayat, R.E., (1994) Chaos and overstability in the thermal convection of viscoelastic fluids, *Journal of Non-Newtonian Fluid Mechanics*, 53, 227–255.

Kim, M. C., Lee, S. B., Kim, S., & Chung, B. J. (2003). Thermal instability of viscoelastic fluids in porous media. *International Journal of Heat and Mass Transfer*, 46, 5065–5072.

Kumar, A., & Bhadauria, B. S. (2011). Double-diffusive convection in a porous layer saturated with viscoelastic fluid using a thermal non-equilibrium model. *Physics of Fluids*, 23(5), 054101.

Lapwood, E. R. (1948). Convection of a fluid in a porous medium. *Mathematical Proceedings of the Cambridge Philosophical Society*, 44, 508–521.

Mahesha, N, Precious Sibanda, Siddheshwar, P. G. and Jayalatha, G. (2013). Linear and nonlinear stability analysis of binary viscoelastic fluidconvection. *Applied Mathematical Modelling*, 37, 8162–8178.

Malashetty, M. S., & Swamy, M. (2007). The onset of convection in a viscoelastic liquid saturated anisotropic porous layer. *Transport in Porous Media*, 67(2), 203–218.

Nield, D. A. (1987). Convective instabilities in porous media with throughflow. *AIChE Journal*, 33, 1222–1224.

Nield, D. A. (1996). The effect of temperature-dependent viscosity on the onset of convection in a saturated porous medium. *ASME Journal of Heat Transfer*, 118, 803–805.

Nield, D. A., & Bejan, A. (2017). *Convection in Porous Media* (5th ed.). Springer, New York.

Rayleigh, L. (1916). On convective currents in a horizontal layer of fluid when the higher temperature is on the underside. *Philosophical Magazine*, 32, 529–546.

Rosenblat, S. (1986). Thermal convection in a viscoelastic liquid. *Journal of Non-Newtonian Fluid Mechanics*, 21, 201–223.

Rudraiah, N., Kaloni, P. N., & Radhadevi, P. V. (1989). Oscillatory convection in a viscoelastic fluid through a porous layer heated from below. *Rheologica Acta*, 28, 48–53.

Sekhar, G. N., & Jayalatha, G. (2009). Elastic effects on Rayleigh–Bénard–Marangoni convection in liquids with temperature-dependent viscosity. Proceedings of ASME IMECE 2009.

Sekhar, G. N., & Jayalatha, G. (2010). Elastic effects on Rayleigh–Bénard convection in liquids with temperature-dependent viscosity. *International Journal of Thermal Sciences*, 49, 67–79.

Sheu, L. J., Tam, L. M., Chen, J. H., Chen, H. K., Lin, K. T., & Kang, Y. (2008). Chaotic convection of viscoelastic fluids in porous media. *Chaos, Solitons & Fractals*, 37(1), 113–124.

Shivakumara I. S. (1999). Boundary and inertia effects on convection in a porous media with throughflow. *Acta Mechanica*, 137, 151–65.

Shivakumara, I. S., Malashetty, M. S., & Chavaraddi, K. B. (2006). Onset of convection in a viscoelastic-fluid-saturated sparsely packed porous layer using a thermal non-equilibrium model. *Canadian Journal of Physics*, 84, 973–988.

Shivakumara, I. S., & Sureshkumar, S. (2007). Convective instabilities in a viscoelastic-fluid-saturated porous medium with throughflow. *Journal of Geophysics and Engineering*, 4, 104–113.

Sokolov, M., & Tanner, R. I. (1972). Convective stability of a general viscoelastic fluid heated from below. *Physics of Fluids*, 15, 534–540.

Straughan, B. (2002). Sharp global non-linear stability for temperature-dependent viscosity. *Proceedings of the Royal Society A*, 458, 1773–1786.

Tan, W., & Masuoka, T. (2007). Stability analysis of a Maxwell fluid in a porous medium heated from below. *Physics Letters A*, 360(3), 454–460.

Torrance, K. E. & Turcotte, D. L. (1971). Thermal convection with large viscosity variations. *Journal of Fluid Mechanics*, 47, 113–125.

Trompert, R. A. & Hansen, U. (1998). On the Rayleigh number dependence of convection with a strongly temperature-dependent viscosity. *The Physics of Fluids*, 10(2), 351–360.

Vanishree, R. K. (2014). Effects of through-flow and internal heat generation on thermo-convective instability in an anisotropic porous medium. *Journal of Applied Fluid Mechanics*, 7(4), 581–590.

Vanishree, R. K., & Siddheshwar, P. G. (2010). Effect of rotation on thermal convection in an anisotropic porous medium with temperature-dependent viscosity. *Transport in Porous Media*, 81(1), 73–87.

Vest, C. M., & Arpaci, V. S. (1969). Overstability of a viscoelastic fluid layer heated from below. *Journal of Fluid Mechanics*, 36(3), 613–623.

Wang, S., & Tan, W. (2008). Stability analysis of double-diffusive convection of Maxwell fluid in a porous medium heated from below. *Physics Letters A*, 372(17), 3046–3050.

Wooding, R. A. (1960). Rayleigh instability of a thermal boundary layer in a flow through a porous medium at large Reynolds or Péclet number. *Journal of Fluid Mechanics*, 9, 183–192.

Yoon, D. Y., Kim, M. C., & Choi, M. C. (2004). The onset of oscillatory convection in a horizontal porous layer saturated with viscoelastic liquid. *Transport in Porous Media*, 55, 275–284.

11 An Analytical Study of Weak Discontinuity Waves in One-Dimensional Flow with Rotational Effects

Nishi Deepa Palo, Paschim vihar

11.1 INTRODUCTION

Shock waves are generated when a disturbance propagates through a medium at a speed exceeding the local acoustic velocity or when the flow undergoes a rapid and substantial contraction in effective cross-sectional area. Such waves are characterized by extremely narrow regions in which macroscopic flow variables, including density, pressure, and velocity, experience abrupt and significant variations. Due to their distinctive physical characteristics and practical importance, shock wave phenomena have been the focus of extensive research over the past several decades.

Shock waves play a crucial role in a wide spectrum of applications ranging from nuclear explosions and high-speed spacecraft propulsion to various astrophysical environments [1]. In industrial engineering, the study of shock dynamics is fundamental to aerodynamics and is widely employed to generate ultra-high pressures in solids and liquids for high-energy-rate material processing. Additional technological applications include explosive demolition, armor penetration systems, and rock fragmentation processes used in drilling, mining, and mineral separation. In the field of high-pressure physics and astrophysics, shock waves are also known to contribute to the heating of cold interstellar clouds embedded within hot, stellar-driven outflows.

Mathematical descriptions of such physical processes typically lead to systems of nonlinear partial differential equations, often involving higher-order nonlinear effects. The presence of nonlinearity results in complex wave behavior, among which finite-time singularity formation and shock development are of particular importance. In recent years, the propagation of shock waves in two-phase flows has attracted significant attention due to its relevance in a variety of engineering and technological applications [2]. In this context, a two-phase flow refers to a mixture consisting of a gaseous phase laden with dispersed solid particles. Recent investigations

DOI: 10.1201/9781003664345-11

have further demonstrated that the inclusion of dust particles can effectively reduce the impact intensity of nuclear explosions [3].

The assumption of an ideal gas is often inadequate for accurately modeling realistic physical environments, necessitating the consideration of more complex and physically representative media. In the present work, a cylindrically symmetric rotating medium containing suspended dust particles is examined. Such flow configurations naturally arise in numerous astrophysical settings, including planetary atmospheres and interstellar gas clouds. Experimental observations indicate that large-scale rotation in outer atmospheres originates from the orbital motion of stars and planets [4–7]. Similar flow conditions may also be reproduced in laboratory experiments, for example, when a cylindrical piston is driven into a shock tube.

From an applied perspective, rotating dusty gas flows are relevant to industrial processes such as the removal of particulate contaminants from semiconductor wafers using shock-induced, non-stationary gas dynamic systems [8]. Moreover, the interaction of shock waves with gas–particle mixtures exhibits close similarities to flow fields encountered in solid rocket engines, where metallic particles such as aluminum are introduced to suppress combustion-induced instabilities.

The evolution of weak discontinuities in fluid media has been extensively investigated using singular surface theory. Grinfeld employed this framework to analyze wave evolution in nonlinear elastic materials [9], while Straughan examined the stability and propagation characteristics of such waves in porous media [10, 11]. Fu and Scott applied singular surface methods to study shock evolution in elastic, non-conducting solids [12, 13]. Singular surface theory is founded on the compatibility conditions originally formulated by Thomas [14]. Within this framework, a wave is represented as a moving surface across which the primary flow variables remain continuous, while their spatial and temporal derivatives may exhibit finite discontinuities. These discontinuities propagate with the wavefront and are required to satisfy a hierarchy of compatibility relations. The lowest-order relations correspond to the Rankine–Hugoniot jump conditions, while higher-order compatibility conditions follow from the assumption of a smooth wavefront.

The governing equations and the corresponding equation of state for the present medium are described in the subsequent sections.

11.1.1 Basic Equations

The mathematical description of the present one-dimensional, unsteady, adiabatic two-phase flow is governed by the following system of partial differential equations [15, 16]. In formulating the model, the following assumptions are adopted: (i) the solid particles suspended in the gas are spherical in shape and occupy less than 5% of the total volume of the mixture; (ii) the particles are sufficiently small and uniformly distributed so that they move in equilibrium with the carrier gas; and (iii) the flow possesses a spatially varying azimuthal velocity component.

$$\rho_t + u\rho_x + \rho u_x + \frac{\rho u}{x} = 0,$$

$$u_t + uu_x + \frac{p_x}{\rho} - \frac{v^2}{x} = 0,$$

$$p_t + up_x + C^2\rho\left(u_x + \frac{u}{x}\right) = 0, \tag{11.1}$$

$$v_t + uv_x + \frac{uv}{x} = 0.$$

Here, x denotes the radial coordinate and t represents time. The variables ρ, u, p, and v correspond to the mixture density, radial velocity, pressure, and azimuthal velocity, respectively. The equilibrium speed of sound in the mixture is defined as

$$C^2 = \frac{\Gamma p}{\rho(1-Z)},$$

where $Z = V_{sp} / V_g$ is the volume fraction of the solid particles, with V_{sp} and V_g denoting the volume of the solid phase and the total mixture volume, respectively [17]. The Grüneisen coefficient Γ is expressed as

$$\Gamma = \frac{\gamma(1+\chi\beta)}{1+\chi\beta\gamma},$$

where γ is the adiabatic index and $\chi = k_p / (1-k_p)$. The mass fraction of solid particles k_p is given by $k_p = m_{sp} / m_T$, where m_{sp} denotes the mass of the solid particles and m_T is the total mass of the mixture. The parameter $\beta = c_{sp} / c_p$ represents the ratio of the specific heat of the solid particles to that of the gas at constant pressure [18]. The density ratio of the solid particles to the gas is defined as $G = \rho_{sp} / \rho_g$. The parameters k_p and G are related through $Z = \theta\rho$, where $\theta = k_p / \rho_{sp}$ and ρ_{sp} denotes the density of the solid particles.

The equation of state for the dusty gas mixture is given by

$$p = \frac{(1-k_p)}{(1-Z)}\rho RT, \tag{11.2}$$

The azimuthal velocity field is assumed to vary linearly with the radial coordinate and is prescribed as

$$v = \Omega x, \tag{11.3}$$

where Ω denotes the angular velocity responsible for the rotational effects in the medium.

The governing system may also be written compactly in matrix form as

$$V_t + AV_x = f.$$

Here,

$$A = \begin{pmatrix} u & \rho & 0 & 0 \\ 0 & u & 1/\rho & 0 \\ 0 & \rho C^2 & u & 0 \\ 0 & v & 0 & u \end{pmatrix},$$

$$V = \left(\rho, \quad u, \quad p, \quad v\right)^{tr},$$

$$F = \left(-\rho u / x, \quad v^2 / x, \quad -C^2 \rho u / x, \quad -vu / x\right)^{tr}.$$

Throughout this work, the superscript *tr* denotes the transpose of a vector or matrix.

11.2 SINGULAR SURFACE

Consider a propagating wavefront represented by the moving surface $x = \Theta(t)$, denoted by Ω. Across this surface, the primary flow variables ρ, u, p, and v are assumed to remain continuous, while their spatial and temporal derivatives may undergo finite discontinuities. When Ω represents a characteristic surface, the associated first- and higher-order discontinuities are required to satisfy the geometrical and kinematical compatibility conditions originally derived by Thomas [14].

$$[\omega_x]| = J, \quad |[\omega_t]| = -VJ, \quad |[\omega_{xx}]| = \bar{J}, \quad |[\omega_{xt}]| = V\left(\frac{\partial J}{\partial x} - \bar{J}\right). \tag{11.4}$$

Here, ω denotes any of the flow variables and V is the speed of propagation of the wavefront Ω. The jump operator $\left\|[y]\right\|$ represents the difference between the values of a quantity immediately behind and ahead of the wave.

For an advancing wave, the propagation speed V is positive. Introducing the notations

$$\left\|[\rho_x]\right\| = \sigma, \left\|[u_x]\right\| = \lambda, \left\|[p_x]\right\| = \eta, \left\|[v_x]\right\| = \omega\alpha, \left\|[\rho_t]\right\| = -V\sigma,$$

$$\left\|[u_t]\right\| = -V\lambda, \quad \left\|[p_t]\right\| = -V\eta, \quad \left\|[v_t]\right\| = -V\bar{\alpha}, \tag{11.5}$$

and evaluating the governing equations on the inner side of the surface Ω, one obtains

$$(V - u_0)\sigma = \rho_0\lambda, \quad \rho_0\lambda(V - u_0) = \eta, \quad (V - u_0)\eta = \lambda C_0\rho_0^2, \quad (V - u_0)\alpha = 0 \tag{11.6}$$

Since the wave propagates with speed $V = u_0 + C_0$ for an advancing disturbance [19, 20], the following relations are obtained:

$$\lambda = \frac{C_0\sigma}{\rho_0} = \frac{\eta}{C_0\rho_0}, \quad \alpha = 0. \tag{11.7}$$

Differentiating the governing equations with respect to x and applying the compatibility relations yields the following system:

$$\left(\frac{\partial\sigma}{\partial x} - \bar{\sigma}\right)V + u_0\bar{\sigma} + \rho_0\bar{\lambda} + \frac{u_0\sigma}{x} + \frac{\rho_0\lambda}{x} + 2\left(\sigma\lambda + \rho_{x0}\lambda + u_{x0}\sigma\right) = 0,$$

$$\left(\frac{\partial\lambda}{\partial x} - \bar{\lambda}\right)V + u_0\bar{\lambda} + \frac{1}{\rho_0}\bar{\eta} + \frac{2v_0\alpha}{x} + 2u_{0_x}\lambda + \lambda^2 - \frac{1}{\rho_0^2}\left(\eta\sigma + \rho_{0_x}\sigma + \rho_{0_x}\eta\right) = 0,$$

$$\left(\frac{\partial\eta}{\partial x} - \bar{\eta}\right)V + \left(\lambda\eta + u_{0_x}\eta + p_{x0}\lambda\right) + u_0\bar{\eta} + \frac{\Gamma}{\left(1-\theta\rho_0\right)}\left(\lambda\eta + u_{0_x}\eta + p_{0_x}\lambda\right) + \frac{\Gamma p_0\bar{\lambda}}{\left(1-\theta\rho_0\right)}$$

$$+ \frac{C_0^2\rho_0}{\left(1-\theta\rho_0\right)}\left(\lambda\theta p_{x0} + u_{0_x}\theta\sigma + \theta\lambda\sigma\right) + \frac{\Gamma u_0\eta}{\left(1-\theta\rho_0\right)x} + \frac{\Gamma p_0\lambda}{x\left(1-\theta\rho_0\right)} - \frac{C_0^2\rho_0 u_0\theta\sigma}{\left(1-\theta\rho_0\right)x} = 0,$$

$$\left(\frac{\partial\alpha}{\partial x} - \bar{\alpha}\right)V + u_0\bar{\alpha} + \frac{v_0\lambda}{x} + \frac{u_0\alpha}{x} + u_{0_x}\alpha + v_{0_x}\lambda + \lambda\alpha = 0. \tag{11.8}$$

Eliminating the higher-order jump terms $\bar{\sigma}$, $\bar{\lambda}$, $\bar{\eta}$, and $\bar{\alpha}$ using the relations obtained earlier, the transport equation governing the amplitude of the acceleration wave is derived as

$$2\frac{d\lambda}{dt} + \frac{\Gamma+1}{\left(1-\theta\rho_0\right)}\lambda^2 + \Omega\lambda = 0, \tag{11.9}$$

where

$$\Omega = u_{0_x}\left(3 + \frac{\Gamma}{\left(1-\theta\rho_0\right)}\right) + \frac{1}{x}\left(C_0 + \frac{u_0\Gamma}{\left(1-\theta\rho_0\right)} + \frac{u_0\theta\rho_0}{\left(1-\theta\rho_0\right)}\right) +$$

$$\rho_{0_x}\left(\frac{\Gamma C_0\theta\rho_0}{\rho_0\left(1-\theta\rho_0\right)} - \frac{C_0}{\rho_0}\right) + \frac{\Gamma\theta p_{0_x}}{\left(1-\theta\rho_0\right)C_0\rho_0}.$$

The above transport equation is consistent with the general result reported in Reference [21], which states that the evolution of a weak discontinuity amplitude is governed by a Bernoulli-type differential equation. Appropriate initial and boundary conditions are required to obtain explicit solutions, which are discussed in the subsequent sections.

11.2.1 Particular Case

For the purpose of solving Equation (11.9) , ahead the shock the flow variable can be characterized as follows:

$$u(x,t)=h(t)x,\ \rho_0=\rho_0(t),\ p_0=p_0(t)\ \text{and}\ v_0=\Omega_a x. \tag{11.10}$$

using Equation (11.10) in the governing equation and integrating, we get

$$\frac{d\rho_0(t)}{dt}=-2\rho_0 h(t),$$

$$\frac{dh(t)}{dt}=\frac{1}{x}\left(\omega_0^2 x^{2\delta-1}-h(t)^2 x\right),$$

$$\frac{dp_0(t)}{dt}=\frac{-2\Gamma p_0 h(t)}{(1-\theta\rho_0)}. \tag{11.11}$$

where $\Omega=\Omega_a=\Omega_0 x^\delta$ where $\delta<1$ [22–23]. Here Ω_0 can be considered as a constant which influences the rotation of the medium. The solution of Equation (11.9) can be obtained by using Equation (11.10). In view of this, the dimensionless form of the flow variables is required, which is given below

$$\tilde{x}=\frac{x}{t_{0_c}a_{0_c}},\ \tilde{t}=\frac{t}{t_{0_c}},\ \tilde{\rho}=\frac{\rho_0}{\rho_{0_c}},\ \tilde{p}=\frac{p_0}{\rho_{0_c}C_{0_c}^2},$$

$$\tilde{\theta}=\theta\rho_{0_c},\ v_0=\frac{C_{0_c}}{t_{0_c}x^{\delta-1}}. \tag{11.12}$$

Here the flow variables with subscript 'c' denote the reference values.

11.2.2 Evolution of Weak Discontinuity and Bernoulli Equation

Using Equations (11.10) and (11.11) in Equation (11.9), we get the following Bernoulli-type transport equation toward the amplitude of acceleration wave.

$$2\frac{d\lambda}{dt}+\frac{(\Gamma+1)}{(1-\theta\rho_0)}\lambda^2+\Lambda\lambda=0,$$

$$\text{where } \Lambda=\left\{\left(3+\frac{2\Gamma}{(1-\theta\rho_0)}+\frac{2\theta\rho_0}{(1-\theta\rho_0)}\right)h(t)+\frac{C_0}{x}\right\} \tag{11.13}$$

The value of the variable 'x' can be computed from the relation $\frac{dx}{dt} = u_0 + C_0$.

Integrating Equation (11.12) analytically, the amplitude of acceleration wave takes the following form

$$\lambda = \frac{\lambda_0 \exp^{-\frac{1}{2}\int_{t0}^{t} \Lambda(s)ds}}{1 + \lambda_0 \frac{(\gamma+1)}{2(1-\theta\rho_0)} \Xi(s)ds}. \tag{11.14}$$

where

$$\Xi(t) = \exp^{-\frac{1}{2}\int_{t0}^{t} \Lambda(s)ds}.$$

While performing numerical integration using RK method of order 4 of Equation (11.13) and depicting the results of amplitude of jump in spatial derivative against $(1/t)$ have been plotted. The graph concludes the following results.

Case 1: If $\lambda_0 > 0$ i.e. the case for rarefaction waves, when t approaches infinity with possible reference values of the dust and rotational parameters k_p, β, G and ω_0 show significant variations.

When any of the dust parameter k_p, β or G increases, the jump in velocity gradient decreases and eventually flattens when time approaches infinity (see Figures 11.1–11.3).

However, an opposite trend has been observed for the angular velocity. When the angular velocity increases there is a decrease in the jump in velocity gradient (see Figure 11.4).

Case 2: The compression wave i.e. $(\lambda_0 < 0)$ can be discussed from Equation (11.13) when the solution breaks down. At some time $t = t_{cr}$ the denominator of Equation (11.13) will tend to zero and the existence of a steepened wave will be confirmed. Here t'_{cr} can be considered as the critical time when the compression wave will be evolved.

11.3 RESULTS AND DISCUSSION

To investigate the functioning of the acceleration wave amplitude under the influence of dust and rotational parameters, the Bernoulli-type transport equation was numerically integrated using the standard fourth-order RK method. The analysis was carried out in a dimensionless structure, and the results were presented as plots of the amplitude of jump in velocity gradient λ versus the time axis $1/t$. The influence of the dust parameters, namely the mass concentration k_p, the specific heat ratio β, the dust index factor G and the rotational parameter Ω_0, was studied separately while keeping the other parameter fixed.

11.3.1 Influence of Dust Parameters k_p

Figure 11.1 shows the evolution of the amplitude of the velocity gradient for the various values of the mass concentration parameters k_p with fixed $G = 1000$, $\beta = 0.6$ and $\Omega_0 = 0.6$. It is observed that an increase in k_p results in monotonic reduction in the amplitude of acceleration wave. Physically, this indicates that a higher dust loading in the medium introduces increased inertia, thereby enhancing wave damping and impeding shock steepening.

11.3.2 Effect of Specific Heat Ratio β

Figure 11.2 illustrates the effect of varying β for fixed values $k_p = 0.4$, $G = 1000$, $\Omega_0 = 0.6$ as β increases the amplitude of the velocity gradient decreases more rapidly with time. This is attributed to the higher energy absorption capacity of the dust phase, which effectively reduces the compressive strength of the wave. The result is consistent with the thermodynamic expectation.

11.3.3 Effect of Dust Gas Density Ratio G

In Figure 11.3 the effect of increasing the density ratio G is examined while keeping $k_p = 0.4$, $\beta = 0.6$ and $\Omega_0 = 0.6$. Similar to the previous two cases, an increase in G leads to suppression of the acceleration wave amplitude λ. This behavior emerges because a higher G value corresponds to heavier solid particles, which leads to a stronger resistance to flow variance and acts as a stabilizing influence on the discontinuity. These three dust-related parameters collectively signify that the presence of a dispersed particulate phase leads to a pronounced attenuation of wave amplitude and acts to suppress the formation of singularities in the flow. The dust thus behaves as a shock-absorbing agent in the medium.

11.3.4 Influence of Rotational Parameters

Figure 11.4 presents the variation in the wave amplitude for different values of the angular velocity parameter Ω_0 with the rest dust parameters held fixed. In contrast

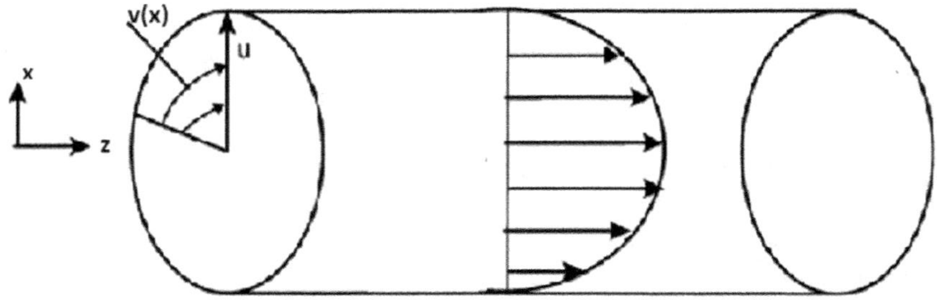

FIGURE 11.1 The component of the velocity vector in rotational flow.

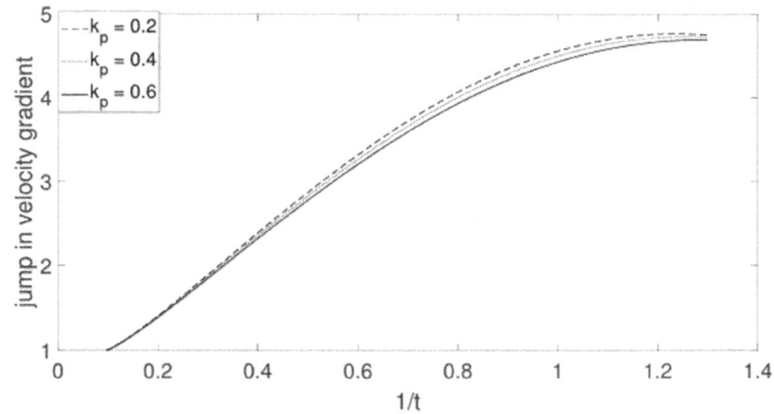

FIGURE 11.2 Variation of the acceleration wave amplitude in a cylindrically symmetric dusty gas flow for different values of the mass fraction of solid particles *kp*, with $G = 1000$, $\Omega_0 = 0.6$, and $\beta = 0.6$.

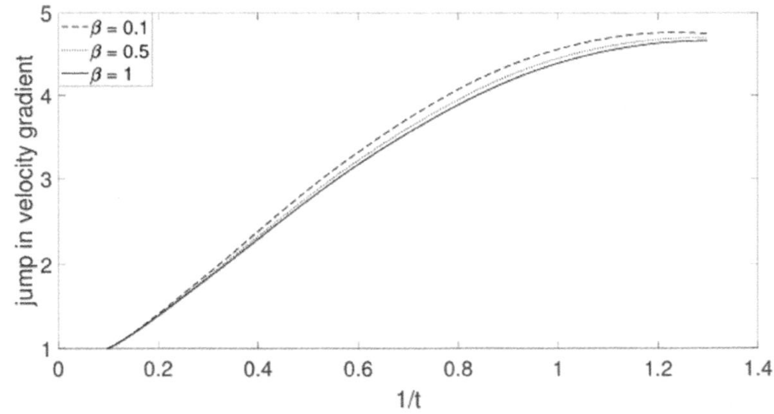

FIGURE 11.3 Effect of the specific heat ratio parameter β on the evolution of acceleration wave amplitude in a rotating dusty gas flow, for $kp = 0.4$, $G = 1000$, and $c = 0.6$.

to the damping effects for observed dust parameters, an increase in Ω_0 leads to a significant enhancement of the amplitude λ of the acceleration wave. Rotational effects promote the growth of discontinuity amplitude potential, accelerating the transition of acceleration waves into fully developed shock waves (Figure 11.5).

11.3.5 Interpretation of Limiting Behaviour

For the case of initial positive amplitude $\lambda_0 > 0$ representing rarefaction waves, it is found that the wave amplitude decays with time asymptotically. The rate of decay is significantly regulated by the choice of parameter values. The presence of dust accelerates this decay while rotational effects delay it or even reverse the trend.

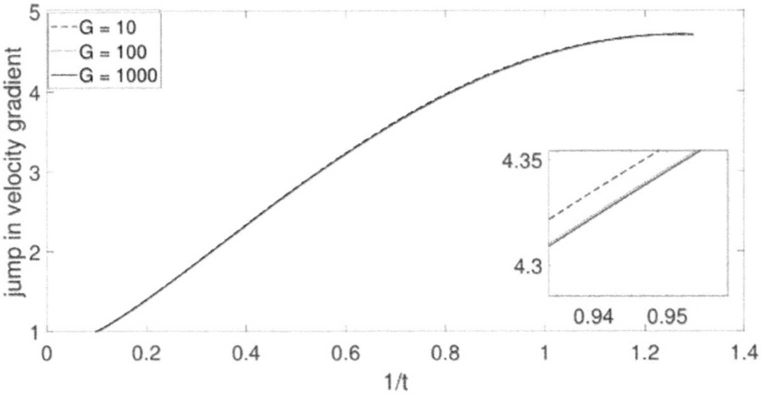

FIGURE 11.4 Influence of the density ratio parameter G on the amplitude variation of acceleration waves in a cylindrically rotating dusty gas, with $kp = 0.4$ and $\Omega_0 = 0.6$.

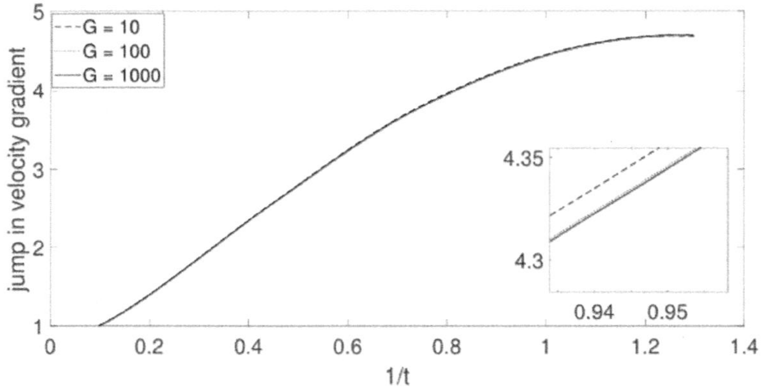

FIGURE 11.5 Influence of the density ratio parameter G on the amplitude variation of acceleration waves in a cylindrically rotating dusty gas, with $kp = 0.4$ and $\Omega_0 = 0.6$.

Conversely, for negative initial amplitude $\lambda_0 < 0$ corresponding to a compression wave, the denominator in the analytical solution for $\lambda(t)$ approaches zero at a finite time, $t = t_{cr}$ concluding a finite time blow-up of the amplitude. This confirms the formation of a shock wave at the critical time t_{cr}, which is inversely connected to the strength of the damping mechanism of medium.

11.4 CONCLUSION

In the present discussion, the evolution of the singular surface influenced by dust and rotational parameters has been discussed. So far, no literature has been developed on the application of singular surfaces in a rotational medium. Here, using a particular case, the jump in velocity gradient has been analyzed. The transport equation that describes the amplitude of weak discontinuity waves satisfied the Bernoulli type of transport equation, as anticipated. In order to discuss the jump in the velocity gradient, the solutions are discussed analytically and numerically. The numerical solutions are depicted using graphs. From Figures 11.1–11.3, it has been observed that with an increase in any of the dust parameters k_p, G and β, the jump in velocity gradient decreases, but a reverse trend has been noticed for the rotational parameter i.e. from Figures 11.4 and 11.5 it is evident that when the angular velocity increases there is an increase in jump in the amplitude of weak discontinuity waves. This concludes that the dust parameters help to subside the impact of the shock wave; on the other hand, the angular velocity influences the enhancement of the amplitude.

11.5 THE ASTROPHYSICAL RELEVANCE AND OUTLOOK FOR FUTURE STUDIES

The study of rotating flows filled with dust is prevalent in many astrophysical situations, including stellar winds, circumstellar disks, and the interstellar medium. These mediums often consist of a gas-dust mixture subjected to rotational effects due to stellar or planetary motions. The present model incorporating a cylindrically symmetric rotating frame with azimuthal velocity captures the essential features of such media. The results demonstrate that the dust parameters k_p, β and G attenuates the amplitude of weak discontinuity waves, suggesting a stabilizing effect of dust on shock fronts. This has direct applications for the shock-induced structures in protostellar clouds, where dust may act as compressive instabilities. On the other hand, the effects of rotational parameters modeled via the azimuthal velocity $v = \Omega x$ enhance the amplitude of acceleration waves, thereby facilitating shock formation in rotating regions such as the inner edge of accretion disks or collapsing protostellar cores. The singular surface theory applied here provides a useful mathematical framework for describing early-stage discontinuity evolution in astrophysical gas-dust flow with rotation. The current study may be extended in several directions, such as (i) inclusion of magnetic fields for modeling magneto-rotational dusty plasmas in astrophysics, (ii) coupling with energy equations for thermal wavefront tracking in dusty environments, (iii) experimental validation using rotating shock tubes with controlled dust injections to validate theoretical predictions.

REFERENCES

1. Whitham, G.B. (1974), *Linear and non linear waves.* International John Wiley and Sons, New York.
2. Pai, S.I. (1976), *Two phase flows.* Braunschweig Vieweg Verlag.
3. Chadha, M., & Jena, J. (2020), Estimation of reduction in damage by a nuclear blast wave with the addition of dust, *J. Phys. Soc. Japan*, 89, 094004. https://doi.org/10.7566/JPSJ.89.094004
4. Sahu, P.K. (2017), Cylindrical shock waves in rotational axisymmetric non-ideal dusty gas with increasing energy under the action of monochromatic radiation, *Phys. Fluids*, 29, 086102. https://doi.org/10.1063/1.4998962

5. Vishwakarma, J.P., & Nath, G. (2012), Similarity solution for a cylindrical shock wave in a rotational axisymmetric dusty gas with heat conduction and radiation heat flux. *Commun. Nonlinear Sci. Numer. Simul*, 17, 154–169. https://doi.org/10.1016/j.cnsns.2011.04.021
6. Nath, G., Sahu, P.K., & Chaurasia, S. (2019), Self-similar solution for the flow behind an exponential shock wave in a rotational axisymmetric non-ideal gas with magnetic field. *Chin. J. Phys.*, 58, 280–293. https://doi.org/10.1016/j.cjph.2019.02.007
7. Levin, V.A. (2004), Detonation wave propagation in rotational gas flows. *J App Mech Tech Phy.*, 45, 457–460. https://doi.org/10.1023/B:JAMT.0000030320.77965.c1
8. Shibasaki, K., Shibasaki, S., et al. (2001), Development of a high-speed cylindrical rotor device for industrial applications of shock waves, Proceedings of 23rd International Symposium on Shock Waves.
9. Grinfeld, M.A. (1978), Ray method of calculating the wavefront intensity in nonlinearly elastic materials, *Prikladnaya Matematika i Mekhanika*, 42, 958–977. http://dx.doi.org/10.1111/j.1365-246X.1994.tb03969.x
10. Straughan, B. (2010), Acoustic waves in Cattaneo–Christov gas, *Physics Letters A.*, 374, 2667–2669.
11. Straughan, B. (2001), *Stability and Wave Motion in Porous Media*, Springer-Verlag, p. 165.
12. Fu Y.B., & Scott N.H. (1989), The evolutionary laws of dilatational spherical and cylindrical weak nonlinear shockwaves in elastic non-conductors, *Arch. Ration. Mech. Anal.*, 108, 11–34.
13. Fu, Y.B., & Scott, N.H. (1990), On one-dimensional shock waves in simple materials with memory, *Proc. R. Soc. Lond. A.*, 428, 547–571.
14. Thomas, T.Y. (1966), The general theory of compatibility conditions, *Int. J. Eng. Sc.*, 4, 207–233. https://doi.org/10.1016/0020-7225(66)90001-2
15. Chadha, M., & Jena, J. (2015), Propagation of weak waves in a dusty, van der Waals gas, *Meccanica*, 51(9), 2145. https://doi.org/10.1007/s11012-015-0354-2
16. Palo, N.D., Jena, J., & Chadha, M. (2022), An analytical approach to study kinematics of shock waves in a dusty, cylindrical gas flow, *ZAMM*, e202200019. https://doi.org/10.1002/zamm.202200019
17. Anand, R.K. (2014), Shock jump relations for a dusty gas atmosphere, *Astrophys. Space Sci.*, 349, 181–195. https://doi.org/10.1007/s10509-013-1638-4
18. Chadha, M., & Jena, J. (2020), Impact of dust in the decay of blast waves produced by a nuclear explosion, *Proc. of Roy. Soc. A., London*, 476. http://dx.doi.org/10.1098/rspa.2020.0105
19. Mittal, S., & Jena, J. (2020), Interaction of a singular surface with a characteristic shock in a relaxing gas with dust particles., *Zeitschrift für Naturforschung A*, 75, 119–129. https://doi.org/10.1515/zna-2019-0217
20. Singh, R., & Jena, J. (2015), One dimensional steepening of waves in non-ideal relaxing gas. *Int. J. Non-Linear Mech.*, 77, 151–161. https://doi.org/10.1016/j.ijnonlinmec.2015.08.001
21. Radha, Ch., Sharma, V.D., & Jeffrey, A. (1993), Interaction of shock waves with discontinuities. *Appl. Anal.*, 50:145–166. https://doi.org/10.1007/s10665-007-9182-2
22. Vishwakarma, J.P., & Nath, G. (2010), Propagation of a cylindrical shock wave in a rotating dusty gas with heat conduction and radiation heat flux, *Physica Scripta*, 81, 045401 (9pp). https://doi.org/10.1088/0031-8949/81/04/045401
23. Vishwakarma, J.P., Maurya, A.K., & Singh, K.K. (2007), Self-similar adiabatic flow headed by a magnetogasdynamic cylindrical shock wave in a rotating non-ideal gas, *Geophys. Astrophys. Fluid Dyn.*, 101(2), 155–168. http://dx.doi.org/10.1080/03091920701298112
24. Müller, I., & Ruggeri, T. (2013), *Rational extended thermodynamics*, Springer, New York, NY, Vol. XV, p. 396. https://doi.org/10.1007/978-1-4612-2210-1

12 Magnetohydrodynamic Effects on Dual-Mode Richtmyer–Meshkov Instability

Liwen Shen, Satyvir Singh, and Manuel Torrilhon

12.1 INTRODUCTION

The Richtmyer–Meshkov instability (RMI) develops when a shock wave impulsively accelerates a perturbed material interface separating fluids of different densities, generating baroclinic vorticity through pressure–density misalignment and triggering interfacial roll-up, nonlinear deformation, and mixing. Since Richtmyer's seminal linear impulsive model and Meshkov's shock-tube observations, RMI has become a canonical problem in compressible-flow instability and shock-driven mixing studies [1, 2]. Comprehensive reviews have highlighted its importance across high-energy-density physics, inertial confinement fusion, and astrophysical flows, and have documented the multiscale pathways from early-time growth to late-time turbulent mixing [3, 4]. Beyond the linear regime, nonlinear theories and experiments have clarified saturation mechanisms, bubble–spike asymmetry, and vortex dynamics that govern morphology and mixing rates [5, 6], while unified theoretical frameworks have systematized linear-to-nonlinear evolution across regimes relevant to shock-driven interfaces [7].

A key challenge in modern RMI research is that realistic interfaces are seldom purely single mode. Multimode or few-mode perturbations introduce mode coupling, spectral energy transfer, and phase-dependent interactions that can strongly alter growth rates and late-time structures compared with single-mode predictions. Simplified perturbation theories and numerical studies have shown that mode competition and harmonic generation can reorganize the spectrum even in weakly nonlinear stages [8]. Recent shock-tube experiments and numerical investigations on dual-mode interfaces have provided controlled evidence that the growth of each constituent mode depends sensitively on phase relations and inter-mode interaction, motivating dedicated studies of dual-mode RMI as a bridge between single-mode physics and fully multimode turbulence [9]. Related investigations in converging geometries further demonstrate that dual-mode coupling can be amplified by wave

DOI: 10.1201/9781003664345-12

focusing and reshock processes, yielding richer interaction pathways for instability growth and mixing [10]. High-resolution simulations of multimode RMI likewise confirm that the interface-to-turbulence transition depends strongly on initial spectral content and mode interaction, reinforcing the need for systematic parameter studies [11, 12].

In many applications of interest, the participating media are electrically conducting plasmas, liquid metals, or magnetized gases, and the instability evolves under the influence of magnetic fields. In the magnetohydrodynamic (MHD) setting, magnetic tension and magnetic pressure introduce additional constraints and anisotropy, modifying vorticity deposition, interface dynamics, and mixing. Early numerical evidence demonstrated that sufficiently strong magnetic fields can suppress classical RMI growth by redistributing baroclinically generated vorticity into magnetosonic structures, thereby reducing interface roll-up and associated mixing [13]. Subsequent theoretical and numerical work established stability properties of impulsively accelerated density interfaces in incompressible and compressible MHD, identifying field-strength thresholds and orientations that govern bounded growth and stabilization [14, 15]. Dedicated studies have quantified critical magnetic-field levels required for suppression and clarified how magnetic tension competes with baroclinic vorticity production during the post-shock evolution [16]. More recent modeling has extended these ideas to cylindrical geometries and incompressible limits, providing reduced descriptions that are useful for interpretation and scaling [17], while two-fluid plasma investigations have revealed magnetization effects beyond ideal single-fluid MHD, including additional dispersive and kinetic-scale influences on the RMI evolution [18]. Hall-MHD analyses further demonstrate that arbitrary field orientation and additional electrodynamic terms can significantly alter growth, anisotropy, and flow topology relative to ideal MHD predictions [19].

Despite this progress, the combined influence of MHD physics and dual-mode initial conditions remains comparatively underexplored, even though dual-mode perturbations represent a minimal configuration in which mode coupling, phase effects, and nonlinear spectral transfer are present. Understanding how magnetic fields regulate dual-mode coupling, suppress or redirect vortex interactions, and reshape mixing morphology is essential for developing predictive models of magnetized shock-driven interfacial mixing. In this work, we perform high-fidelity simulations of magnetohydrodynamic dual-mode RMI using the Athena++ framework, which provides robust Godunov-type solvers and scalable performance for compressible MHD problems [20]. We systematically examine the effects of magnetic-field strength and orientation on interface evolution, mode-amplitude history, vorticity and current structures, and mixing measures, and we identify regimes in which magnetic forces primarily damp mode coupling versus regimes in which they redirect energy transfer and yield strongly anisotropic morphology. The present results aim to provide a controlled reference for magnetized, few-mode RMI and to support future reduced-order and data-driven modeling of MHD shock-driven mixing.

The remainder of the study is organized as follows. Section 12.2 describes the computational methodology, including the governing equations for compressible MHD flows, numerical method, and validation. Section 12.3 presents problem configuration. Section 12.4 illustrates the results and discusses the impact of magnetic fields on interface evolution, mode coupling, and flow structures. Finally, Section 12.5 summarizes the key findings and outlines directions for future research in magnetized shock-driven interfacial instabilities.

12.2 COMPUTATIONAL METHODOLOGY

12.2.1 Governing Equations for Compressible MHD Flows

To investigate the MHD effects on the evolution of dual-mode RMI, we solve the two-dimensional equations of compressible, inviscid, adiabatic, ideal MHD in conservative form. This formulation captures the coupled dynamics of shock propagation, baroclinic vorticity generation, magnetic tension, and magnetic-pressure effects that govern the growth, interaction, and suppression of perturbed material interfaces in electrically conducting media.

The governing equations describe the conservation of mass, momentum, total energy, and magnetic flux, and are written for a two-dimensional Cartesian domain as

$$\frac{\partial \rho}{\partial t}+\frac{\partial}{\partial x}(\rho u)+\frac{\partial}{\partial y}(\rho v)=0, \tag{12.1}$$

$$\frac{\partial(\rho u)}{\partial t}+\frac{\partial}{\partial x}\left(\rho u^2-\frac{B_x^2}{\mu}+P^*\right)+\frac{\partial}{\partial y}\left(\rho uv-\frac{B_xB_y}{\mu}\right)=0, \tag{12.2}$$

$$\frac{\partial(\rho v)}{\partial t}+\frac{\partial}{\partial x}\left(\rho uv-\frac{B_xB_y}{\mu}\right)+\frac{\partial}{\partial y}\left(\rho v^2-\frac{B_y^2}{\mu}+P^*\right)=0, \tag{12.3}$$

$$\frac{\partial E}{\partial t}+\frac{\partial}{\partial x}\left[\left(E+P^*\right)u-\frac{B_x\left(B_xu+B_yv\right)}{\mu}\right]+\frac{\partial}{\partial y}\left[\left(E+P^*\right)v-\frac{B_y\left(B_xu+B_yv\right)}{\mu}\right]=0, \tag{12.4}$$

$$\frac{\partial B_x}{\partial t}-\frac{\partial}{\partial y}\left(vB_x-uB_y\right)=0, \tag{12.5}$$

$$\frac{\partial B_y}{\partial t}+\frac{\partial}{\partial x}\left(vB_x-uB_y\right)=0. \tag{12.6}$$

Here, ρ denotes the mass density, u and v are the velocity components, p is the gas pressure, E is the total energy density, and B_x and B_y are the magnetic-field components. The magnetic permeability is set to $\mu=1$ without loss of generality. The total pressure P^* incorporates both gas and magnetic contributions and is defined as

$$P^* = p + \frac{B^2}{2\mu}, \tag{12.7}$$

where $B^2 = B_x^2 + B_y^2$. The total energy density is decomposed into internal, kinetic, and magnetic components according to

$$E = \frac{p}{\gamma - 1} + \frac{1}{2}\rho\left(u^2 + v^2\right) + \frac{B_x^2 + B_y^2}{2\mu}, \tag{12.8}$$

where γ is the ratio of specific heats. This formulation ensures consistent coupling between compressibility, shock dynamics, and magnetic forces, which is essential for accurately resolving post-shock interface evolution and mode interaction in dual-mode RMI.

To quantify the influence of magnetic fields on dual-mode RMI growth and mode coupling, we introduce the inverse plasma-beta parameter

$$\beta^{-1} = \frac{B^2}{2p_0}, \tag{12.9}$$

which represents the ratio of magnetic pressure to gas pressure in the pre-shock state. By varying β^{-1}, we systematically examine regimes ranging from purely hydrodynamic evolution to magnetically dominated behavior, enabling direct assessment of magnetic suppression, anisotropic growth, and redirection of energy transfer between interacting interface modes.

12.2.2 Numerical Method

In this study, all simulations presented are performed using the Athena++ computational framework, a high-performance, grid-based code designed for solving the equations of compressible hydrodynamics (HD) and MHD [20]. Athena++ employs a finite-volume Godunov-type formulation and is particularly well suited for shock-driven flows due to its robust Riemann solvers and constrained-transport capabilities for maintaining the divergence-free condition of the magnetic field [20]. The ideal MHD equations described in Section 12.2 are solved using an approximate Riemann solver within the unsplit corner-transport-upwind (CTU) integrator [20]. Magnetic-field evolution is handled using the constrained transport (CT) method, which ensures that the solenoidal constraint $\nabla \cdot \mathbf{B} = 0$ is satisfied to machine precision throughout the simulation. This is essential for accurately capturing magnetic tension effects and avoiding spurious numerical artifacts in magnetized RMI simulations [13, 15].

Spatial discretization is achieved using a second-order accurate piecewise-linear reconstruction of the conserved variables within each computational cell. Characteristic-based reconstruction is employed to limit numerical dissipation while preserving stability near shocks and material interfaces. Fluxes at cell faces are computed using an approximate Riemann solver appropriate for compressible MHD, providing a balance between accuracy and robustness in the presence of strong shock waves [20]. Time integration is performed using an explicit, second-order accurate time-stepping scheme. The time step is dynamically adjusted according to a Courant–Friedrichs–Lewy (CFL) condition based on the maximum local signal speed, including contributions from fluid velocity, sound speed, and Alfvén speed. This ensures stable integration of both HD and MHD wave families during shock interaction and subsequent instability growth [20].

12.2.3 Validation

Figure 12.1 presents a validation of the numerical methodology by comparing the present MHD dual-mode RMI simulations with previously published reference results. Density-field snapshots at selected nondimensional times are shown for two representative magnetic-field strengths, corresponding to the hydrodynamic limit ($\beta^{-1} = 0$) and a magnetized case ($\beta^{-1} = 0.5$) at shock Mach number $M_s = 2.0$. The reference results reported by Samtaney [13] are displayed in Figure 12.1(a), while the corresponding simulations obtained in the present work using the Athena++ framework are shown in Figure 12.1(b). In the absence of a magnetic field ($\beta^{-1} = 0$), both datasets exhibit the classical evolution of shock-accelerated interfaces, including the formation of pronounced spike and bubble structures, subsequent roll-up, and the onset of nonlinear deformation driven by baroclinic vorticity deposition. When a magnetic field is present ($\beta^{-1} = 0.5$), the instability growth is markedly altered, with reduced interface roll-up and smoother interface morphology, reflecting the stabilizing influence of magnetic tension and magnetic pressure. The present simulations accurately reproduce these key features, including the suppression of small-scale structures and the modification of post-shock flow patterns observed in the reference

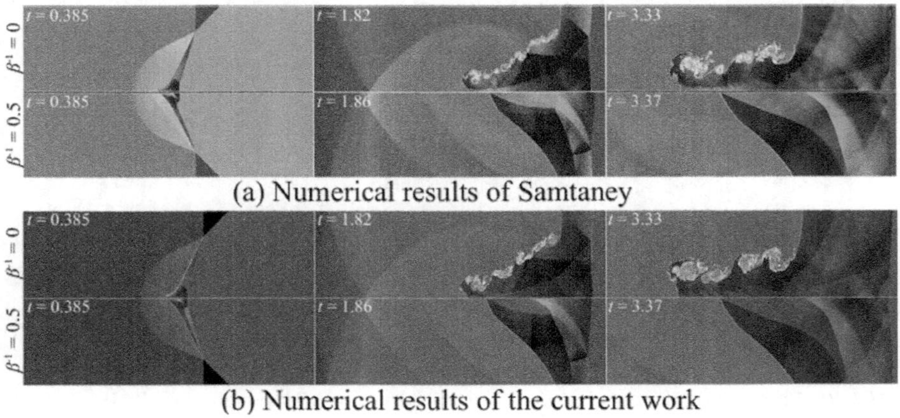

FIGURE 12.1 Comparison of density-field evolution at selected nondimensional times for two magnetic-field strengths, $\beta-1 = 0$ and $\beta-1 = 0.5$: (a) reference numerical results reported by Samtaney [13], (b) the corresponding results obtained using the Athena++ framework.

study. The close qualitative agreement between the two sets of results in terms of shock structure, interface evolution, and magnetic-field–induced stabilization demonstrates that the numerical approach adopted in this study reliably captures the essential physics of magnetized RMI.

12.3 PROBLEM CONFIGURATION

Figure 12.2 illustrates the physical and computational configuration employed to investigate MHD effects on dual-mode RMI at a heavy fluid layer. A two-dimensional rectangular computational domain extends over $\left[0,0.08\right]$ mm in the streamwise (x) direction and $\left[0,0.04\right]$ mm in the transverse (y) direction. In this study, we consider the shock interaction with a dual-mode perturbed SF_6 heavy-gas layer. A finite-thickness SF_6 layer of thickness $d = 0.02$ mm is embedded within a lighter ambient gas and subjected to a uniform magnetic field. A planar shock wave with Mach number $M_s = 1.25$ is initially located 50 mm from the left boundary and propagates from left to right, impulsively accelerating the perturbed interfaces of the heavy-gas layer.

In a Cartesian coordinate system, the dual-mode interfacial perturbation is prescribed as a superposition of two cosine modes,

$$y = a_1\cos\left(k_1x + \phi_1\right) + a_2\cos\left(k_2x + \phi_2\right), \tag{12.10}$$

where a_i and k_i denote the initial amplitudes and wavenumbers of the individual modes, and ϕ_i is the phase shift between them. In the present simulations, the interface profile is specified as

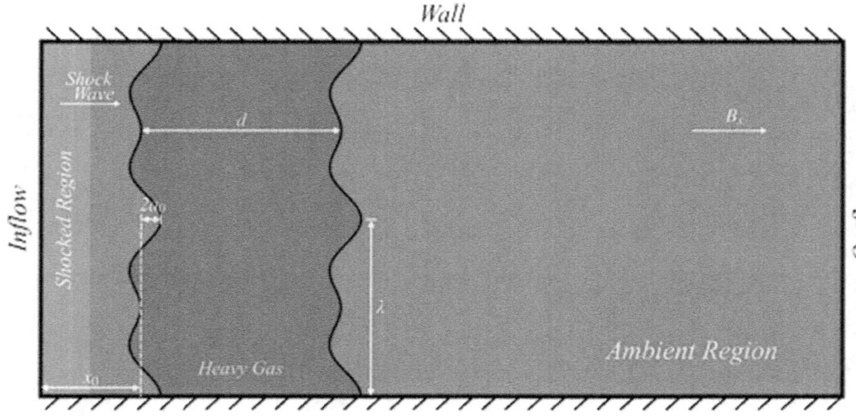

FIGURE 12.2 Schematic of the computational setup for MHD dual-mode RMI. A planar shock accelerates a finite-thickness, dual-mode perturbed heavy-gas layer embedded in a lighter ambient medium. The initial interface perturbations are characterized by amplitude a_0, wavelength λ and layer thickness d, and a uniform magnetic field Bx is imposed parallel to the shock direction.

$$y = x_0 + a_0\cos(kx + \pi) + a_0\cos\left(\frac{k}{2}x\right), \tag{12.11}$$

with initial amplitude $a_0 = 0.001$ mm and mean interface position $x_0 = 0.01$ mm. The fundamental wavenumber is defined as $k = 4\pi/\lambda$, corresponding to a wavelength $\lambda = 20$ mm. The imposed phase shift of π between the two modes produces a characteristic dual-mode interface geometry, as illustrated in Figure 12.1.

All gases are initialized at a uniform pressure $P_0 = 101325$ Pa and temperature $T_0 = 293$ K. The left boundary is prescribed as an inflow boundary, the right boundary as an outflow boundary, and the top and bottom boundaries are treated as solid walls. A uniform magnetic field is imposed throughout the computational domain, with its strength and orientation varied systematically. The magnetic-field influence is quantified using the inverse plasma beta parameter β^{-1}, with cases including $\beta^{-1} = 0$ and $\beta^{-1} = 0.0005$, and the magnetic field aligned along the streamwise (x) direction.

12.4 NUMERICAL RESULTS AND DISCUSSION

This section presents and discusses the numerical results obtained from high-fidelity simulations of shock-accelerated dual-mode RMI under hydrodynamic and magnetohydrodynamic conditions. The analysis focuses on the evolution of interface morphology, vorticity dynamics, and mixing characteristics, with particular emphasis on the role of magnetic-field strength in regulating mode coupling and nonlinear growth. All simulations are performed on a uniform Cartesian grid with a resolution of 3000×1500 grid points, which is sufficient to accurately resolve shock structures, interface deformation, and small-scale vortical features throughout the domain.

Figure 12.3 illustrates the flow physics governing the evolution of shock-accelerated dual-mode RMI in the hydrodynamic limit ($\beta^{-1} = 0$). Immediately after shock impact ($t = 0$–100), the impulsive acceleration of the dual-mode perturbed interfaces generates baroclinic vorticity due to the misalignment of pressure and density gradients. During this early linear stage, the interface deformation closely follows the imposed dual-mode perturbation, and growth remains relatively smooth with weak coupling between the constituent modes. As time progresses into the weakly nonlinear regime ($t \approx 200$–400), differential growth rates of the interacting modes become apparent. The heavier fluid forms pronounced spike structures penetrating into the lighter ambient medium, while lighter-fluid bubbles rise in the opposite direction. Mode coupling intensifies, leading to asymmetric spike–bubble development and the emergence of localized shear layers along the interface. The shadowgraph images reveal the associated wave interactions and reflected shock structures generated by interface deformation. At later times ($t \geq 500$), the flow enters the fully nonlinear regime characterized by strong interface roll-up, vortex pairing, and complex vortex interactions. The initially distinct modal features merge through nonlinear interactions, producing larger coherent vortical structures and enhanced mixing across the interface. Kelvin–Helmholtz instabilities (KHI) develop along the spike edges due to

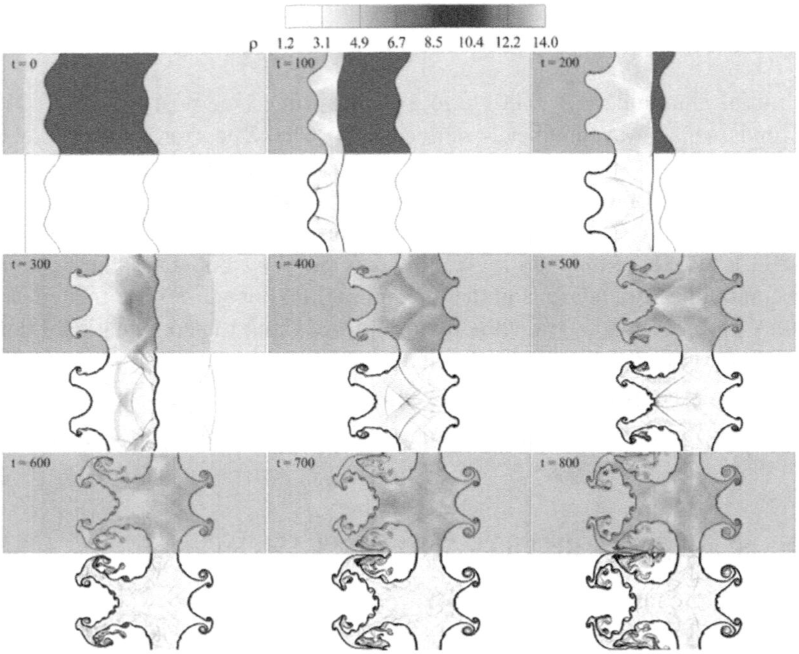

FIGURE 12.3 Time evolution of density fields (upper-half) and corresponding shadowgraphs (lower-half) for shock-accelerated dual-mode Richtmyer–Meshkov instability in the hydrodynamic limit (β–1 = 0). Snapshots at successive nondimensional times illustrate the transition from the early linear growth to the nonlinear regime characterized by interface roll-up, vortex interaction, and enhanced mixing in the absence of magnetic-field effects.

velocity shear, further increasing small-scale structure and interfacial complexity. In the absence of magnetic tension, there is no restoring mechanism to suppress vorticity growth, allowing sustained roll-up and vigorous mixing.

Figure 12.4 illustrates the flow evolution of shock-accelerated dual-mode RMI in the presence of a weak magnetic field ($\beta^{-1} = 0.0005$). Following shock impact, baroclinic vorticity is deposited at the dual-mode perturbed interfaces due to pressure–density misalignment, initiating interface deformation similar to the hydrodynamic case. During the early linear stage ($t \lesssim 100$), the interface growth closely follows the imposed dual-mode perturbation, indicating that the magnetic field is initially too weak to significantly alter the impulsive vorticity generation. As the flow evolves into the weakly nonlinear regime ($t \approx 200$–400), differences from the hydrodynamic limit become apparent. The presence of the magnetic field introduces a weak magnetic tension that resists interface curvature, leading to reduced growth of small-scale undulations and a delay in the onset of strong vortex roll-up. While the dominant spike–bubble structures continue to develop, the associated shear layers are smoother and less prone to secondary instability compared with the unmagnetized case. At later times ($t \geq 500$), nonlinear interactions between the two interface

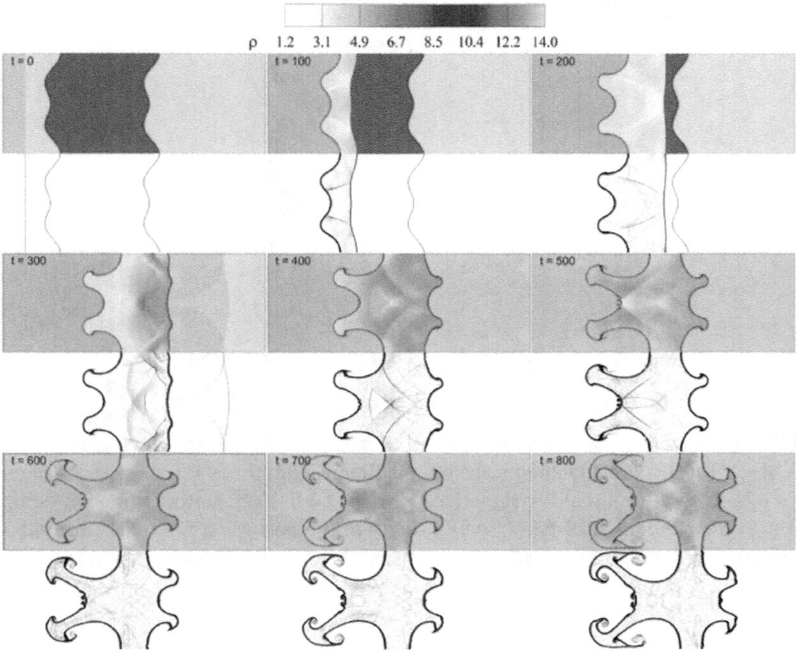

FIGURE 12.4 Time evolution of density fields (upper-half) and corresponding shadowgraphs (lower-half) for shock-accelerated dual-mode RMI at a weak magnetic-field strength, β–1 = 0.0005. Snapshots at successive nondimensional times illustrate the early development of baroclinic vorticity and subsequent nonlinear interface evolution under the influence of a weak magnetic field. Compared to the hydrodynamic case, the magnetic field mildly suppresses small-scale interface roll-up and delays vortex interaction while preserving the overall dual-mode growth and large-scale spike–bubble morphology.

modes intensify, and large-scale vortical structures emerge. However, the weak magnetic field partially suppresses vortex pairing and limits the generation of fine-scale structures, resulting in a comparatively smoother interface and reduced mixing efficiency. The shadowgraph images reveal weakened wave–interface interactions and diminished shock-induced compression features relative to the hydrodynamic case.

Figure 12.5 illustrates the time evolution of spanwise vorticity contours for shock-accelerated dual-mode RMI, comparing the hydrodynamic case ($\beta^{-1} = 0$) with a weakly magnetized case ($\beta^{-1} = 0.0005$). Immediately after shock impact ($t = 100$), strong baroclinic vorticity is deposited along the perturbed interfaces in both cases as a result of pressure–density misalignment. At this early stage, the vorticity distribution closely follows the imposed dual-mode interface geometry, and magnetic effects remain weak. As the flow evolves ($t = 200$–400), the hydrodynamic case exhibits rapid amplification of vorticity concentrated along the spike and bubble flanks, leading to the formation of coherent vortex sheets and the onset of vortex roll-up. Nonlinear mode coupling becomes increasingly pronounced, and vorticity

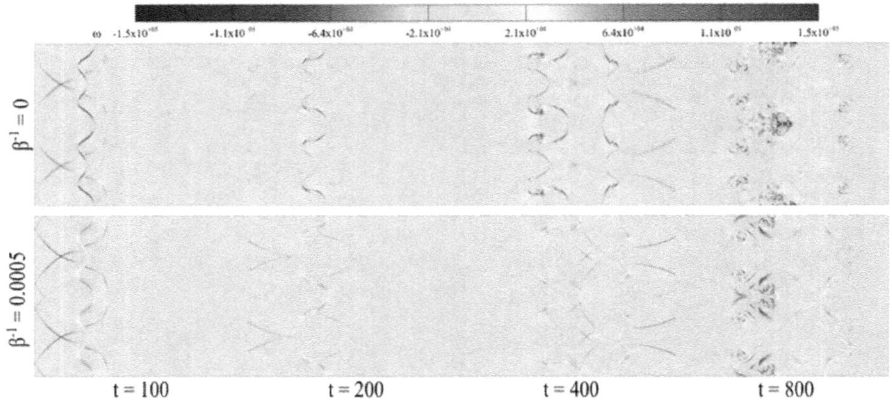

FIGURE 12.5 Time evolution of spanwise vorticity contours for shock-accelerated dual-mode RMI. Results are shown at selected nondimensional times for the hydrodynamic case (β–1 = 0, top row) and a weakly magnetized case (β–1 = 0.0005, bottom row). The comparison highlights the generation of baroclinic vorticity at the perturbed interfaces, the subsequent growth and interaction of vortical structures, and the progressive suppression and redistribution of vorticity induced by magnetic tension in the presence of a weak magnetic field.

structures begin to interact and merge, signaling the transition toward vortex-dominated dynamics. In contrast, for $\beta^{-1} = 0.0005$, the presence of a weak magnetic field modifies the vorticity evolution by redistributing vorticity along magnetic field lines. Magnetic tension resists sharp interface curvature, resulting in smoother vorticity filaments and delayed vortex pairing compared with the unmagnetized case. At later times ($t = 800$), the hydrodynamic flow is characterized by intense vorticity clustering, strong vortex interaction, and the development of fine-scale vortical structures associated with KHI. The weakly magnetized case, however, shows reduced peak vorticity and more elongated, organized vortical patterns. While large-scale vortices still form due to dual-mode interaction, small-scale vorticity amplification is suppressed, and vortex merging occurs more gradually. This demonstrates that even a weak magnetic field can significantly influence vorticity dynamics by damping small-scale rotational motion and altering energy transfer pathways.

Figure 12.6 quantifies the influence of a weak magnetic field on the evolution of extreme vorticity in shock-accelerated dual-mode RMI by tracking the temporal histories of the maximum and minimum spanwise vorticity. In both cases, sharp peaks in ω_{max} and ω_{min} appear immediately after shock interaction, corresponding to impulsive baroclinic vorticity deposition at the perturbed interfaces. These early-time extrema reflect the rapid generation of shear layers along spike and bubble flanks during the linear growth phase. In the hydrodynamic case ($\beta^{-1} = 0$), the magnitudes of both maximum and minimum vorticity continue to grow and exhibit strong temporal fluctuations as the flow transitions into the nonlinear regime. These oscillations are associated with intense interface roll-up, vortex pairing, and intermittent amplification of vorticity due to vortex–vortex interactions and secondary

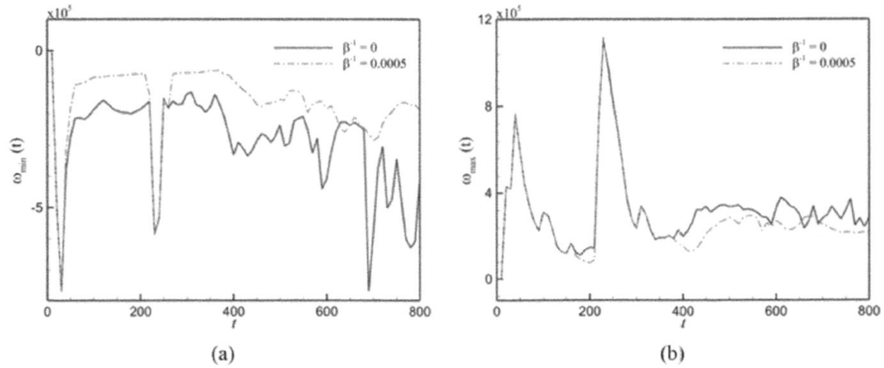

FIGURE 12.6 Temporal evolution of vorticity extrema in the shock-accelerated dual-mode RMI for the hydrodynamic case (β–1 = 0) and a weakly magnetized case (β–1 = 0.0005): (a) maximum vorticity (ω_{max}) and (b) minimum vorticity (ω_{min}). The comparison highlights the attenuation and temporal smoothing of vorticity peaks induced by the magnetic field, reflecting the role of magnetic tension in moderating vorticity amplification and nonlinear vortex interactions.

KHI. The increasingly irregular behavior at later times reflects the development of complex, multiscale vortical dynamics and enhanced mixing. By contrast, in the weakly magnetized case ($\beta^{-1} = 0.0005$), the peak vorticity levels are systematically reduced and display smoother temporal evolution. Magnetic tension acts to resist sharp interface curvature and suppress the formation of highly concentrated vortex cores, thereby limiting extreme vorticity amplification. While the overall timing of major vorticity peaks remains similar to the hydrodynamic case—indicating that the magnetic field does not significantly alter the initial baroclinic vorticity generation—it moderates subsequent nonlinear growth and damps rapid fluctuations associated with vortex interaction.

The overall intensity of vortical activity in the shock-accelerated dual-mode RMI is quantified using the enstrophy, defined as

$$\Omega(t) = \int_D \omega^2 dxdy. \tag{12.12}$$

where ω denotes the spanwise vorticity and D represents the computational domain. Enstrophy provides a global measure of vorticity amplification and is particularly sensitive to the development of small-scale structures arising from interface roll-up, vortex interaction, and secondary instabilities during the nonlinear evolution of dual-mode RMI.

Figure 12.7 illustrates the temporal evolution of enstrophy Ω, providing a global measure of vorticity intensity and small-scale activity in shock-accelerated dual-mode RMI. Immediately after shock impact, a sharp rise in enstrophy is observed in both cases, reflecting the impulsive deposition of baroclinic vorticity at the perturbed interfaces. This early-time peak corresponds to the rapid formation of shear layers along the spike and bubble flanks during the linear growth stage. As the flow

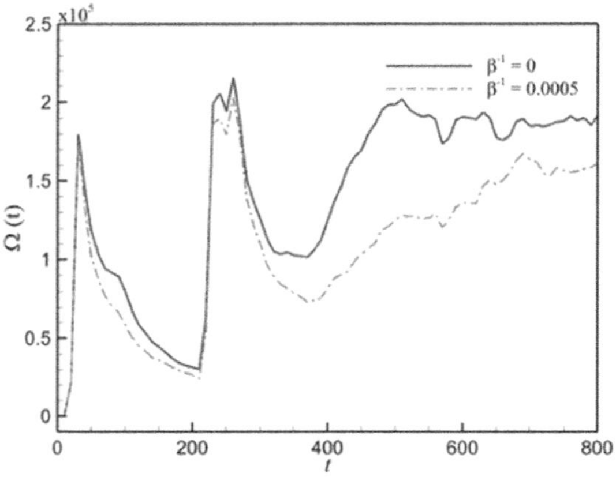

FIGURE 12.7 Temporal evolution of enstrophy Ω in the shock-accelerated dual-mode RMI for the hydrodynamic case (β–1 = 0) and a weakly magnetized case (β–1 = 0.0005). The enstrophy growth reflects the generation and amplification of vorticity following shock impact, while the reduced levels and smoother temporal evolution in the magnetized case indicate the damping of small-scale vortical activity and suppression of nonlinear vortex interactions by magnetic tension.

transitions into the weakly nonlinear regime, the enstrophy exhibits pronounced growth associated with interface roll-up, vortex pairing, and mode coupling between the two imposed perturbation modes. In the hydrodynamic case ($\beta^{-1} = 0$), enstrophy continues to increase and displays strong temporal fluctuations, indicating sustained vorticity amplification and the development of multiscale vortical structures driven by nonlinear vortex interactions and secondary instabilities. In contrast, for the weakly magnetized case ($\beta^{-1} = 0.0005$), the enstrophy level is consistently lower and evolves more smoothly in time. The magnetic field introduces magnetic tension that resists sharp interface curvature and suppresses the formation of intense small-scale vortices, thereby reducing the rate of enstrophy growth. While the overall trend of enstrophy increase remains similar – showing that magnetic effects do not significantly alter the initial baroclinic vorticity generation – the subsequent nonlinear amplification is moderated, leading to reduced fluctuations and delayed saturation.

12.5 CONCLUSIONS

This chapter has examined the influence of magnetic fields on shock-accelerated dual-mode RMI using high-fidelity numerical simulations based on Athena++. In the hydrodynamic limit, the dual-mode interface exhibits rapid baroclinic vorticity generation, strong nonlinear mode coupling, interface roll-up, and enhanced mixing. The introduction of a magnetic field modifies this evolution by redistributing vorticity, suppressing small-scale structures, and moderating nonlinear vortex interactions through magnetic tension. Quantitative analyses based on vorticity extrema and enstrophy demonstrate that even weak magnetization significantly smooths the temporal evolution of vortical activity while leaving the initial baroclinic vorticity deposition largely unchanged. These results highlight the regulatory role of magnetic fields in dual-mode RMI and provide a controlled numerical reference for magnetized shock-driven interfacial instabilities relevant to high-energy-density and astrophysical flows.

ACKNOWLEDGMENTS

The authors acknowledge funding through the German Research Foundation within the research unit DFG–FOR5409.

REFERENCES

1. R. D. Richtmyer, Taylor instability in shock acceleration of compressible fluids, *Commun. Pure Appl. Math.*, 13 (1960), pp. 297–319.
2. E. E. Meshkov, Instability of the interface of two gases accelerated by a shock wave, *Fluid Dyn.*, 4 (1969), pp. 101–104.
3. M. Brouillette, The Richtmyer–Meshkov instability, *Annu. Rev. Fluid Mech.*, 34 (2002), pp. 445–468.
4. Y. Zhou, Rayleigh–Taylor and Richtmyer–Meshkov instability induced flow, turbulence, and mixing. I, *Phys. Rep.*, 720–722 (2017), pp. 1–136.
5. A. L. Velikovich, and G. Dimonte, Nonlinear perturbation theory of the incompressible Richtmyer–Meshkov instability, *Phys. Rev. Lett.*, 76 (1996), pp. 3112–3115.
6. O. Sadot, L. Erez, D. Oron, G. Erez, G. Ben-Dor, U. Alon, L. A. Levin, and D. Shvarts, Study of nonlinear evolution of single-mode and two-bubble interaction in Richtmyer–Meshkov instability, *Phys. Rev. Lett.*, 80 (1998), pp. 1654–1657.
7. K. Nishihara, J. G. Wouchuk, C. Matsuoka, R. Ishizaki, and V. V. Zhakhovsky, Richtmyer–Meshkov instability: theory of linear and nonlinear evolution, *Phil. Trans. R. Soc. A*, 368 (2010), pp. 1769–1807.
8. M. Vandenboomgaerde, S. Gauthier, and C. Mügler, Nonlinear regime of a multimode Richtmyer–Meshkov instability: a simplified perturbation theory, *Phys. Fluids*, 14 (2002), pp. 1111–1122.
9. X. Luo, L. Liu, Y. Liang, J. Ding, and C.-Y. Wen, Richtmyer–Meshkov instability on a dual-mode interface, *J. Fluid Mech.*, 905 (2020), p. A5.
10. Z. Zhou, J. Ding, Z. Zhai, W. Cheng, and X. Luo, Mode coupling in converging Richtmyer–Meshkov instability of dual-mode interface, *Acta Mech. Sin.*, 36 (2020), pp. 356–366.
11. B. Thornber and Y. Zhou, Numerical simulations of the two-dimensional multimode Richtmyer–Meshkov instability, *Phys. Plasmas*, 22 (2015), p. 032309.
12. G. Dimonte and P. Ramaprabhu, Simulations and model of the nonlinear Richtmyer–Meshkov instability, *Phys. Fluids*, 22 (2010), p. 014104.
13. R. Samtaney, Suppression of the Richtmyer–Meshkov instability in the presence of a magnetic field, *Phys. Fluids*, 15 (2003), pp. L53–L56.
14. V. Wheatley, D. I. Pullin, and R. Samtaney, Stability of an impulsively accelerated density interface in magnetohydrodynamics, *Phys. Rev. Lett.*, 95 (2005), p. 125002.
15. V. Wheatley, R. Samtaney, and D. I. Pullin, The Richtmyer–Meshkov instability in magnetohydrodynamics, *Phys. Fluids*, 21 (2009), p. 082102.
16. T. Sano, K. Nishihara, C. Matsuoka, and A. Inoue, Critical magnetic field strength for suppression of the Richtmyer–Meshkov instability, *Phys. Rev. Lett.*, 111 (2013), p. 205001.

17. A. Bakhsh and R. Samtaney, Incompressible models of magnetohydrodynamic Richtmyer–Meshkov instability in cylindrical geometry, *Phys. Rev. Fluids*, 4 (2019), p. 063906.
18. D. Bond, V. Wheatley, Y. Li, R. Samtaney, and D. I. Pullin, The magnetised Richtmyer–Meshkov instability in two-fluid plasmas, *J. Fluid Mech.*, 903 (2020), p. A41.
19. N. Shen, D. I. Pullin, V. Wheatley, and R. Samtaney, Magnetohydrodynamic Richtmyer–Meshkov instability under an arbitrarily oriented magnetic field, *Phys. Plasmas*, 27 (2020), p. 062101.
20. J. M. Stone, T. A. Gardiner, P. Teuben, J. F. Hawley, and J. B. Simon, The Athena++ adaptive mesh refinement framework: Design and magnetohydrodynamic solvers, *Astrophys. J. Suppl. Ser.*, 249 (2020), p. 4.

13 Biological Fluids
Study of Casson fluids in porous media

Deepak Kumar Maurya, Satya Deo, and Pankaj Kumar Maurya

13.1 INTRODUCTION

Biological fluids transfer nutrients, gases, hormones, and wastes, maintain homeostasis, cushion organs, and support immunity. Life requires biological fluids to carry, protect, and regulate body systems (Spagnolie, 2015). Cellular function, communication, and defense depend on them, which make up 50–60% of healthy people's body weight. Blood is sometimes modeled as a Casson fluid, a non-Newtonian fluid that flows only when a critical stress is exceeded. Due to suspended cells (such as erythrocytes) and plasma proteins, blood flow properties are complex, making this rheological model essential in biomedical engineering, hemodynamics, and clinical diagnostics.

The study of Casson fluids in porous media aims to replicate biological fluids, such as blood, where non-Newtonian characteristics and porous architectures, including tissues, capillaries, or artificial scaffolds, significantly affect flow, thermal transfer, and mass transport (Bhaskar *et al.*, 2025; Maurya *et al.*, 2026). In recent decades, numerous researchers have been drawn to studying Casson fluid, one of the non-Newtonian fluids, due to its numerous uses in food processing, metallurgy, bioengineering, and drilling operations (Casson, 1959; Ali *et al.*, 2023). A classical book describes the heat transfer performance of several heat exchangers employing various fluids (Sheikholeslami and Ganji, 2017). A comprehensive review of the characteristics of generalized Newtonian and non-Newtonian fluids is presented in the influential book (Bird *et al.*, 1987). The Soret and Dufour effects on a Casson fluid around the stagnation point on a stretching sheet have been addressed by Kameshwaran *et al.* (2014). In their study, Venkatesan *et al.* (2013) examined the blood's flow through a small artery with a bell-shaped stenosis, treating the blood as Casson fluid. Employing the non-linear temperature distribution, the mathematical model for magnetohydrodynamic flow of a Casson fluid over a stretching sheet was reported by Mustafa and Khan (2015). Kumar *et al.* (2019) addressed the flow of a Casson fluid in two horizontal coaxial rotating cylinders, taking into account the effect of a magnetic field. Following Casson's stress–strain relation, Batra and

DOI: 10.1201/9781003664345-13

Das (1992) investigated the flow properties of the time-independent non-Newtonian fluid in the annular space between two coaxially rotating cylinders when the outer cylinder rotates at a constant speed and the inner cylinder is at rest. MHD micropolar Casson fluid flow across a linearly extending sheet with a heat source/sink effect in a porous medium was numerically solved by Kumar *et al.* (2024). Nandeppanavar and Jagadevappa (2025) used the Lie similarity transformation to study the flow, heat, and mass transfer characteristics of MHD Casson fluid through a porous stretching sheet with effects of varying viscosity and uniform heat source. Chaudhary *et al.* (2025) used the BVP4C method to study the magnetohydrodynamic boundary layer flow of Casson fluid driven by a moving permeable plate with Soret and Dufour impacts. Jat *et al.* (2024a, 2024b) extended their understanding of the non-Newtonian characteristics of Casson fluid flow through two concentric cylinders in a porous medium under torsional motion. Brownian motion and the Darcy–Forchheimer equation of porosity have been used to numerically model Casson fluid flow based on a curved surface. The effect of simultaneously including radiation absorption and chemical reactions on Casson fluid flow across an exponentially expanding surface under the impact of different exponential order flow parameters was studied by Nandhini *et al.* (2023).

In the presence of a magnetic field, the flow of micropolar fluid through a cylindrical tube containing an impermeable core coated with a porous layer was studied by Deo *et al.* (2020). The generalization of this problem is being addressed by Deo *et al.* (2021). Maurya *et al.* (2021) compared the flow patterns for two different kinds of boundary value problems and investigated the Stokesian flow of a micropolar fluid passing through a porous circular cylinder. The hydromagnetic flow of a micropolar fluid layer between two Newtonian fluid layers via a composite porous channel and cylindrical shells was examined by Deo and Maurya (2022). The magnetohydrodynamic flow of two immiscible micropolar fluids across horizontal coaxial cylindrical cells, consisting of a hypothetical cylindrical cell containing a solid impermeable cylindrical core wrapped in a porous layer, was explored by Maurya (2025). Deo *et al.* (2024) presented a comprehensive study of multiphase fluid flow through cylindrical cavities and porous channels. Incorporating the cell models, Selvi *et al.* (2023) investigated the flow of an incompressible couple stress fluid within an incompressible Reiner–Rivlin liquid sphere surrounded by an additional porous sphere (Selvi *et al.*, 2024). Maurya *et al.* (2023) employed the separation of variables method to investigate the Stokesian flow of an axisymmetric, incompressible couple stress fluid within a porous medium around a solid sphere, subjected to a uniform magnetic field.

13.1.1 Single-Phase Flow and Related Phenomenon

In fluid mechanics, the term 'single-phase flow' describes the flow of a fluid that is completely in one physical state (either a liquid or a gas) without changing phases or combining with other phases (Nield and Bejan, 2006). By treating the fluid as a continuous medium under the continuum hypothesis and assuming uniformity in physical characteristics like density, viscosity, and temperature, it is a fundamental idea that makes the analysis of fluid behavior easier. The continuity equation, the Navier–Stokes equations, and the energy equation are commonly used to explain the basic conservation laws of mass, momentum, and energy that govern such types of flow. Single-phase flow can be categorized based on its features into laminar or turbulent flow, determined by the Reynolds number, and into compressible or incompressible flow, based on the significance of density fluctuations. The examination of single-phase flow is essential across various engineering and scientific domains, encompassing pipeline, pump, and turbine design; blood flow analysis in biomedical engineering; and groundwater and river system modeling in environmental research. Single-phase flow models in porous media characterize the movement of fluids such as water or oil through soil or rock formations, frequently utilizing Darcy's equation to correlate flow rate with pressure gradient and permeability. Comprehending single-phase flow facilitates the formulation of control strategies for fluid systems, the forecasting of flow patterns, and the evaluation of energy losses attributable to friction and turbulence. Single-phase flow constitutes a vital area in fluid mechanics that connects fundamental theory with practical applications, providing a coherent framework for the analysis and design of systems where fluid motion is paramount. A rough sketch for single-phase flow models through channel, cylindrical, and spherical configurations can be seen from Figure 13.1.

13.1.2 Certain Single-Phase Flow Models Based on Casson Fluid

Chakrabarti and Gupta (1979) derived a similarity solution for the velocity and heat transfer properties of Casson fluid in a flow with uniform suction at the stretched wall. Polymer technology, which involves stretching plastic sheets, and metallurgy, which uses hydromagnetic methods, may leverage this challenge. Many metallurgical methods cool continuous strips or filaments by drawing them through a quiescent fluid, which sometimes stretches the strips. Dash and Mehta (1996) employed the Brinkman model to simulate the flow of a Casson fluid in a tube filled with a homogeneous porous media, accounting for Darcy resistance. The findings suggest that this study may simulate cardiovascular blood flow when cholesterol plaques and blood clots form in the coronary artery lumen. When the Casson fluid is considered to be viscous and incompressible, Mukhopadhyay (2009) examined the impact of thermal radiation on the unstable boundary layer mixed convection heat transfer

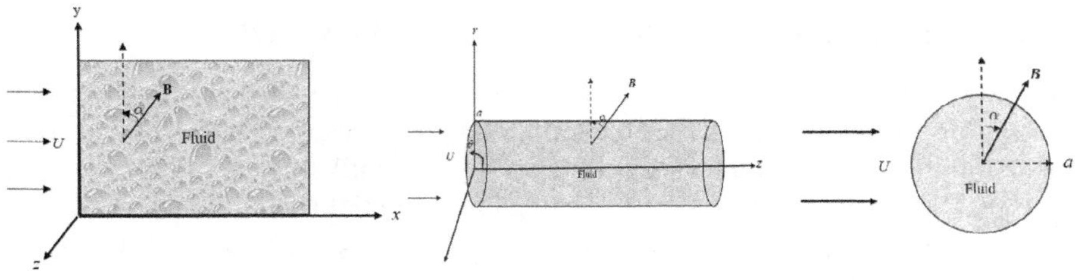

FIGURE 13.1 Single-phase flow models in various geometries.

problem from a vertical porous stretching surface immersed in porous medium. Employing a homotopic methodology, Mustafa *et al.* (2011) investigated the unsteady flow and thermal transfer of a Casson fluid over a moving flat plate under the influence of a parallel free stream. Mukhopadhyay (2011) examined the unstable mixed convective boundary layer flow and heat transfer over a porous stretched vertical surface with slip effects. Mukhopadhyay *et al.* (2013) applied similarity transformations to transform partial differential equations into ordinary differential equations for unsteady two-dimensional Casson fluid flow over a stretching surface with a prescribed temperature. Mukhopadhyay (2013) reported the boundary layer analysis in non-Newtonian fluid flow and heat transfer over a nonlinearly enhancing surface, using the Casson fluid model to characterize the fluid behavior. Venkatesan *et al.* (2013) examine the blood flow via a constricted artery exhibiting bell-shaped stenosis, modeling blood as a Casson fluid. Pramanik (2014) addressed the boundary layer flow of Casson fluid with heat transfer concerning an exponentially stretching surface, incorporating thermal radiation into the temperature field equation. Hussanan *et al.* (2017) investigated the exact solutions for unsteady magnetohydrodynamic flow of a Casson fluid over an infinite vertical oscillating plate immersed in a porous medium in the presence of Newtonian heating. In the scenario consisting of thermal radiation and a magnetic field, Naqvi *et al.* (2020) explored the electrically conducting Casson fluid combined with metallic particles of nanoscale size associated with a moving boundary cylinder. Rasool *et al.* (2020) used the slip conditions to determine the temperature, velocity, and concentration of the nanoparticles while incorporating the Darcy–Forchheimer relation to discuss the properties of a Casson-type incompressible viscous fluid maintained to flow through porous medium over a non-linear stretching surface from the perspective of heat and mass transfer developments. Maurya *et al.* (2023) investigated the role of magnetic field on a steady flow of Casson viscous fluid over a porous stretching plate. Maurya and Sharma (2025) employed Schauder's fixed point theorem to investigate the hydromagnetic effects on the constant flow of Casson fluid through two coaxial horizontal rotating cylinders which revolved about their axes at varying angular velocities. In the presence of Stefan blowing and slip conditions, Hanaya *et al.* (2024) investigated the flow and heat transfer of Casson fluids over a solid sphere saturated in an isotropic porous medium. Pushpalatha *et al.* (2016) investigated numerical solutions of Casson fluid flow at the stagnation point region of a sphere with cross diffusion effects using Runge-Kutta and Newton's technique.

13.2 MECHANISM OF CASSON FLUID FLOW MODEL

A Casson fluid is a non-Newtonian fluid that demonstrates distinct flow behavior, marked by a yield stress, which is the minimal force necessary for the fluid to commence movement. Casson fluids behave as a solid until this energy threshold is reached, analogous to a heavy box that requires an initial push to move. A Casson fluid is characterized by two flow parameters: yield stress $\left(\tau_y\right)$, associated with the attractive forces between fluid particles, and plastic viscosity $\left(\mu_p\right)$, which pertains to the asymmetry of the fluid particles (Abdus and Sharif, 2015). The Casson fluid model effectively characterizes the rheological behavior of specific complex fluids that display yield stress alongside shear-thinning characteristics, indicating that they function as solids until a critical shear stress is surpassed, after which they flow with a viscosity that diminishes as the shear rate escalates. The Casson fluid flow model was initially introduced by Casson (1959) in connection to printing inks to account for the phenomenon that these fluids exhibit resistance to flow below a specific stress threshold, and thereafter, once activated, they adhere to a non-linear relationship between stress (τ) and shear rate (e). The constitutive equations for Casson fluids combine a finite yield stress with shear-thinning behavior and are often represented implicitly as

$$\sqrt{\tau} = \sqrt{\tau_y} + \sqrt{e\mu_p}. \tag{13.1}$$

13.2.1 Reduction of Casson Fluid into Newtonian Fluid

Casson fluids fit many biological and industrial fluids that resist initial deformation but thin under shear better by adding a threshold stress for flow initiation and a shear-dependent viscosity to the Newtonian framework. The Casson model reduces to a Newtonian fluid when its yield stress is negligible or when shear rates are significantly high (Figure 13.2). A zero yield stress indicates that the fluid presents no threshold resistance and exhibits linear flow in response to applied stress. At extremely high shear rates, even fluids with finite yield stress exhibit almost Newtonian behavior due to the predominance of the viscous term. In cylindrical polar coordinates, Deo and Maurya (2019) introduced the generalized stream function solution for the Brinkman equation (Maurya and Deo, 2024), and deduced the solutions reported by Pop and Cheng (1992) as a specific case (Deo *et al.*, 2016). The comprehensive stream function solution for the Brinkman and Stokes equations in parabolic cylindrical coordinates, as derived by Maurya and Deo (2020). Tagg *et al.* (2023) used the simplified Navier–Stokes equations to study the viscous fluid flow problem through rotating coaxial cylinders. Analytical solutions for Newtonian fluid flow through porous media governed by the Brinkman equation are also available in tangent-sphere coordinates, cardioid coordinates, bispherical polar coordinates, and toroidal coordinates (Maurya and Sharma, 2024; Yadav and Maurya, 2025).

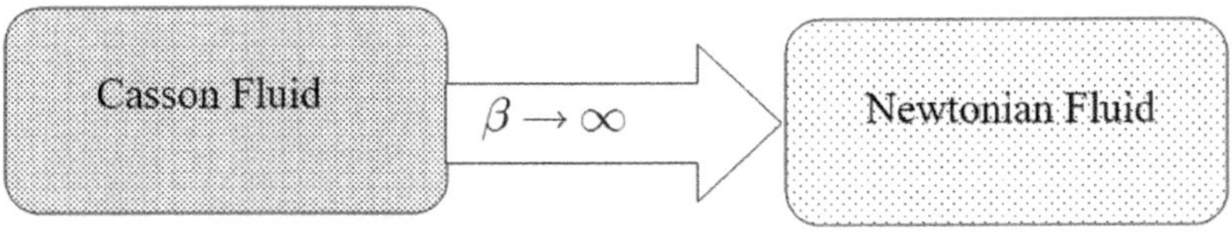

FIGURE 13.2 Reduction of Casson fluid to Newtonian fluid.

13.3 GOVERNING EQUATIONS OF CASSON FLUID FLOWS

The governing equations for Casson fluid flow are derived from the Navier–Stokes equations, modified to incorporate the Casson constitutive relation, given in Equation (13.1). The governing equations of Casson fluid consist of conservation of mass, linear momentum equation, and energy equation. These equations in vector form for an incompressible Casson fluid model can be expressed as follows:

$$\nabla \cdot \mathbf{v} = 0, \tag{13.2}$$

$$\rho \frac{d\mathbf{v}}{dt} = \mathbf{f} - \nabla p + \nabla \cdot \tau, \tag{13.3}$$

$$\rho c_p \frac{dT}{dt} = \kappa \nabla^2 T + \Phi, \tag{13.4}$$

where $\mathbf{v}$ is fluid velocity at any time t, ρ is fluid density, $\mathbf{f}$ is body force, p is fluid pressure, c_p is specific heat at constant pressure, T is temperature, κ is thermal conductivity and Φ is the viscous dissipation term and τ is shear stress which is defined in Equation (13.1) for Casson model.

13.3.1 For Unsteady Porous Stretching Surface

In the absence of body force and magnetic field, the governing equations for laminar boundary layer, two-dimensional flow, and heat transfer of an incompressible, conducting non-Newtonian Casson fluid over an unsteady porous stretching sheet/surface (Mukhopadhyay *et al.*, 2013), are given by

$$\frac{\partial u}{\partial x} + \frac{\partial v}{\partial y} = 0, \tag{13.5}$$

$$\frac{\partial u}{\partial t} + u\frac{\partial u}{\partial x} + v\frac{\partial u}{\partial y} = -\frac{\nu}{k}u + \nu\left(1 + \frac{1}{\beta}\right)\frac{\partial^2 u}{\partial y^2}, \tag{13.6}$$

TABLE 13.1
Characteristics of Casson and Newtonian Fluid

Characteristics	Newtonian Fluid	Casson Fluid
Constitutive equation	$\tau = e\mu$	$\sqrt{\tau} = \sqrt{\tau_y} + \sqrt{e\mu_p}$
Yield stress	0	>0
Casson parameter	∞	Finite
Viscosity	Almost constant	Decreases with e
Examples	Water, air, thin oil, etc.	Blood, chocolate, thick paint, etc.

$$\frac{\partial T}{\partial t}+u\frac{\partial T}{\partial x}+v\frac{\partial T}{\partial y}=\kappa\frac{\partial^2 T}{\partial y^2}, \tag{13.7}$$

where (u, v) are velocity components along horizontal x and vertical y directions, ν is kinematic viscosity, k is permeability, and β is Casson parameter.

13.3.2 For Steady Stretching Cylinder under Magnetic Field

The governing equations (Mukhopadhyay, 2013) for steady axi-symmetric flow of incompressible Casson fluid along a stretching cylinder in the presence of magnetic field can be expressed

$$\frac{\partial(rv_r)}{\partial r}+\frac{\partial(rv_z)}{\partial z}=0, \tag{13.8}$$

$$v_r\frac{\partial v_z}{\partial r}+v_z\frac{\partial v_z}{\partial z}=\nu\left(1+\frac{1}{\beta}\right)\left(\frac{\partial^2 v_z}{\partial r^2}+\frac{1}{r}\frac{\partial v_z}{\partial r}\right)-\frac{\sigma B^2}{\rho}v_z, \tag{13.9}$$

$$v_r\frac{\partial T}{\partial r}+v_z\frac{\partial T}{\partial z}=\kappa\left(\frac{\partial^2 T}{\partial r^2}+\frac{1}{r}\frac{\partial T}{\partial r}\right), \tag{13.10}$$

where (v_r, v_z) are velocity components along radial coordinate r and z of cylindrical coordinate (r, θ, z), B is magnitude of magnetic field intensity, and σ is electrical conductivity.

13.4 SINGLE-PHASE FLOW OF CASSON FLUID THROUGH A ROTATING ANNULUS

This single-phase flow problem considers the steady motion of an incompressible, viscous, fully developed Casson fluid between two infinitely long co-axial horizontal rotating cylinders (Maurya and Sharma, 2025). A uniform magnetic field of strength $\mathbf{B}$ is applied perpendicular to fluid flow, with $|\mathbf{B}|=B$. The inner and outer cylinders are considered to rotate with constant angular velocity ω_1 and ω_2, respectively, with curved arrows near the boundary of the cylinder (Figure 13.3). Inner and outer cylinder radii are written as a and b, respectively. As a is the distance from the axis of coaxial cylinders to the solid core boundary, therefore the radius of the porous cylinder will be $b-a$. Curved arrows indicate the direction of fluid particle movement in a porous cylinder as a Casson fluid flows with velocity (v_θ).

Circular Couette flow allows us to disregard the pressure gradient along variable θ. Thus, Casson fluid governing equations are such as

$$\frac{\partial v_\theta}{\partial \theta}=0, \tag{13.11}$$

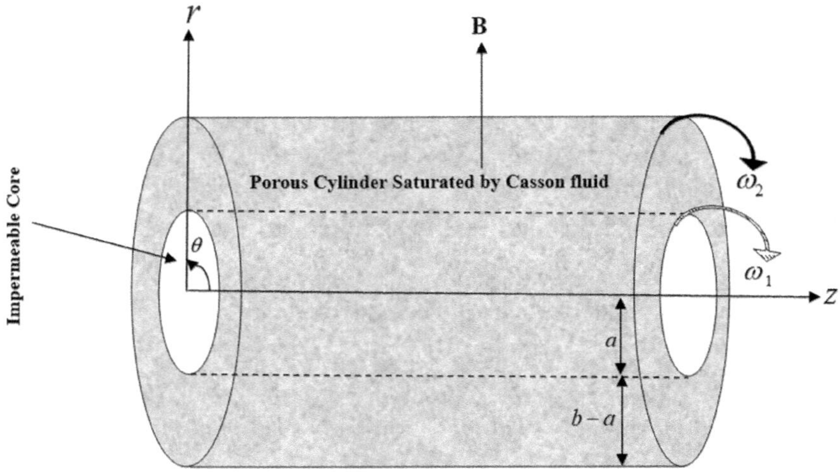

FIGURE 13.3 Representation of the physical model.

$$\mu\left(1+\frac{1}{\beta}\right)\left(\frac{\partial^2 v_\theta}{\partial r^2}+\frac{1}{r}\frac{\partial v_\theta}{\partial r}-\frac{v_\theta}{r^2}\right)=\sigma B^2 v_\theta+\frac{\mu}{k}v_\theta, \tag{13.12}$$

with the boundary conditions (BCs): $v_\theta(a)=a\omega_1$ and $v_\theta(b)=b\omega_2$.

Introducing the following dimensionless quantities to pursue the further analysis:

$$\widetilde{v_\theta}=\frac{v_\theta}{a\omega_1},\quad \tilde{r}=\frac{r}{a},\quad \tilde{R}=\frac{b}{a},\quad \tilde{\omega}=\frac{\omega_2}{\omega_1},\quad \tilde{M}=\tilde{H}^2+\frac{a^2}{k}\quad \text{and}\quad \tilde{H}=aB\sqrt{\frac{\sigma}{\mu}}. \tag{13.13}$$

We omit the tilde (~) sign throughout the manuscript for broader understanding. Equations (13.11) and (13.12) and BCs can be represented as a boundary value problem (BVP) as follows:

$$\left(1+\frac{1}{\beta}\right)\left(\frac{d^2 v_\theta}{dr^2}+\frac{1}{r}\frac{dv_\theta}{dr}-\frac{v_\theta}{r^2}\right)-Mv_\theta=0 \quad \text{with BCs} \quad v_\theta=\begin{cases}1 & \text{at} \quad r=1\\ R\omega & \text{at} \quad r=R\end{cases}. \tag{13.14}$$

We can convert the above boundary value problem (BVP) into a modified Bessel differential equation of order one by arranging its terms. So, the general solution of the BVP will be

$$v_\theta(r)=A_1 I_1\left(r\sqrt{M\beta/(1+\beta)}\right)+A_2 K_1\left(r\sqrt{M\beta/(1+\beta)}\right), \tag{13.15}$$

where
$$A_1=\frac{K_1\left(R\sqrt{M\beta/(1+\beta)}\right)-R\omega K_1\left(\sqrt{M\beta/(1+\beta)}\right)}{\begin{vmatrix} I_1\left(\sqrt{M\beta/(1+\beta)}\right) & I_1\left(R\sqrt{M\beta/(1+\beta)}\right)\\ K_1\left(\sqrt{M\beta/(1+\beta)}\right) & K_1\left(R\sqrt{M\beta/(1+\beta)}\right)\end{vmatrix}},$$

and $$A_2 = -\frac{I_1\left(R\sqrt{M\beta/(1+\beta)}\right) - R\omega I_1\left(\sqrt{M\beta/(1+\beta)}\right)}{\begin{vmatrix} I_1\left(\sqrt{M\beta/(1+\beta)}\right) & I_1\left(R\sqrt{M\beta/(1+\beta)}\right) \\ K_1\left(\sqrt{M\beta/(1+\beta)}\right) & K_1\left(R\sqrt{M\beta/(1+\beta)}\right) \end{vmatrix}}.$$

Here, $I_1(\cdot)$ and $K_1(\cdot)$ are modified Bessel functions of order one, of first and second kinds, respectively.

The formula for calculating the skin friction coefficient (C_f) was previously described by Maurya and Sharma (2025). Therefore, the mathematical expression of C_f will be

$$C_f = \left(\frac{1+\beta}{\beta}\right)\left[c_1\left(\frac{\alpha}{2}\left(I_0(\alpha r) + I_2(\alpha r)\right) - \frac{1}{r}I_1(\alpha r)\right)\right.$$
$$\left. - c_2\left(\frac{\alpha}{2}\left(K_0(\alpha r) + K_2(\alpha r)\right) + \frac{1}{r}K_1(\alpha r)\right)\right], \quad (13.16)$$

where $\alpha = a\sqrt{\frac{\beta}{1+\beta}\left(\frac{\sigma B^2}{\mu} + \frac{1}{k}\right)}$.

13.5 SPECIAL CASES

13.5.1 Casson Fluid Reduces to Newtonian Fluid (i.e. $\beta \to \infty$)

When $\beta \to \infty$, then the present problem of Casson fluid reduces to the problem of Newtonian fluid

$$v_\theta = \frac{I_1(r\alpha)\left(K_1(R\alpha) - R\omega K_1(\alpha)\right)}{I_1(\alpha)K_1(R\alpha) - K_1(\alpha)I_1(R\alpha)} + \frac{K_1(r\alpha)\left(I_1(R\alpha) - R\omega I_1(\alpha)\right)}{K_1(\alpha)I_1(R\alpha) - I_1(\alpha)K_1(R\alpha)}, \quad (13.17)$$

where $\alpha = a\sqrt{\left(\frac{\sigma B^2}{\mu} + \frac{1}{k}\right)}$.

This finding confirms the velocity profile of Newtonian fluid as reported in the book by Deville (2022).

13.5.2 Permeability Tends to Infinity (i.e. $k \to \infty$)

When $k \to \infty$, then the expression of velocity profile will be of the form

$$v_\theta = \frac{I_1(r\alpha)\left(K_1(R\alpha) - R\omega K_1(\alpha)\right)}{I_1(\alpha)K_1(R\alpha) - K_1(\alpha)I_1(R\alpha)} + \frac{K_1(r\alpha)\left(I_1(R\alpha) - R\omega I_1(\alpha)\right)}{K_1(\alpha)I_1(R\alpha) - I_1(\alpha)K_1(R\alpha)}, \quad (13.18)$$

$$\text{where } \alpha = Ba\sqrt{\frac{\sigma\beta}{\mu(1+\beta)}}.$$

The above result of the velocity profile matches with the velocity profile as reported in the research work by Kumar *et al.* (2019).

13.6 GRAPHICAL REPRESENTATION OF VELOCITY PROFILE

This section explains the graphical variation of velocity profiles for several flow parameters, including permeability k, Hartmann number H, Casson parameter β, rotation parameter ω, and aspect ratio R of cylinder radii. Here, the permeability parameter (k) varies from zero to infinity, Hartmann number (H) varies from 0 to ∞, Casson parameter (β) is always a non-negative real number, rotation parameter $\omega \in \mathbb{R}$ and $R \geq 1$. Figures 13.4 and 13.5 explore the impact of the viscous parameter β on the radial velocity profile under similar global conditions $(R = 10, \omega = 1, H = 0.5)$ with and without permeability. When β rises in Figure 13.4, the velocity decreases and the radial gradient steepens, showing that viscous stress reduces core flow and compresses the shear layer. However, Figure 13.4 shows that slip reduces viscous resistance and expands the high-velocity region compared to Figure 13.5. Figures 13.6 and 13.7 perform the same comparison for the magnetic Hartmann number H at $\beta = 1, R = 10, \omega = 1$. As Hartmann number H increases in Figure 13.6, velocity profiles decrease and the region of significant velocity narrows due to Lorentz forces opposing motion and thinned conductive boundary layer.

13.6.1 Graphical Representation of Skin Friction Coefficient

The authors examined the skin friction coefficient (C_f) vs radius (r) for different values of rotation parameter (ω) and aspect ratio (R). We compared the data of

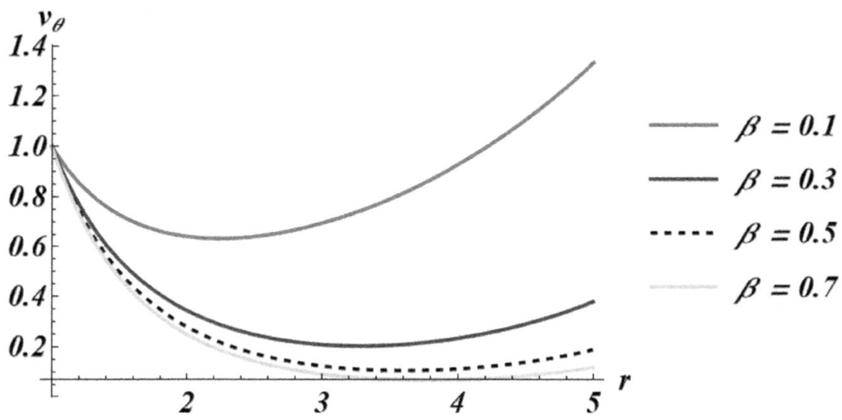

FIGURE 13.4 Velocity profile when $a = 1, R = 10, \omega = 1, k = 0.5, H = 0.5$.

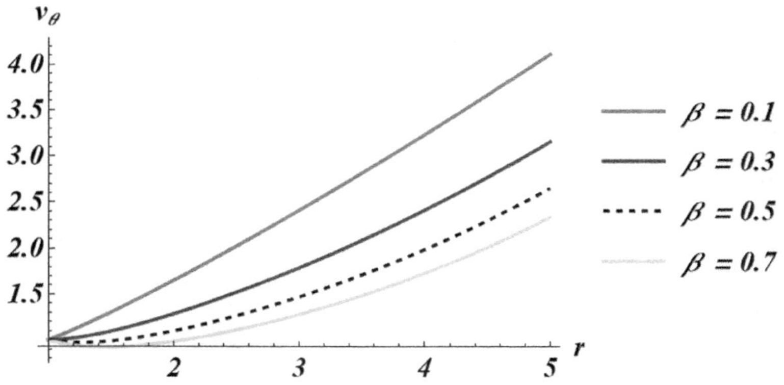

FIGURE 13.5 Velocity profile (Kumar *et al.*, 2019) when $R = 10$, $\omega = 1$, $H = 0.5$.

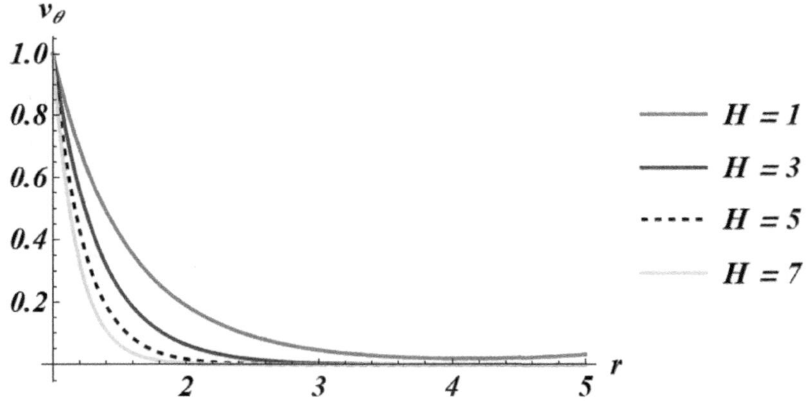

FIGURE 13.6 Velocity profile when $R = 10$, $\omega = 1$, $k = 0.5$, $\beta = 1$.

skin friction coefficient with results reported by Kumar *et al.* (2019). Permeability parameter is present in Figure 13.8, but its absence is shown in Figure 13.9. Figures 13.5–13.8 show how the skin friction coefficient C_f varies with respect to the radial coordinate r under various fluid flow analysis parameters. Figure 13.8 shows that rising ω increases C_f, indicating increased shear stress at the boundary due to stronger rotational effects. In Figure 13.9, the base parameters remain unchanged, but a and k are removed to focus on the effect of ω. The trends remain consistent, reinforcing the role of rotation in modulating frictional forces. Figure 13.10 shows that raising R decreases C_f, implying that stronger inertial forces reduce boundary shear compared to viscous forces. Figure 13.11 investigates the influence of R under comparable conditions, but without specifying a and k, and verifies the inverse relationship between R and C_f.

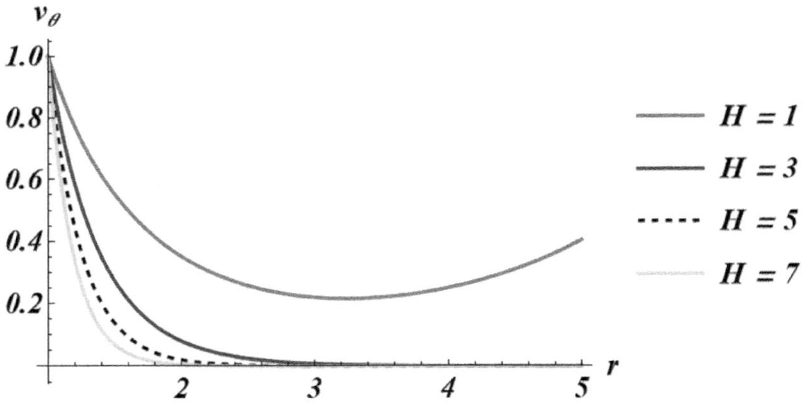

FIGURE 13.7 Velocity profile (Kumar *et al.*, 2019) when $R = 10$, $\omega = 1$, $\beta = 1$.

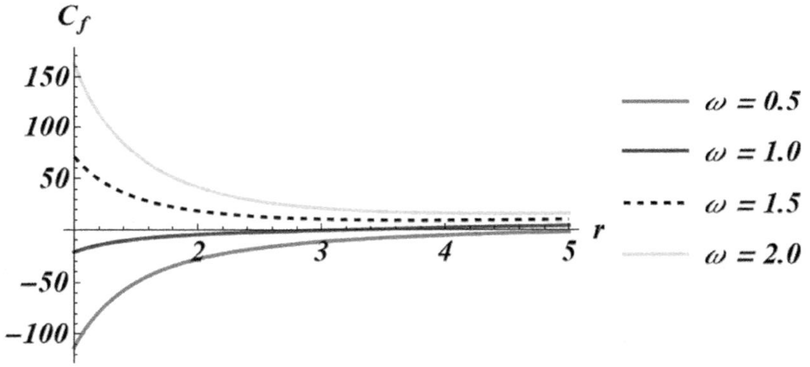

FIGURE 13.8 Skin friction coefficient when $a = 1$, $\beta = 0.01$, $R = 10$, $H = 0.5$, $k = 1.5$.

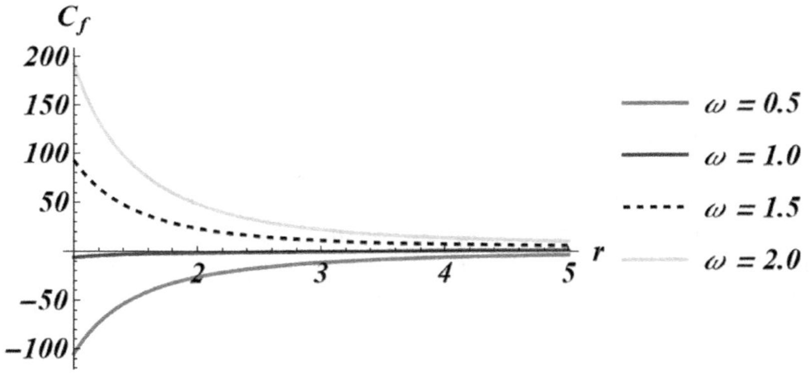

FIGURE 13.9 Skin friction coefficient (Kumar *et al.*, 2019) when $\beta = 0.01$, $R = 10$, $H = 0.5$.

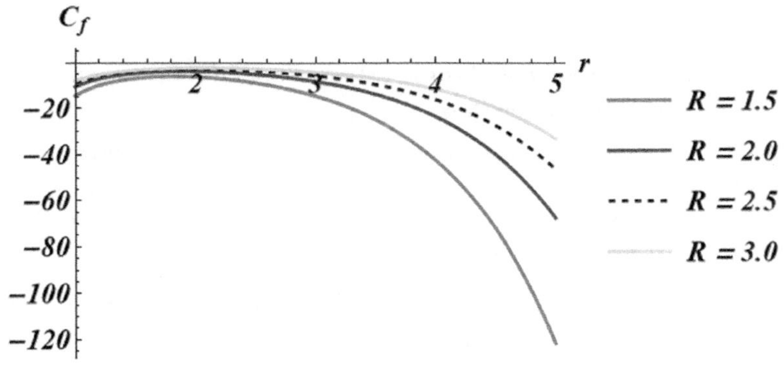

FIGURE 13.10 Skin friction coefficient when $a = 1, \beta = 0.01, \beta = 0.01, H = 0.5, k = 1.5$.

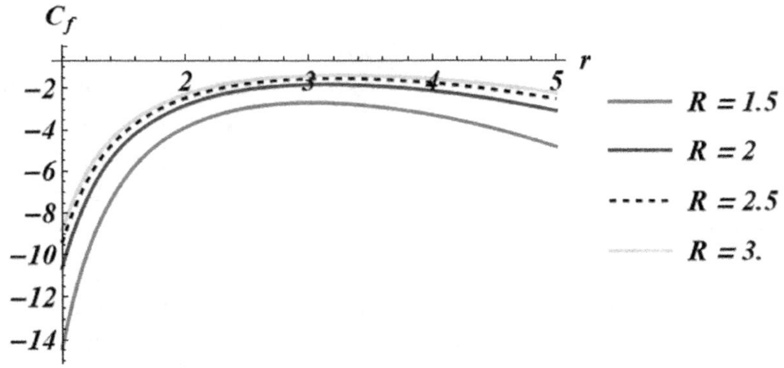

FIGURE 13.11 Skin friction coefficient (Kumar *et al.*, 2019) when $\beta = 0.01$, $\beta = 0.01$, $H = 0.5$.

13.7 APPLICATIONS

Blood flow modeling in biomedical engineering uses the Casson model. At low shear rates, red blood cells agglomerate, causing yield stress in blood. This is critical for understanding circulation in narrow blood arteries and capillaries, where flow requires more effort. The flow of liquid chocolate during molding and coating can be predicted. Pumping and packaging equipment can be designed for tomato paste, sauces, and condensed milk, which exhibit Casson fluid behavior. The Casson model drives the paint and coatings business. Paint should spread smoothly and have a yield stress to prevent dripping. Formulators and shampoo manufacturers employ the Casson model to attain this balance. Casson fluid is highly effective in lubricating systems with permeable surfaces, in the chemical, pharmaceutical, and food sectors where filtration via permeable plates is required, and in the flow through porous tissues or membranes, as in dialysis or drug administration systems. There are some dairy products that have Casson fluid properties. Pharmaceutical suspensions and

TABLE 13.2
Numerical Values of Velocity Profile (v) and Skin Friction Coefficient (C_f) at $r = 1.5$ for Different Parameters

Assigned Parameters		Comparative Study of Proposed Model with Kumar *et al.* (2019) and Tagg *et al.* (2023)					
		Proposed Casson Model		Kumar *et al.*, (2019)		Tagg *et al.*, (2023)	
		v	C_f	v	C_f	v	C_f
β	0.1	0.7266	−8.6589	1.3027	−2.2443	0.3649	−10.3501
	0.3	0.5445	−4.0701	1.0946	−1.8124	0.3649	−4.0773
	0.5	0.4972	−2.8817	0.9880	−1.5812	0.3649	−2.8227
	0.7	0.4723	−2.3451	0.9240	−1.4376	0.3649	−2.2850
H	1	0.4158	−1.9226	0.5499	−1.8720	0.3266	−1.8311
	3	0.2428	−1.6518	0.2707	−1.7229	0.1508	−1.3149
	5	0.1264	−1.1919	0.1354	−1.2391	0.0595	−0.7408
	7	0.0639	−0.7780	0.0672	−0.8042	0.0226	−0.3691
R	1.5	−1.5000	−10.0020	−1.5000	−9.6450	−1.5000	−10.7858
	2.0	−0.6609	−7.0107	-0.7149	−7.1007	−0.5612	−6.7775
	2.5	−0.3478	−5.8944	-0.4663	−6.2951	−0.1807	−5.1528
	3.0	−0.1525	−5.1983	-0.3425	−5.8937	0.0408	−4.2067

emulsions, which need careful mixing to preserve stability and efficacy, can profit from this study. In the context of reactor design, non-viscous fluids are used in many chemical processes. With important applications in industrial processes, energy systems, environmental engineering, material science, and the aerospace and automotive industries, where knowledge of heat transfer mechanisms in complex systems can improve efficiency, performance, and safety, this innovative method offers fresh perspectives on the behavior of Casson fluids. Applications for Casson fluid models are extensive and include energy systems, automotive, aerospace, and environmental engineering, where heat transfer phenomena are utilized to improve performance, safety, and efficiency.

13.8 CONCLUSIONS AND FUTURE SCOPE

The literature on single phase Casson fluid flow models has been briefly reviewed in this chapter. This chapter also discusses the situation where Casson fluid flow models exhibit behavior similar to that of Newtonian models, and an explanation of relevant research works on Newtonian models is reported here. Additionally, the fundamental ideas, governing equations, and corresponding mechanisms for the mathematical modeling of Casson fluid for both magnetohydrodynamic steady stretching cylinders and unsteady porous stretching surfaces have been explored. The flow of Casson fluid within rotating co-axial cylinders containing a solid core, subjected to an external magnetic field, is investigated by a single phase flow model. In the proposed research problem, the fluid velocity demonstrates graphical characteristics for many parameters, including Hartmann number (H), Casson parameter (β), etc. Graphical sketches are also drawn for different values of ω and R. A comparison of flow fields is conducted with and without the permeability term of porous media. In the absence of the permeability term, the skin friction coefficient exhibits lesser reactivity to variations in the Casson parameter. The main findings of the single phase flow problem can be encapsulated as follows:

The radial velocity of the Casson fluid is decelerated, and the boundary layers are sharpened as the Casson parameter β and the magnetic field strength H increase.

The radial velocity profile of Casson fluid is elevated and broadened as the rotation parameter ω is increased, resulting in an intensification of skin friction C_f. In addition, a larger aspect ratio R reduces the peak shear stress and redirects the location of friction decay.

The skin friction coefficient demonstrates a nonlinear relationship with r, characterized by unique profiles influenced by the interaction of rotational and inertial components.

FUNDING INFORMATION

Deepak Kumar Maurya expresses gratitude to the Council of Science & Technology, U.P. India, for the financial support arranged through the research project (Grant ID: CST/D-1517).

BIBLIOGRAPHY

Abdus, S.S. and Sharif, U., *MHD Boundary Layer Casson Fluid over a Nonlinear Stretching Sheet*, Lambert Academic Publishing, 2015.

Ali, H.M., Hassan, A. and Wahab, A., *Nanofluids for Heat Exchangers*, Springer, 2023.

Batra, R.L. and Das, B., Flow of a Casson Fluid Between two Rotating Cylinders, *Fluid Dyn. Res.*, vol. 9, pp. 133–141, 1992.

Bhaskar, M., Babu, K.R., Khan, S.U., Lodhi, R.K., Oudina, F.M. and Sohail, M., Multiscale Modeling of Jeffrey–Casson Immiscible Fluid Dynamics Under Electrokinetic and Magnetohydrodynamic Effects, *Multiscale Multidiscip. Model. Exp. Des.*, vol. 8, no. 383, pp. 1–13, 2025.

Bird, R.B., Armstrong, R.C. and Hassager, O., *Dynamics of Polymeric Liquids*, Wiley, New York, 1987.

Casson, N., A Flow Equation for Pigment–oil Suspensions of the Printing-ink Type, in *Rheology of Disperse Systems*, Pergamon Press, pp. 84–104, 1959.

Chakrabarti, A. and Gupta, A.S., Hydromagnetic Flow and Heat Transfer over a Stretching Sheet, *Q. Appl. Math.*, vol. 37, pp. 73–78, 1979.

Chaudhary, H., Singh, N., Roshan, M., Yadav, P.K., and Saleem, S., Modeling of Magnetohydrodynamic Casson Fluid Flow Due to a Moving Permeable Plate with Soret and Dufour Impacts, *J. Korean Phys. Soc.*, vol. 86, pp. 745–768, 2025.

Dash, R.K. and Mehta, K.N., Casson Fluid Flow in a Pipe Filled with a Homogeneous Porous Medium, *Int. J. Eng. Sci.*, vol. 34, pp. 1145–1156, 1996.

Deo, S. and Maurya, D.K., Generalized Stream Function of the Brinkman Equation in Cylindrical Polar Coordinates, *Spec. Top. Rev. Porous Media*, vol. 10, pp. 421–428, 2019.

Deo, S. and Maurya, D.K., Investigation of MHD Effects on Micropolar–Newtonian Fluid Flow Through a Composite Porous Channel, *Microfluid. Nanofluid.*, vol. 26, no. 64, pp. 1–16, 2022.

Deo, S., Ansari, I.A., and Srivastava, B.G., On the General Stream Function Solution of the Brinkman Equation in Cylindrical Polar Coordinates, *Adv. Theor. Appl. Mech.*, vol. 9, pp. 21–30, 2016.

Deo, S., Maurya, D.K. and Filippov, A.N., Influence of Magnetic Field on Micropolar Fluid Flow in a Cylindrical Tube Enclosing an Impermeable Core Coated with a Porous Layer, *Colloid J.*, vol. 82, pp. 649–660, 2020.

Deo, S., Maurya, D.K. and Filippov, A.N., Effect of Magnetic Field on Hydrodynamic Permeability of Biporous Membrane Relative to Micropolar Liquid Flow, *Colloid J.*, vol. 83, pp. 662–675, 2021.

Deo, S., Maurya, D.K. and Patel, P.K., Multiphase Flow Through Porous Channels and Cylindrical Shells: A Review, *BGP Conf. Proc.*, vol. 71, pp. 112–125, 2024.

Deville, M.O., *An Introduction to the Mechanics of Incompressible Fluids*, Springer, 2022.

Hanaya, A.A., Rashed, Z.Z. and Ahmed, S.E., MHD Casson Flow over a Solid Sphere Surrounded by Porous Material in the Presence of Stefan Blowing and Slip Conditions, *Alexandria Eng. J.*, vol. 103, pp. 402–412, 2024.

Hussanan, A., Salleh, M.Z., Khan, I. and Tahar, R.M., Heat Transfer in Magnetohydrodynamic Flow of a Casson Fluid with Porous Medium and Newtonian Heating, *J. Fluids*, vol. 6, pp. 1–10, 2017.

Jat, K., Sharma, K., Choudhary, P., Sonia, P., Ali, R. and Ganesh, M., Significance of Darcy–Forchheimer Casson Fluid Flow Past a Non-Permeable Curved Stretching Sheet with the Impacts of Heat and Mass Transfer, *Case Stud. Therm. Eng.*, vol. 61, no. 104907, pp. 1–18, 2024a.

Jat, K., Sharma, K., Choudhary, P., Thamizharasi, R., Abbas, M. and Devi, V.R., Computational Study on Torsional Casson Fluid Flow Through Concentric Cylinders in a Porous Medium, *Alexandria Eng. J.*, vol. 108, pp. 400–414, 2024b.

Kameshwaran, P.K., Shaw, S. and Sibanda, P., Dual Solutions of Casson Fluid Flow over a Stretching or Shrinking Sheet, *Sādhanā*, vol. 39, pp. 1573–1583, 2014.

Kumar, D., Ramesh, K. and Chandok, S., Influence of an External Magnetic Field on the Flow of a Casson Fluid in a Micro-Annulus Between two Concentric Cylinders, *Bull. Braz. Math. Soc. (N.S.)*, vol. 50, pp. 515–531, 2019.

Kumar, P., Singh, B., Goyal, S., Kumar, K., Nisar, K.S., Alkhazaleh, S.A., and Aty, A.H., Numerical Solution of MHD Micropolar Casson Fluid Flow over Porous Linearly Stretching Sheet with Heat Source/Sink, *Appl. Math. Inf. Sci.*, vol. 18, pp. 183–190, 2024.

Maurya, D.K., Role of Transverse Magnetic Field on Flow of Immiscible Micropolar Fluids Through Coaxial Horizontal Cylindrical Membrane Using Cell Models, *Phys. Fluids*, vol. 37, no. 023137, pp. 1–13, 2025.

Maurya, D.K. and Deo, S., Effect of Magnetic Field on Newtonian Fluid Sandwiched Between non-Newtonian Fluids Through Porous Cylindrical Shells, *Spec. Top. Rev. Porous Media*, vol. 13, pp. 75–92, 2022.

Maurya, D.K. and Deo, S., Stream Function Solution of the Brinkman Equation in Parabolic Cylindrical Coordinates, *Int. J. Appl. Comput. Math.*, vol. 6, no. 167, 2020.

Maurya, D.K. and Deo, S., Brinkman Equation in nanofluids: Generalized Stream Function Solution, in Nanofluids Technology for Thermal Sciences and Engineering, *CRC Press*, vol. 14, pp. 260–271, 2024.

Maurya, D.K. and Sharma, V.K., Brinkman Equation in Tangent Sphere and Cardioid Coordinate Systems: Stream Function Solution, *Jnanabha*, vol. 54, pp. 329–337, 2024.

Maurya, D.K. and Sharma, V.K., Hydromagnetic Effects on Steady Flow of Casson Nanofluid Through Rotating Coaxial Cylinders, *Phys. Fluids*, vol. 37, no. 013116, pp. 1–8, 2025.

Maurya, D.K., Deo, S. and Khanukaeva, D.Y., Analysis of Stokes Flow of Micropolar Fluid Through a Porous Cylinder, *Math. Methods Appl. Sci.*, vol. 44, pp. 6647–6665, 2021.

Maurya, D.K., Deo, S. and Maurya, P.K., Hydromagnetic Flow of Casson Fluid over a Stretching Plane Through a Porous Medium, *Z. Angew. Math. Mech.*, vol. 104, no. e202300675, pp. 1–15, 2024.

Maurya, D.K., Yadav, J. and Sharma, V.K., Poiseuille Flow of Newtonian–Couple Stress Fluids Through Coaxial Porous Tubes Enclosing an Impermeable Core: Role of Inclined Magnetic Field, *J. Mar. Sci. Technol.*, vol. 31, pp. 40–53, 2026.

Maurya, P.K., Deo, S. and Maurya, D.K., Couple Stress Fluid Flow Enclosing a Solid Sphere in a Porous Medium: Effect of Magnetic Field, *Phys. Fluids*, vol. 35, no. 072006, pp. 1–8, 2023.

Mukhopadhyay, S., Effect of Thermal Radiation on Unsteady Mixed Convection Flow and Heat Transfer over a Porous Stretching Surface in a Porous Medium, *Int. J. Heat Mass* Transfer, vol. 52, pp. 3261–3265, 2009.

Mukhopadhyay, S., Effects of Slip on Unsteady Mixed Convective Flow and Heat Transfer Past a Porous Stretching Surface, *Nucl. Eng. Des.*, vol. 241, pp. 2660–2665, 2011.

Mukhopadhyay, S., Casson Fluid Flow and Heat Transfer over a Nonlinearly Stretching Surface, *Chin. Phys.*, vol. B-22, no. 074701, pp. 1–6, 2013.

Mukhopadhyay, S., MHD Boundary Layer Slip Flow Along a Stretching Cylinder, *Ain Shams Eng. J.*, vol. 4, pp. 317–324, 2013.

Mukhopadhyay, S., De, P.R., Bhattacharya, K. and Layek, G.C., Casson Fluid Flow over an Unsteady Stretching Surface, *Ain Shams Eng. J.*, vol. 4, pp. 933–938, 2013.

Mustafa, M. and Khan, J.A., Model for Flow of Casson Fluid past a Non-linearly Stretching Sheet Considering Magnetic Field Effects, *AIP Adv.*, vol. 5, no. 077148, pp. 1–11, 2015.

Mustafa, M., Hayat, T., Pop, I. and Aziz, A., Unsteady Boundary Layer Flow of a Casson Fluid due to an Impulsively Started Moving Flat Plate, *Heat Transf. Asian Res.*, vol. 40, pp. 563–576, 2011.

Naqvi, S.M., Muhammad, T. and Asma, M., Hydromagnetic Flow of Casson Fluid over a Porous Stretching Cylinder with Newtonian Heat and Mass Conditions, *Physica A*, vol. 550, no. 123988, 2020.

Nandhini, C.A., Jothimani, S. and Chamkha, A.J., Combined Effect of Radiation Absorption and Exponential Parameter on Chemically Reactive Casson Fluid over an Exponentially Stretching Sheet, *Partial Differ. Equ. Appl. Math.*, vol. 8, no. 100534, pp. 1–14, 2023.

Nandeppanavar, M.M. and Jagadevappa, V., Mathematical Analysis of Flow, Heat and Mass Transfer Variations of MHD Casson Fluid due to Variable Physical Properties over a Stretching Sheet, *Int. J. Appl. Comput. Math.*, vol. 11, pp. 1–22, 2025.

Nield, D.A. and Bejan, A., *Convection in Porous Media*, Springer, New York, 2006.

Pop, I. and Cheng, P., Flow Past a Circular Cylinder Embedded in a Porous Medium Based on the Brinkman Model, *Int. J. Eng. Sci.*, vol. 30, pp. 257–262, 1992.

Pramanik, S., Casson Fluid Flow and Heat Transfer Past an Exponentially Porous Stretching Surface in the Presence of Thermal Radiation, *Ain Shams Eng. J.*, vol. 5, pp. 205–212, 2014.

Pushpalatha, K., Sugunamma, V., Reddy, J.V.R. and Sandeep, N., Dufour and Soret Effects on Unsteady Flow of a Casson Fluid in the Stagnation-Point Region of a Rotating Sphere, *Middle East J. Sci. Res.*, vol. 24, pp. 1141–1150, 2016.

Rasool, G., Chamkha, A.J., Muhammad, T., Shafiq, A. and Khan, I., Darcy–Forchheimer Relation in Casson-Type MHD Fluid Flow over a Nonlinear Stretching Surface, Propuls. *Power Res.*, vol. 9, pp. 159–168, 2020.

Selvi, R., Maurya, D.K. and Shukla, P., Analytical Solution of a Couple Stress Fluid Saturated in a Porous Medium Through a Reiner–Rivlin Liquid Sphere, *Phys. Fluids*, vol. 35, no. 073106, pp. 1–12, 2023.

Selvi, R., Maurya, D.K., Shukla, P. and Chamkha, A.J., Analysis of a Reiner–Rivlin Liquid Sphere Enveloped by a Permeable Layer, *Phys. Fluids*, vol. 36, no. 023303, pp. 1–11, 2024.

Spagnolie, S.E., *Complex Fluids in Biological Systems: Experiment, Theory, and Computation*, Springer, 2015.

Sheikholeslami, M. and Ganji, D.D., *Applications of Fluid for Heat Transfer Enhancement*, Elsevier, Amsterdam, 2017.

Tagg, J., Hawkins, R. and Danby, J., Fluid Flow Between Rotating Concentric Cylinders, *J. Appl. Eng. Math.*, vol. 10, pp. 1–5, 2023.

Venkatesan, J., Sankar, D.S., Hemalatha, K. and Yatim, Y., Mathematical Analysis of Casson Fluid Model for Blood Rheology in Stenosed Narrow Arteries, *J. Appl. Math.*, vol. 2013, no. 583809, pp. 1–12, 2013.

Yadav, J. and Maurya, D.K., Brinkman Equation in Bispherical Polar and Toroidal Coordinates: Stream Function Solution, *Spec. Top. Rev. Porous Media*, vol. 16, pp. 15–25, 2025.

14 Optimization in Fluid Dynamics

Divya Mahajan and Hritik

14.1 INTRODUCTION

For engineering and fluid dynamics, the search for optimization is a necessity. The increasing demands for cleaner energy, faster transportation, and sustainable industrial systems have turned the optimization of fluid systems into an extremely urgent task. Fluid motion plays a leading part in major technological applications. From aerospace propulsion and renewable energy to biomedical microdevices, large-scale chemical processing, and advanced thermal management systems, fluid motion is needed (Zhou et al. (2023)). If these systems aren't finely tuned, the outcome is severe. There is unnecessary energy wastage, unsteady or hard-to-predict flow patterns, and reduced product lifespan. These inefficiencies are not just throwing money down the drain, they're also causing environmental strain. All this makes optimization an absolute must-have.

Optimization in fluid dynamics is well known as the systematic boost to the design and operation of fluid systems. It uses mathematical formulas and state-of-the-art computational tools to improve on existing designs. It doesn't rely on trial-and-error methods, which were always a gamble on what might work. Optimization is a great tool for resolving conflicts between competing objectives. Take aerodynamics for example. Here, reducing drag can be at odds with the need for lift. Also, in piping networks, lowering pressure drop should be balanced with the requirement for sufficient heat transfer or structural resilience. In these situations, optimization gives engineers a logical way to navigate the trade-offs and create systems that are not just functional, but truly exceptional (Carbas et al. (2021)).

Thanks to the rapid advancement of computing capabilities, the frontiers of fluid optimization have been pushed forward. Moreover, high-fidelity computational fluid dynamics simulations, which were once inaccessible and reserved for academia, are now being used in industrial design. To alleviate this challenge, surrogate models and reduced-order approaches have been developed. These provide accurate approximations of fluid behavior at a fraction of the computational cost. These methods enable the exploration of vast and intricate design landscapes, making optimization both practical and scalable in real-world engineering contexts.

At the same time, data-driven approaches are reshaping the field in profound ways (Mendez et al. (2023)). The integration of artificial intelligence and machine

DOI: 10.1201/9781003664345-14

learning with physics-based models has given rise to hybrid methods. These methods combine the interpretability of governing equations with the adaptability of data-driven learning. Such approaches make it possible to perform real-time optimization, adaptive control, and predictive maintenance. For example, fluid systems can now adjust themselves dynamically in response to fluctuating operating conditions, thereby enhancing reliability, improving efficiency, and extending system lifespans. This shift represents a transformation from static and rigid designs toward systems that are intelligent, adaptive, and capable of evolving with their environments.

An equally important aspect of modern optimization frameworks is the consideration of uncertainty and robustness. Real systems are never free from variations. This is due to fluctuating operating conditions, manufacturing tolerances, or nonlinear instabilities. Designs that appear optimal in controlled simulations may fail when subjected to such variability. Robust optimization, therefore, emphasizes solutions that deliver consistently strong performance across a range of conditions rather than simply maximizing performance under ideal assumptions. This perspective bridges the gap between theoretical elegance and practical reliability. This also ensures that optimized designs succeed not just in theory but in application.

The purpose of this chapter is to weave together these various threads—mathematical principles, computational techniques, algorithmic strategies, and data-driven innovations—into a unified discussion on optimization in fluid dynamics. This chapter not only presents the theoretical foundations but also illustrates their application through case studies such as pipe flow systems, microfluidic devices, and large-scale heat exchangers. In doing so, it highlights how optimization simultaneously advances scientific understanding and drives technological innovation. For developing efficient, robust, and sustainable engineering solutions, incorporating optimization into the design and operation of fluid systems is essential. This establishes a foundation for the future generation of intelligent fluid technologies.

14.2 MATHEMATICAL FOUNDATIONS OF FLUID DYNAMICS

The study of how fluids flow is based on a few key rules. These rules tell how fluids move, share speed, and pass on power. The math looks simple, but the rules are thick and twist into each other, making them hard to work out in real work spots. So, knowing this math is key. It helps catch how fluids work and sets the limits for making things better. Following, we discuss some mathematical foundations involved in fluid dynamics.

14.2.1 Conservation of Mass

The continuity equation incorporates the mass conservation principle, ensuring that matter does not suddenly enter the system. For incompressible flows, the continuity equation is given by

$$\nabla \cdot \mathbf{u} = 0, \tag{14.1}$$

where **u** is the velocity field. This condition reports the balance between incoming and outgoing fluid in any given control volume. Figure 14.1 depicts the continuity equation for an incompressible fluid. Since no system in the real world violates the fundamental principle of mass conservation, the continuity equation helps ensure that proposed solutions are physical in the context of design optimization.

14.2.2 Conservation of Momentum

The Navier–Stokes equations embody the principles of fluid mechanics, and the equations' foundation lies in the conservation of momentum. These equations can be framed in the following form,

$$\rho\left(\frac{\partial u}{\partial t}+u\cdot\nabla u\right)=-\nabla\rho+\mu+\nabla^2 u+f \tag{14.2}$$

where ρ is the density of the fluid, p is the pressure, μ is the dynamic viscosity, and **f** is the body force. The term $u\cdot\nabla$u is a nonlinear convective term which makes the system analytically difficult and is responsible for turbulence, vortex shedding, and flow instabilities. In optimization problems, it is crucial to control the influence of this term. This is because in the case of small geometric changes or in the case of small operating variations, the flow behavior can be extremely different and unpredictable.

14.2.3 Conservation of Energy

Fluid systems are seldom controlled by momentum alone; thermal actions can be equally important. The energy equation expresses the retention and transport of heat in the fluid and can be written as

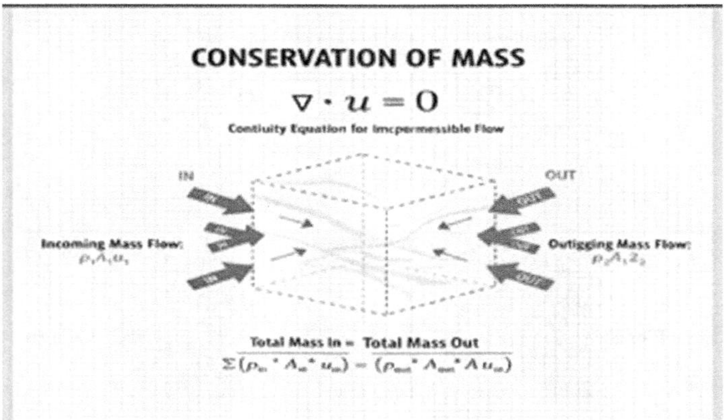

FIGURE 14.1 Conservation of mass.

$$\rho c_p\left(\frac{\partial T}{\partial t}+u\cdot VT\right)=k\nabla^2 T+\phi \tag{14.3}$$

where T is the temperature, c_p is specific heat capacity at constant pressure, k is thermal conductivity, and Φ is rate of internal heat generation per unit volume. This equation illustrates the interplay of thermal and momentum transport and explains why heat exchangers, cooled channels, and combustion systems pose multiphysics optimization problems. Such systems require the engineer's dilemma of maintaining hydrodynamic efficiency with optimal thermal performance, which significantly complicates the optimization problem (Figure 14.2).

14.2.4 Nonlinearity and Boundary Conditions

Even though the above governing equations are quite simple, they are nonlinear. Only certain idealized situations allow for closed-form analytical solutions with an actual derivation. In addition to the issues brought about by natural flow, the boundary conditions that must be rendered to solve for any given situation make matters worse. Boundary conditions exist as physical restraints that affect the ultimate solution. This can be seen in examples like the no-slip condition at a solid wall, arbitrary inlet velocities, or fixed outlet pressures. These boundary conditions exist as restraints and opportunities for adjustment and optimization. They constrain what can be worked with but also provide variables that can be manipulated for optimal efficiency.

14.2.5 Analytical and Numerical Approaches

Because the governing equations of fluid mechanics are nonlinear and coupled, there is usually no analytical solution for real-world applications. Instead, numerical solutions for practical application are found through finite volume, finite element, and spectral methods. These render the engineer capable of solving a flow field about

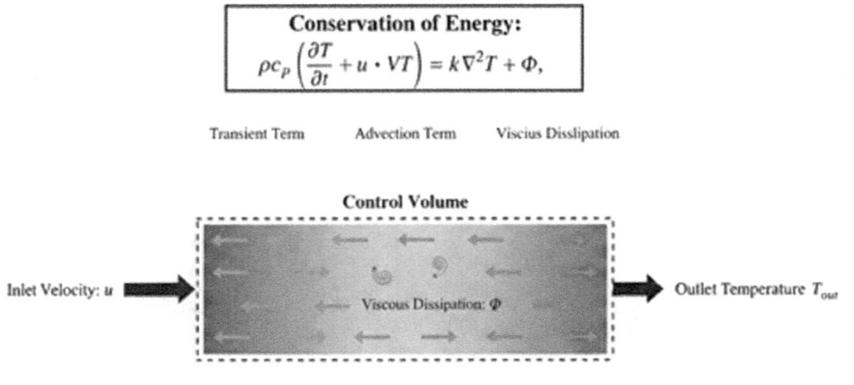

FIGURE 14.2 Diagram illustrating the principle of energy conservation.

an otherwise too complicated geometry, turbulent regime, or multiphase interaction. But even so, computational costs are intensive—more so when hundreds, if not thousands, of simulations need to be completed across varying design parameters for optimal exploration. Thus, surrogate models, reduced-order models, and machine learning approaches are employed to ease and accelerate design space exploration.

14.2.6 Role in Optimization

In optimization, the governing equations are the fluid's physics and the constraints that all feasible solutions must honor. Therefore, regardless of the design variables adjusted during the optimization process—geometry, boundary conditions, operating conditions—the governing equations will forever be present to guarantee that whatever is solved is still physically valid. The overlap of this conservation equation and the later solving procedure for fluid optimization fosters engineering advancements. This further translates what is learned theoretically into practically minded solutions for efficient performance, safety, and conservation (Figure 14.3).

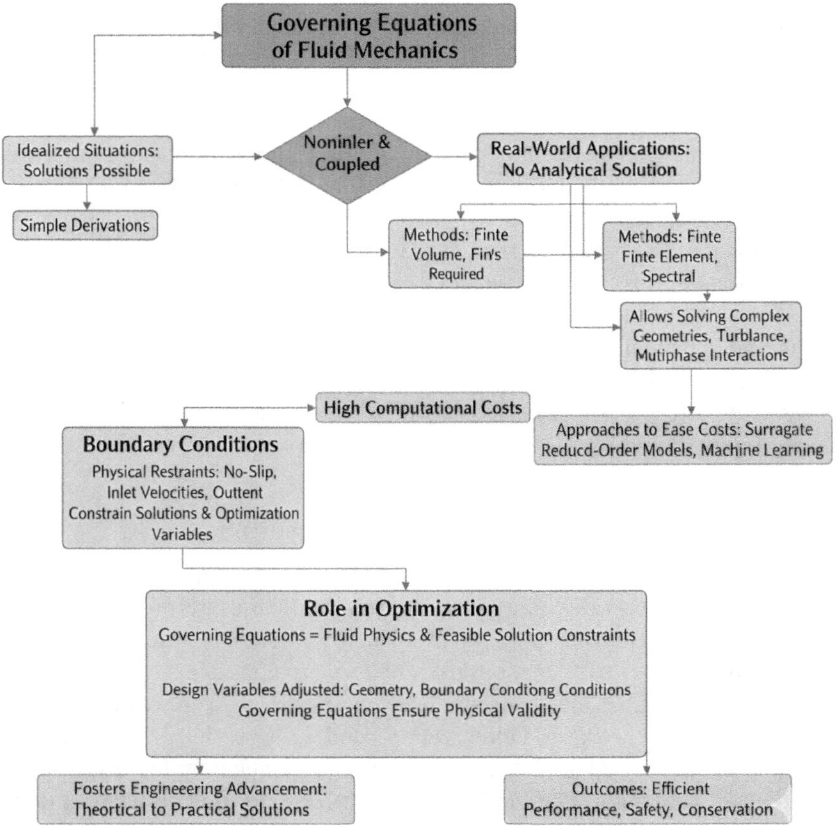

FIGURE 14.3 Governing equations of fluid mechanics.

14.3 OPTIMIZATION TECHNIQUES IN FLUID DYNAMICS

Optimization of fluid systems is not reliant upon one specific method. Numerous computational methods are available to determine solutions relative to the intricacies of the governing equations. These methods span from gradient-based applications to metaheuristic, evolutionary, and hybrid strategies (Mirjalili et al. (2023)). Each class of method operates uniquely and renders unique contributions relative to one another, creating an extensive toolbox for fluid system optimization. Following this, we discuss some of these methods.

14.3.1 Gradient-Based Methods

Gradient-based methods are among the most well-established optimization techniques. They perform well within a relatively tame, continuous problem. These techniques rely upon derivative information to help engineers progressively reach better designs. By determining sensitivities—how performance metrics change based on variations in design variables—engineers effectively find a path of least resistance upon which to access better results. As such, gradient-based optimization is also extremely effective in large design variable problems since there exists no random wandering as occurs with heuristic methods. Adjoint methods are a special case of these approaches. Adjoint formulations allow for sensitivity relative to thousands of design variables to be determined at a numerical cost. These variables are virtually independent of the number of design variables. In other words, the ability to determine what happens when shape changes, topology changes, or flow control strategies are applied that otherwise require exorbitant amounts of computational resources to determine are now feasible to be assessed. The range of applications in this category includes aerodynamic surface optimizations, drag reduction, and thermodynamic-fluid systems for energy-efficient use. However, these methods become problematic in turbulent, chaotic solutions where the solution is highly non-linear and gradients are not as reliable for convergence. When this happens, convergence occurs slowly or not at all, and local optima are selected instead of global ones.

14.3.2 Metaheuristic and Evolutionary Algorithms

Where the gradient-based approach falters, metaheuristic and evolutionary algorithms emerge successfully as a stable alternative. These are processes based on natural or social phenomena. Since no gradient information is needed, they are typically effective for objective functions that are discontinuous or excessively nonlinear. Some widely applicable metaheuristic or evolutionary algorithms include Genetic Algorithms, Particle Swarm Optimization, and Differential Evolution. Genetic Algorithms operate based on selection, crossovers, and mutations to create successive generations of a solution. Particle Swarm Optimization mimics socially motivated occurrences as particles interact with one another to stay on course within the search space. Differential Evolution allows solutions to be deduced from a large number of possibilities through recombination and mutation influence. These methods work well in global and complicated design spaces without premature convergence to local minima. They're commonly used in turbulence modeling, multi-parameter calibration, and flow separation control. The downside to such advantages, however, comes at a price. These metaheuristic approaches are highly convergent but require countless evaluations to settle on a solution, rendering them expensive in a computational sense when developed concurrently with high-fidelity CFD solutions. In addition, their viability often hinges on setting parameters (population size, mutation index, etc.) which complicate those applying the algorithms.

14.3.3 Hybrid Approaches

Recognizing the strengths and weaknesses of both gradient-based and metaheuristic methods, hybrid approaches have emerged as a promising middle ground. These strategies combine the precision and efficiency of gradient-based techniques with the global search capability of metaheuristics. For instance, a metaheuristic algorithm might be used to broadly explore the design space and identify promising regions. After this, a gradient-based method refines solutions to achieve higher precision. This layered approach makes optimization more robust and efficient, especially in challenging fluid dynamics problems where the design landscape is rugged or highly sensitive to small changes. A key enabler of hybrid methods is the use of surrogate modeling. Surrogate models—constructed using techniques such as Kriging (Gaussian processes), radial basis functions, or neural networks—approximate the behavior of complex fluid systems at a fraction of the computational cost. By replacing expensive CFD evaluations with faster surrogates, optimization can proceed more rapidly while still retaining sufficient accuracy for decision-making. In many frameworks, surrogates act as filters that guide the search, reserving high-fidelity simulations for final validation. Multi-fidelity approaches extend this concept further by combining models of varying resolution, balancing speed and accuracy dynamically during the optimization process.

14.3.4 A Unified Perspective

Taken together, these optimization techniques provide complementary strategies for improving fluid systems. Gradient-based methods excel when the problem is smooth and sensitivities are reliable. Metaheuristic and evolutionary algorithms are most valuable when the search space is rugged, nonlinear, or discontinuous. Hybrid approaches, supported by surrogate and multi-fidelity modeling, combine the advantages of both to achieve practical efficiency and robustness. In modern engineering practice, the choice is rarely between one method and another. Effective optimization frameworks often integrate multiple approaches to exploit their respective strengths. This convergence of methods reflects the broader evolution of the field, where flexibility and adaptability are just as important as mathematical rigor (Figure 14.4).

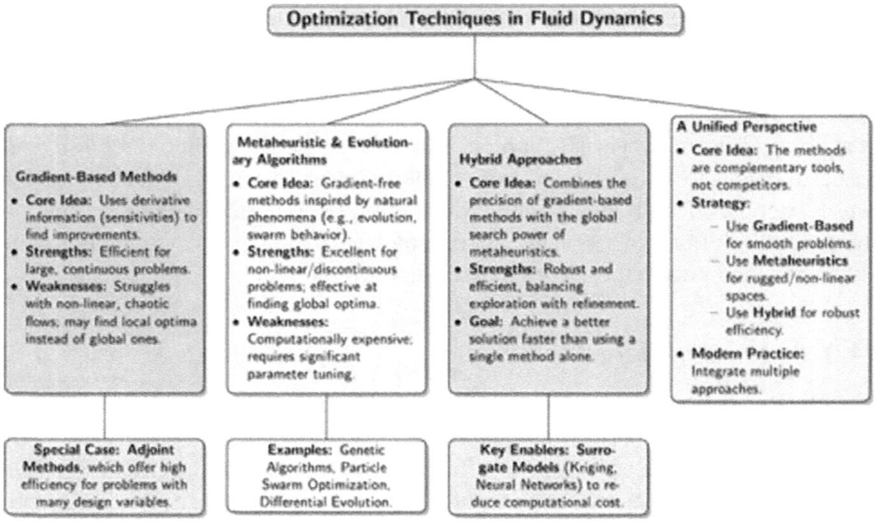

FIGURE 14.4 Optimization techniques in fluid dynamics.

14.4 SIMULATION FRAMEWORKS FOR OPTIMIZATION

The success of optimization in fluid dynamics depends heavily on the quality of the underlying simulation frameworks. These frameworks act as the bridge between abstract mathematical formulations and real engineering applications. They provide the computational environment in which fluid systems can be modeled, tested, and refined under different operating conditions. Optimization involves repeated evaluations across potentially thousands of design configurations. Due to this, the choice of simulation framework directly impacts feasibility, accuracy, and efficiency. Modern approaches to simulation can be broadly grouped into three categories: high-fidelity computational fluid dynamics (CFD) solvers, physics-informed machine learning (PIML) and reduced order modeling, and high-performance computing (HPC) infrastructures. Each of these plays a unique role, and together they form a comprehensive foundation for optimization workflows. These three categories are discussed below.

14.4.1 Computational Fluid Dynamics Solvers

Traditional CFD solvers remain the cornerstone of fluid simulation for optimization purposes (Joshi et al. (2022)). Well-established platforms such as OpenFOAM, ANSYS Fluent, and COMSOL provide robust environments for solving the Navier–Stokes equations and associated conservation laws across a wide variety of geometries and flow regimes. These solvers allow engineers to incorporate turbulence models, multiphysics interactions, and complex boundary conditions with a high degree of fidelity. Because of their maturity and validation across countless case studies, they are often used as the reference standard against which surrogate or reduced-order models are compared. However, their biggest limitation lies in their computational intensity. High-fidelity CFD simulations, particularly those involving turbulence or transient behavior, can require significant computational resources and long runtimes. This makes them challenging to integrate directly into optimization loops without additional strategies.

14.4.2 Physics-Informed Machine Learning and Reduced-Order Modeling

In recent years, data-driven methods have significantly expanded the landscape of simulation frameworks (Brunton et al. (2022)). PIML seeks to combine the predictive power of machine learning with the reliability of physical laws, embedding governing equations directly into neural network architectures. This results in models that are both flexible and physically consistent. Furthermore, these models can make fast predictions while respecting conservation principles. Alongside this, reduced-order modeling techniques compress high-dimensional CFD data into simplified but representative models. Methods such as Proper Orthogonal Decomposition (POD), Dynamic Mode Decomposition (DMD), and autoencoder-based neural networks allow engineers to capture the dominant flow structures and dynamics with fewer degrees of freedom. When integrated into optimization, these approaches dramatically reduce computational cost, making it possible to explore vast design spaces that would otherwise be inaccessible.

14.4.3 High-Performance Computing and GPU Acceleration

Even with advanced solvers and surrogate models, the computational demands of optimization remain immense. HPC has therefore become a critical enabler. Parallelization across large-scale clusters allows CFD simulations to be distributed, reducing runtimes, and enabling the analysis of highly complex systems. GPU-accelerated solvers further enhance this capability. They provide substantial speed-ups in cases where thousands of design evaluations are required. The integration of HPC with optimization algorithms has shifted CFD from being merely a diagnostic tool to an active component of design workflows. In practical terms, this means engineers can now perform optimization studies that were once considered infeasible. For example, optimizing large-scale networks of pipes, turbines, or heat exchangers with full 3D unsteady simulations becomes feasible.

These three directions—high-fidelity solvers, PIML and reduced-order models, and HPC—are not isolated. They increasingly converge into integrated frameworks. For instance, surrogate models are often trained on CFD data generated using HPC resources, and then deployed to accelerate optimization loops. Similarly, physics-informed neural networks can be embedded into solver environments, blending data-driven and physics-based approaches. This convergence reflects a broader trend in fluid dynamics optimization, the recognition that no single tool is sufficient. Progress lies in creating ecosystems where traditional CFD, machine learning, and advanced hardware complement each other. Such integration provides the scalability, adaptability, and precision needed to address modern engineering challenges (Figure 14.5).

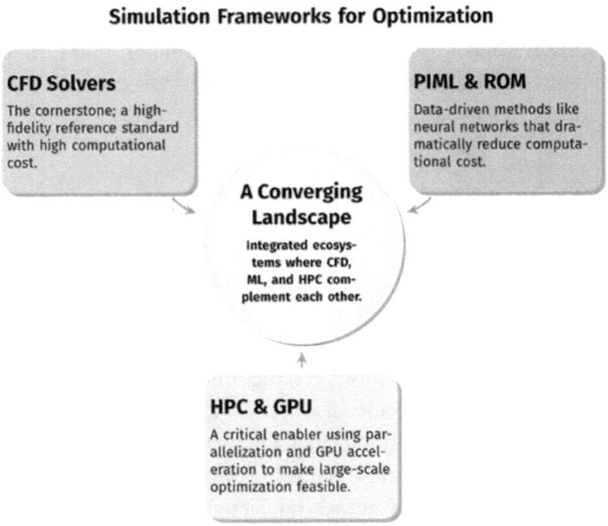

FIGURE 14.5 Simulation frameworks for optimization.

14.5 CASE STUDIES

To illustrate the role of optimization in fluid dynamics, this section presents three practical case studies. These cases span different scales and applications, from large-scale pipeline systems to microfluidic biomedical devices and industrial heat exchangers. Each case links to published Scopus-indexed studies and demonstrates how optimization bridges theoretical formulations with engineering practice.

14.5.1 Pipe Flow Systems

Pipelines are the backbone of water distribution, oil and gas networks, and process industries (Lal et al. (2023)). Their efficiency is strongly influenced by pressure losses, which are affected by pipe diameter, surface roughness, and routing. If the pipe diameter is too small, pumping costs rise due to frictional losses; if too large, material and installation costs escalate. Optimization provides a rational balance between these competing factors.

Scopus Study: Gunawan et al. (2021) reported a three-dimensional pipeline design optimization that reduced drag resistance and pressure drop by approximately 13%. Their findings emphasize that optimal diameter selection must also consider bends, crossings, and layout to achieve significant energy savings.

Worked Example: Consider a pipe of length $L = 100\,\text{m}$, volumetric flow rate $Q = 0.05\text{m}^3/\text{s}$, roughness $\epsilon = 5.0\times10^{-5}$ m, and water as the working fluid. The pressure drop can be modeled using Darcy–Weisbach, and pumping power is given by $P(D) = \Delta p(D)Q$. Defining a cost function

$$J(D) = C_e \frac{P(D)T}{1000} + C_p DL,$$

with energy cost $C_e = 0.08\$/\text{kWh}$, lifetime $T = 5$ years, and material cost coefficient $C_p = 1500$, a coarse grid search indicates an optimum at $D^* \approx 0.22$ m.

Interpretation: This example highlights the central trade-off (refer Figure 14.6), showing larger diameters reduce pressure losses but increase material costs. The Scopus study reinforces that in real-world settings, complete optimization should include geometry, routing, and operational parameters for maximum efficiency.

14.5.2 Microfluidic Networks

At the microscale, fluidic systems behave differently from macroscale flows. Microfluidic devices are crucial for applications such as targeted drug delivery and lab-on-a-chip diagnostics, where precision and reliability are vital. Because flows are laminar and dominated by viscous effects, channel geometry must be optimized to achieve uniform distribution, efficient mixing, and minimal pressure requirements.

Scopus Study: Santhanes et al. (2024) demonstrated that careful control of microfluidic flow conditions significantly enhanced the transfection capability of nanoparticle systems. Their work shows how tuning channel dimensions and flow ratios translates directly into biomedical performance improvements.

Worked Example: Referring to Figure 14.7, imagine a Y-junction with one inlet and two outlets of equal length. For equal flow division, channel widths w_1 and w_2 must satisfy

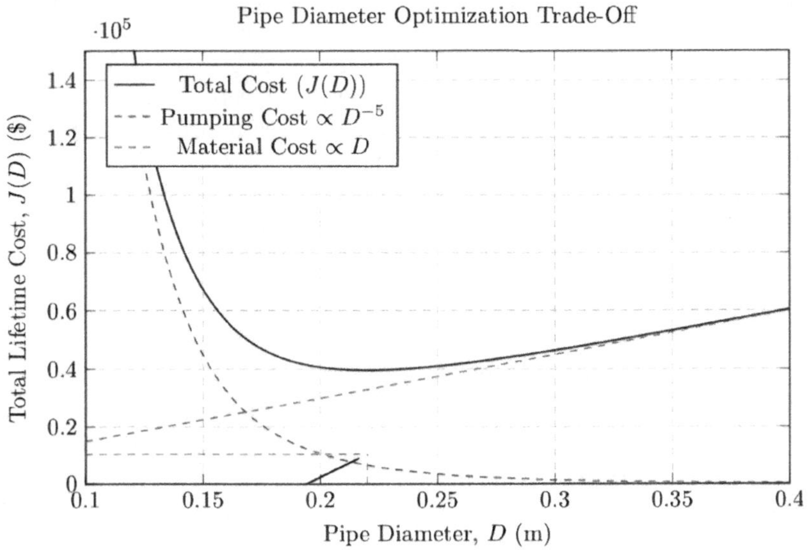

FIGURE 14.6 Pipe flow systems interpretation.

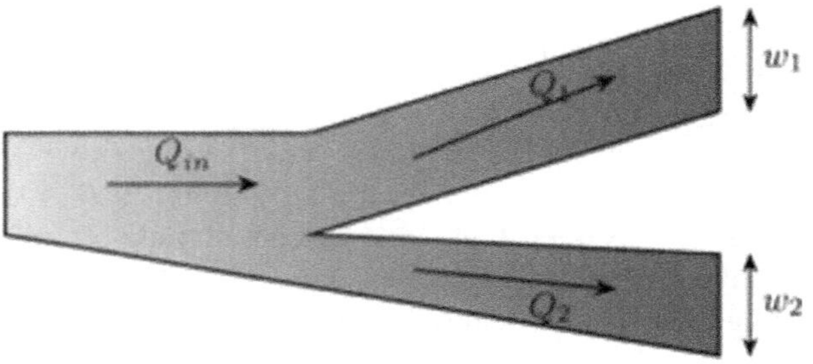

FIGURE 14.7 Microfluidic Y-junction.

$$\frac{L}{w_1^3} \approx \frac{L}{w_2^3}$$

Assuming fabrication limits between $50 \text{ and } 200\ \mu\text{m}$, optimization yields $w_1^* \approx w_2^* \approx 140\ \mu\text{m}$, which balances the flows and ensures precise delivery.

Interpretation: This example underscores how sensitive microfluidic systems are to geometric variations. Optimization provides a systematic path to achieve precise flow control, which is critical for biomedical reliability. The Scopus study confirms that such optimization can have a direct impact on therapeutic outcomes.

14.5.3 Heat Exchangers

Heat exchangers are key components in energy systems, refrigeration, and chemical processes. Designers must balance two conflicting objectives: maximizing heat transfer for thermal efficiency and minimizing pumping power to reduce operational costs. This inherently creates a multi-objective optimization problem.

Scopus Study: Ravanbakhsh et al. (2022) performed a multi-objective optimization of fins geometries in domestic water-heater exchangers. Their results identified Pareto-optimal designs, simultaneously improving heat transfer (via higher Nusselt numbers) while keeping pressure drops manageable.

Worked Example: Consider a shell-and-tube exchanger with tube diameter d, number of tubes N, and fin spacing s. Define an objective function:

$$J(d,N,s) = w_1 \frac{P_{pump}}{P_{ref}} - w_2 \frac{U}{U_{ref}}$$

where U is the overall heat transfer coefficient, and P_{pump} is pumping power. Using weights $w_1 = 0.4$ and $w_2 = 0.6$, an optimization search in practical ranges yields $d^* = 0.03,\quad N^* = 35,\quad s^* = 12\text{ mm}$. Figure 14.8 gives the trade-off curve between heat transfer and normalized pumping power.

Interpretation: Heat exchanger optimization does not result in one universally "best" solution, but rather a trade-off curve (Pareto front). The Scopus study emphasizes that engineers must choose based on application-specific priorities, whether

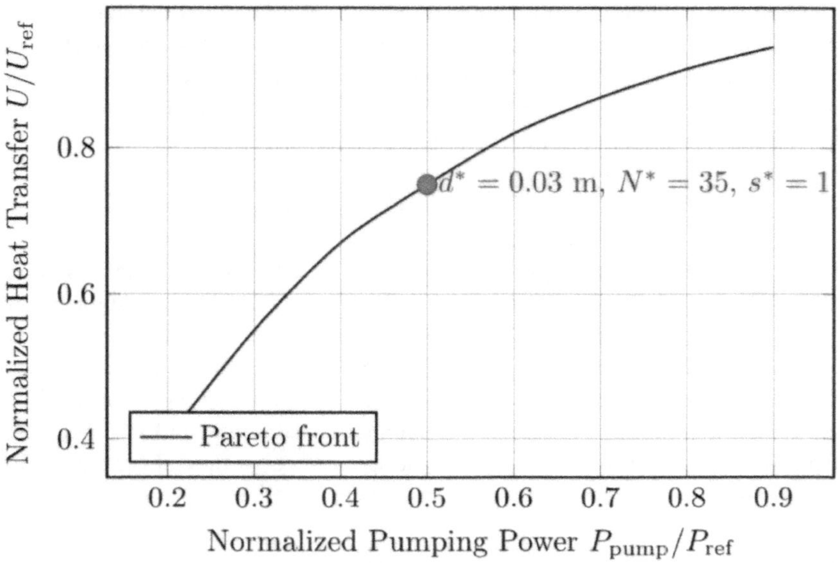

FIGURE 14.8 Trade-off curve (Pareto front) for heat exchanger optimization.

it is higher efficiency or lower pumping cost. Optimization thus makes trade-offs explicit and actionable.

14.6 PRACTICAL CHALLENGES AND REAL-WORLD CONSTRAINTS

When moving from theory and numerical examples into real-world optimization of fluid systems, engineers must grapple with several practical challenges. These arise because idealized models often ignore material limits, manufacturing realities, uncertain conditions, and complex flow behaviors. This section discusses three of the most critical constraints: material and manufacturability issues, uncertainty quantification and robust optimization, and flow instabilities and turbulence control. Each is a major factor that shapes what designs are feasible, not just what is theoretically optimal.

14.6.1 Material Restrictions, Manufacturability, and Cost Functions

Every optimization is bounded by what can actually be built. Materials have limits—strength, corrosion resistance, thermal expansion, surface finish—that affect not just performance but longevity. A pipe with a very smooth internal surface might reduce friction loss, but the material or treatment needed might be expensive, hard to maintain, or degrade rapidly in field conditions. Manufacturability includes geometric tolerances, fabrication methods, machining cost, and maintenance access. Cost functions used in optimization must include not just raw material cost, but fabrication, assembly, maintenance, and the cost of deviation from ideal geometry.

Example Scenario. Suppose an optimization of a large duct system for HVAC considers minimizing pressure drop by choosing very smooth internal surfaces (low roughness) and certain cross-sectional shapes. However, to achieve a very low roughness, special coatings or polishing may be required, which greatly increase cost. An adjusted cost function might thus include a penalty term:

$$J_{adj} = J_{hydraulic} + \alpha C_{material\ or\ fabrication}$$

where α reflects how strongly manufacturability or surface finish cost impacts the total design. If α is large, optimization may select a shape or roughness that is slightly less efficient hydraulically but much cheaper to manufacture and maintain.

Interpretation. Incorporating manufacturability and realistic material costs into optimization prevents designs that are ideal only in simulation but infeasible in practice. The best "optimal" design in a theoretical sense might fail early or cost too much. Real-world constraints shrink the design space, but they also ensure reliability, lower lifecycle cost, and greater industrial applicability.

14.6.2 Uncertainty Quantification (UQ) and Robust Optimization under Variable Operating Conditions

Fluid dynamics systems rarely operate under perfectly controlled conditions. Inflow rates, temperature, fluid properties, loadings, and boundary conditions can vary. Manufacturing imperfections and wear over time introduce additional uncertainty. Designs optimized only for a single "nominal" scenario may perform poorly under variation. To mitigate this, UQ methods are used to assess how sensitive performance metrics are to variations. Also, the robust optimization techniques seek solutions that perform well across a range of plausible conditions.

Relevant Book Resource. A useful reference is *Uncertainty Quantification of Stochastic Defects in Materials* (Chu, 2022), which, while focused on material defects, presents many of the same approaches: polynomial chaos, Monte Carlo, and stochastic modeling. These tools can be adapted to quantify uncertainty in fluid geometry, boundary conditions, or operating parameters.

Example Scenario. Consider a pipe flow optimization where inlet flow rate Q may vary by $\pm10\%$ and roughness ϵ is uncertain due to surface wear or scale buildup. Suppose we define the cost function as before, but want $J(D)$ to be small not just for nominal Q, ϵ, but on average across possible variations. One robust optimization formulation might be:

$$\min E\left[J(D;Q,\varepsilon)\right] + \beta\, Var\left[J(D;Q,\varepsilon)\right]$$

where $E[\cdot]$ is expectation, $Var[\cdot]$ is variance, and β sets how much we penalize unstable/high-variance designs. Optimized diameter under this formulation may shift away from the one selected by nominal-only optimization in favor of more conservative (less sensitive) designs.

Interpretation. Robust optimization builds in margin—it accepts slightly worse performance in the nominal case in exchange for greater stability across uncertainty. In many industrial fluid systems, that trade-off yields fewer failures, lower maintenance, and more predictable performance over time.

14.6.3 Flow Instabilities and Turbulence Intensity Control

In many applications, flows are not steady or laminar. Turbulence, flow separation, and instabilities can degrade performance, cause noise, vibration, or failure. But many optimization approaches assume simple physics (steady, laminar, ideal boundary conditions). To get realistic and usable designs, one must account for turbulent behavior, unsteady flows, and how control of turbulence or suppression of separation affects performance and durability.

Example scenario. For an airfoil or turbine blade, optimizing shape not only for aerodynamic drag in "clean" flow, but also controlling turbulence intensity near the surface, delaying separation, or reducing unsteady eddies, may lead to different optimal geometries. Suppose a cost objective includes both drag D_g and a turbulence penalty T_p which is proportional to turbulence intensity I. Then one might write:

$$J = w_1 D_g + w_2 T_p(I), \qquad T_p(I) = kI^2$$

where w_1, w_2 are weights. Optimizing shape under this expanded cost penalizes shapes that produce high turbulence (or suffer from early separation), leading to designs that are more reliable under real flow conditions.

Interpretation. Without accounting for turbulence or instabilities, a design that looks good in steady simulation may perform poorly in reality—increased drag, vibration, noise, and perhaps early structural failure. Including turbulence intensity or stability criteria makes optimization more aligned with what is physically and operationally realistic. It may reduce theoretical maximum performance, but increase robustness, service life, and real-world utility.

14.7 FUTURE DIRECTIONS

As optimization in fluid dynamics continues to mature, several emerging trends are poised to redefine what is possible. The following are likely to be critical in the coming years: real-time integration with AI, multi-scale and reduced-order modeling for very large or complex systems, and the rising role of digital twins for predictive optimization of fluid systems. Each of these directions holds promise but also raises new challenges. These are discussed in detail below.

14.7.1 Integration with AI and Machine Learning for Real-Time Optimization

The fusion of artificial intelligence (AI) and machine learning (ML) with fluid dynamics opens the door to real-time or near real-time optimization. Physics-informed neural networks (PINNs), for example, allow models that embed governing equations (e.g. Navier–Stokes) to learn flow dynamics directly from data. This enables predictions that satisfy physical constraints while handling noisy or incomplete information. In the future, optimization workflows might adapt dynamically as operating conditions evolve (e.g. changing flow rates, temperatures) with AI components guiding decision-making on the fly. This could be transformative for systems like smart HVAC, process plants, or adaptive aerodynamic surfaces.

14.7.2 Multi-Scale and Reduced-Order Modeling for Large-Scale Problems

Many practical fluid systems span multiple scales—spatially (from micro-channels to full systems), temporally (from fast transients to slow cycles), or in parameter space (e.g. varying fluid properties, loads, boundary conditions) (Iske et al. (2023)). Full high-fidelity simulation for all these scales is usually infeasible. Reduced-order models (ROMs) and multi-scale modeling provide tools to compress complexity. This helps in capturing essential dynamics with fewer degrees of freedom, coupling coarse and fine models, or using hierarchical model fidelity. As compute capability and data capture improve, these methods will allow optimization of huge systems (city-scale water networks, large energy plants, whole-wing aerodynamics) with speed and acceptable accuracy.

14.7.3 Digital Twins for Predictive Optimization of Fluid Systems

Digital twins are virtual representations of physical fluid systems that are continuously updated with data (Fitzgerald et al. (2024)). They combine live sensor input, simulation, and optimization to predict system behavior, alert to malfunctions or inefficiencies, and suggest adjustments. In fluid dynamics, digital twins could monitor flow networks, heat exchangers, pipelines, or even biomedical devices to proactively optimize for energy use, maintenance, or performance. The future points toward fluid systems that learn and adapt over time, reducing downtime, improving efficiency, and extending service life.

14.8 CONCLUSION

This chapter has shown that optimization in fluid dynamics provides a structured way to improve efficiency, stability, and design reliability. Starting from mathematical foundations and advanced algorithms to simulation frameworks, real-world case studies are discussed. The discussion highlighted how optimization moves beyond trial-and-error toward systematic decision-making. The examples of pipe networks, microfluidic devices, and heat exchangers demonstrated its versatility across scales. At the same time, challenges such as material limits, uncertainty, and turbulence control remind us that optimization must account for robustness and manufacturability, not just theoretical performance. Looking ahead, emerging tools such as machine learning, reduced-order models, and digital twins promise greater scalability and adaptability. In summary, optimization is a bridge between fluid dynamics theory and engineering practice—making fluid systems not only more efficient, but also more resilient and relevant to modern industry.

REFERENCES

Brunton, S. L., & Kutz, J. N. (2022). *Data-Driven Science and Engineering: Machine Learning, Dynamical Systems, and Control* (2nd ed.). Cambridge University Press.

Carbas, S., Toktas, A., & Ustun, D. (Eds.). (2021). *Nature-Inspired Metaheuristic Algorithms for Engineering Optimization Applications*. Springer Singapore.

Chu, L. (2022). *Uncertainty Quantification of Stochastic Defects in Materials*. CRC Press.

Fitzgerald, J., Gomes, C., & Larsen, P. G. (Eds.). (2024). *The Engineering of Digital Twins*. Springer Cham.

Gunawan, Y., Utomo, A. S. A., Fariz, F., & Lambang, S. A. (2021). *Reduction of Drag Resistance by Pressure Drop in Pipeline with 3-dimensional Design Optimization*. Scopus.

Iske, A., & Rung, T. (Eds.). (2023). *Modeling, Simulation and Optimization of Fluid Dynamic Applications*. Lecture Notes in Computational Science and Engineering, Vol. 148. Springer Cham.

Joshi, Y., & Kumar, S. (Eds.). (2022). *Computational Fluid Dynamics and Heat Transfer: Recent Advances and Applications*. Springer.

Lal, B., Bavoh, C. B., & Sayani, J. K. S. (2023). *Machine Learning and Flow Assurance in Oil and Gas Production*. Springer Cham.

Mendez, M. A., Ianiro, A., & Noack, B. R. (2023). *Data-Driven Fluid Mechanics*. Cambridge University Press.

Mirjalili, S., & Gandomi, A. H. (Eds.). (2023). *Comprehensive Metaheuristics: Algorithms and Applications*. Elsevier.

Ravanbakhsh, M., Deymi-Dashtebayaz, M., & Rezapour, M. (2022). Multi-objective optimization of three different fins of the heat exchangers used in domestic gas-fired water heaters: A hydrothermal performance and entropy generation analysis. *Journal of Thermal Analysis and Calorimetry*. Scopus.

Santhanes, D., Zhang, H., Wilkins, A., et al. (2024). Precise control of microfluidic flow conditions is critical for harnessing the in vitro transfection capability of pDNA-loaded lipid-Eudragit nanoparticles. *Drug Delivery and Translational Research*. Scopus.

Zhou, Q., Zhao, M., Hu, J., & Ma, M. (2023). *Multi-fidelity Surrogates: Modeling, Optimization and Applications*. Springer Singapore.

15 Thermal Optimization of a Magnetohydrodynamic U-Channel

Rupchand Malo, Krishno D. Goswami, and Anirban Chattopadhyay

15.1 INTRODUCTION

Thermogravitational convection has captivated the scientific community for decades owing to its myriad relevance, including solar energy, nuclear energy, electronic equipment efficiency, and aeronautics. Despite the diverse methodologies employed, the ultimate objective remains constant: managing fluid flow to enhance thermal efficiency for cooling or insulating purposes [1]. numerous approaches have been developed to upgrade devices' cooling performance, such as utilizing fins [2], obstacles [3], and magnetic fields. Researchers claimed that until the thermal conductivity of coolants is less, all these factors will fail to achieve optimum efficiency. To overcome such hurdles, they advocate for the adoption of nanofluids instead of usual heat transfer liquids (water, oil, ethylene glycol, etc.).

Nowadays, hybrid nanofluid is a trendy field among investigators in the application of engineering in almost every line of business, like microstructural, microelectric, medical, propulsion, mechanical, defense, aerospace, shipping, and shipbuilding. Suresh [4] first indicated the idea of hybrid nanofluids to encourage the significant characteristic of simple nanoparticles. In modern technology realms such as microelectronics and thermal control systems, there's an increasing demand for efficient energy transport and regulation. To address this, the integration of multiple thin baffles alongside hybrid nanofluids has emerged as a promising strategy. Thin baffles play a pivotal role in enhancing thermal management within confined spaces, making them indispensable in microelectronic and heat exchange applications. Consequently, the exploration of hybrid nanofluids [5] in conjunction with thin baffles has garnered significant attention, epitomizing a burgeoning research frontier.

Motivated by this, our study delves into the influence of multiple thin baffles on thermal regulation within enclosures filled with hybrid nanoliquid. Bilgen [6] explored thermal convection in a domain with a slim fin on the hot wall, noting a decrease in the Nusselt number with increased baffle length. Abedini et al. [7] examined the influence of MHD on thermal convection in a baffled C-structured device filled with Fe_3O_4-water nano composite, highlighting positive outcomes with

DOI: 10.1201/9781003664345-15

increased aspect ratio or particle concentration. Rahimi et al. [8] explored entropy generation in an H-structured domain using SiO_2-TiO_2/water-ethylene glycol nano composite with square inner blocks, while Armaghani et al. [9] worked free and entropy optimization in an L-type domain filled with Al_2O_3-water nanofluids using baffles. Pandit et al. [10] delved into the area of thermogravitational convection within a multi-baffled enclosure suffused with magneto-hybrid nanofluid.

Among the significant studies in this domain is the precious research made by Ma et al. [11], who thoroughly analyzed the innermost aspects of magnetohydrodynamic (MHD) nanofluid natural convection in a U-shaped cavity with baffles. Their detailed discussion revealed how changes in Rayleigh number and solid concentration influence mean Nusselt numbers. Similarly, Al-Farhany et al. [12] explored the impact of MHD on natural convection in a U-shaped cavity with saturated nanocomposite porous media, the insightful role of Rayleigh number, Darcy number, solid–fluid volume ratio, and Hartmann number on thermal convection. Keramat et al. [13] worked on thermal convection of alumina-water composite in an H-shaped cavity with a V-type baffle, establishing improvement in heat transfer to high Rayleigh numbers and solid–fluid volume fraction. Similarly, Zemani et al. [14] found the role of dimension ratio and baffle structure on thermal transportation within a U-type cavity filled with CuO water-based nanocomposite, showcasing the notable improvement of mean Nusselt numbers with rising Rayleigh numbers, solid–fluid concentration ratio, and aspect ratios, particularly in the presence of a T-shaped baffle. Moreover, Shao et al. [15] explored the entropy optimization of nanocomposite non-Newtonian fluid free convection in U-type geometry with a hot tree-type baffle, establishing interdynamics between natural convection dynamics and entropy generation.

Iftikhar et al. [16] analyzed through heatlines in magnetohydrodynamic (MHD) convection of ferrofluid in a U-shaped domain. Their findings exposed good improvement in streamline patterns and heat transfer in the presence of ferroparticles. Additionally, Pasha et al. [17] embarked on a multifaceted exploration, probing entropy and hydrothermal dynamics in nano-encapsulated phase change materials within a U-shaped enclosure replete with diverse baffle structures and corrugated walls. Their discerning analysis uncovered intriguing phenomena, such as Hartmann number-induced variations in streamline density, isotherms, and velocity profiles, while elucidating the nuanced impact of baffle length and corrugated wall height on local heat transfer rates, particularly in the magnetic field's presence. Similarly, Nabwey et al. [18] meticulously analyzed MHD thermal transportation in an inclined U-type geometry with Cu-water nanofluid, revealing intriguing trends in mean Nusselt number modulation with varying heat source parameters and Hartmann numbers, shedding light on the intricate interplay of transport phenomena in complex geometries.

Simultaneously, the notion of viscous dissipation arises as a fundamental element within the domain of fluid dynamics, taking on the function of an energy reservoir crucial for comprehending intricate fluid behaviors in various systems. Regardless of its fundamental relevance, viscous dissipation is frequently obscured in larger-scale geometries; nonetheless, its effects become clearly observable in domains marked by steep gradients. In a more profound inquiry, Makinde et al. [19] examined the influence of magnetohydrodynamic (MHD) flow in a permeable medium, observing a decline in heat convection rates attributed to both the magnetic field and the Eckert number. In a detailed investigation, Pandit et al. [20] analyzed the behavior of magnetic nanofluid flow within a rectangular framework, noted for its isotropic thermal diffusion and viscous dissipation. Their insightful examination revealed barriers to thermal diffusion resulting from viscous dissipation, along with a significant reduction in heat transfer rates as the values of viscous dissipation increased. In parallel, Ghalambaz et al. [21] explored the complex domain of nanofluid natural convection within a conventional apparatus filled with porous media, examining the simultaneous effects of radiation and viscous dissipation, which revealed fascinating variations in Nusselt number distributions along the walls of the apparatus, dependent on the Eckert number. Chandra Shekar et al. [22] analyzed the elaborate interactions of natural convection within a porous medium infused with nanofluid, influenced by magnetohydrodynamics (MHD) and viscous dissipation. Their research elucidated the profound impacts of magnetic fields, Brownian motion, and viscous dissipation on the intricate transport phenomena manifesting within the apparatus.

Meanwhile, Alsabery et al. [23] scrutinized the synergistic consequences of dissipation and radiation on flow physics within a slanting porous enclosure subjected to uniform heat flux. Their numerical analysis revealed that amplifying the magnetic force led to a diminution in the intensity of streamlines and isotherm patterns. Furthermore, they noted a significant enhancement in overall heat transmission with increasing values of viscous dissipation and radiation parameters. Recently, Miri et al. [24] conducted a recent investigation into entropy optimization and thermal transmission rates in MHD thermal convection of nanofluid, with a particular focus on the nuanced impacts of viscous dissipation, especially in the presence of a magnetic field. In a parallel endeavor, Tanveer et al. [25] investigated the territory of heat transfer analysis using a sophisticated Heat Transfer Neural Analysis methodology applied to a lid-driven rectangular enclosure filled with Bingham-Papanastasiou fluid. Their innovative approach, leveraging artificial neural networks, yielded precise predictions based on simulation data, underscoring the potential of advanced computational techniques in fluid dynamics research. At the same time, Divya et al. [26] embarked on a numerical exploration of MHD naturally convective flow through an exponentially accelerating plate. Their study revealed temperature variations driven by changes in the Eckert number, Dufour number, and Prandtl number, shedding light on the multifaceted nature of heat transfer phenomena in complex fluid systems.

The escalating scholarly interest in hybrid nanofluid natural convection across diverse geometrical configurations of devices, in conjunction with the ramifications of thin baffles and the energy dissipation attributable to viscosity, emphasizes a complex and multifaceted research landscape. Nonetheless, in the midst of this academic enthusiasm, a discernible lacuna in research endures, particularly regarding the implications of viscous dissipation in hybrid nanofluids contained within U-shaped devices that are equipped with corner baffles. While antecedent inquiries have primarily concentrated on the arrangement of baffles along the walls or within the flow field, our investigation presents a pioneering methodology: corner baffles deliberately situated at the lower corners of the U-shaped device. This avant-garde configuration signifies a notable deviation from traditional research frameworks and provides an intriguing pathway for further investigation within the scientific community. Although preceding studies have thoroughly scrutinized various dimensions of nanofluid dynamics, the comprehension of viscous dissipation influences within partially active U-shaped devices, characterized by corner baffles, remains strikingly underexplored.

This research endeavor seeks to fill this void by examining MHD natural convection thermal conduction within a U-typed-corner-baffled geometry, employing a water-based TiO_2-$CoFe_2O_4$ hybrid-nanofluid under varying orientations of the magnetic field ($0^0 \leq \gamma \leq 90^0$). We explore the effects of physical factors including Rayleigh number (Ra), Hartmann number (Ha), solid–fluid volume concentration (ϕ), and Eckert number (Ec) in the presence of baffles, examining their influence on streamlines, isotherms, and average Nusselt number. The outcomes of this investigation hold promise for informing the structure of electronic equipment and advancing our understanding of heat transfer phenomena in confined geometries.

15.2 MATHEMATICAL FORMULATION

15.2.1 Problem Statement

Our investigation delves into the intricacies of a 2D U-shaped device, delineated in Figure 15.1. The device contains a water-based TiO_2-$CoFe_2O_4$ hybrid-nanofluid, with strategically positioned baffles at the left-bottom and right-bottom positions. We explore four distinct geometrical configurations, indicated as Case-I, Case-II, Case-III, and Case-IV, each representing different arrangements of baffle placement at the bottom corners. The lower boundary is subjected to partial heating with non-dimensional heater length b ($= 0.25, 05, 0.75$ and 1), while the vertical walls, including the inner section of the U-shape and the baffles, remain maintained at a cool temperature. Additionally, an adiabatic condition characterizes the top wall, excluding the inner portion of the U-shape. Importantly, we treat the hybrid-nanofluid as a single-phase, Newtonian fluid, distinguished by its incompressible nature.

15.2.2 Attributes of the Fluid Medium

The nanoliquid deployed in our work comprises three primary components: water serving as the base fluid, and nano-sized particles of TiO_2 and $CoFe_2O_4$. The constant physicothermal properties [27] of these constituents are outlined in Table 15.2. Furthermore, Table 15.3 presents theoretical correlations [27, 28] relevant to the utilized fluid medium, providing a comprehensive overview of its key characteristics and behavior.

15.3 CONTROLLING EQUATIONS IN USUAL FORM

The fundamental physics governing the two-dimensional natural convection phenomenon, under the stipulated assumptions, are comprised of the Navier-Stokes equations and the heat equation. Articulated in the subsequent manner:

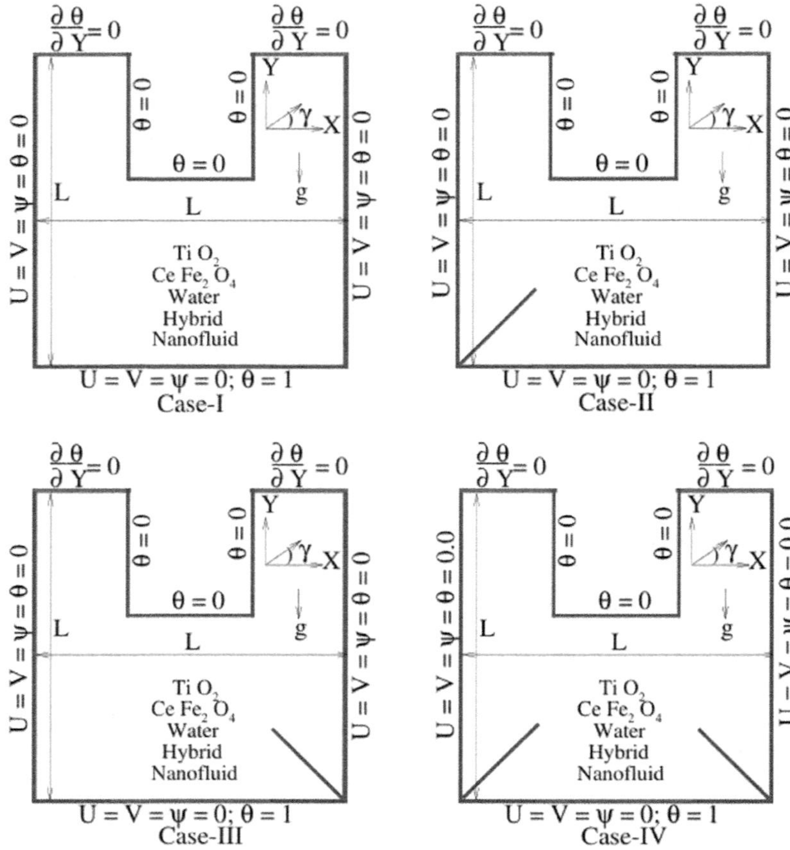

FIGURE 15.1 Physical model: Schematic representations illustrating the dimensional boundary conditions of the device filled with a water-based TiO-CoFeO hybrid-nanofluid.

$$\frac{\partial u}{\partial x}+\frac{\partial v}{\partial y}=0, \tag{15.1}$$

$$u\frac{\partial u}{\partial x}+v\frac{\partial u}{\partial y}=-\frac{1}{\rho_{\mathrm{hnf}}}\frac{\partial p}{\partial x}+\frac{\mu_{\mathrm{hnf}}}{\rho_{\mathrm{hnf}}}\nabla^2 u+\frac{\sigma_{\mathrm{hnf}}B^2}{\rho_{\mathrm{hnf}}}\left(v\sin\gamma\cos\gamma-u\sin^2\gamma\right), \tag{15.2}$$

$$u\frac{\partial v}{\partial x}+v\frac{\partial v}{\partial y}=-\frac{1}{\rho_{\mathrm{hnf}}}\frac{\partial p}{\partial y}+\frac{\mu_{\mathrm{hnf}}}{\rho_{\mathrm{hnf}}}\nabla^2 v+\frac{\sigma_{\mathrm{hnf}}B^2}{\rho_{\mathrm{hnf}}}\left(u\sin\gamma\cos\gamma-v\cos^2\gamma\right)$$

$$+\frac{(\rho\beta)_{\mathrm{hnf}}}{\rho_{\mathrm{hnf}}}g\left(T-T_c\right), \tag{15.3}$$

TABLE 15.1
Summary of Existing Literature and Contribution of the Present Study

Author(s)	Focus of Study	Key Contribution/Findings
Suresh [4]	Concept of hybrid nanofluids	First introduction of hybrid nanofluid idea to enhance thermal properties of conventional nanofluids.
Bilgen [6]	Convection in cavity with fin	Longer baffle length decreases Nusselt number.
Abedini et al. [7]	MHD convection in baffled C-device with Fe_3O_4-water	Heat transfer improved with higher aspect ratio and particle concentration.
Rahimi et al. [8]	Entropy in H-shaped cavity with hybrid nanofluid	Highlighted entropy generation trends with SiO_2-TiO_2/water-EG nanofluid.
Armaghani et al. [9]	Free/entropy optimization in L-shaped cavity with Al_2O_3 -water nanofluid	Role of baffles in enhancing thermal and entropy characteristics.
Pandit et al. [10]	Thermogravitational convection in multi-baffled enclosure with magneto-hybrid nanofluid	Effect of magnetic field and multiple baffles on convection.
Ma et al. [11]	MHD natural convection in U-shaped cavity with baffles	Rayleigh number and solid concentration significantly alter Nusselt numbers.
Al-Farhany et al. [12]	MHD convection in U-shaped cavity with porous nanocomposite	Impact of Rayleigh, Darcy, Hartmann number, and solid–fluid ratio on convection.
Keramat et al. [13]	Alumina-water in H-shaped cavity with V-baffle	Heat transfer improves at high Ra and solid–fluid fractions.
Zemani et al. [14]	CuO-water in U-shaped cavity with T-baffle	Aspect ratio and concentration strongly affect Nusselt number enhancement.
Shao et al. [15]	Entropy optimization with tree-type baffle	Interplay of entropy and natural convection.
Iftikhar et al. [16]	Heatlines in MHD ferrofluid convection	Ferro-particles improve streamlines and heat transfer.
Pasha et al. [17]	Nano-PCM in U-cavity with baffles and corrugations	Magnetic field and baffle/corrugation geometry affect streamline density and heat transfer.
Nabwey et al. [18]	MHD Cu-water nanofluid in inclined U-cavity	Heat source and Hartmann number strongly regulate Nusselt number.
Present study	Hybrid nanofluid convection in enclosure with multiple thin baffles and viscous dissipation	Examines combined influence of multiple thin baffles + hybrid nanofluid + viscous dissipation on heat transfer and entropy generation – a novel integrated analysis extending previous works.

$$u\frac{\partial T}{\partial x}+v\frac{\partial T}{\partial y}=\alpha_{\text{hnf}}\nabla^2 T+\frac{\mu_{\text{hnf}}}{(\rho C_p)_{\text{hnf}}}\phi, \tag{15.4}$$

where

$$\phi=\left\{2\left[\left(\frac{\partial u}{\partial x}\right)^2+\left(\frac{\partial v}{\partial y}\right)^2\right]+\left(\frac{\partial v}{\partial x}+\frac{\partial u}{\partial y}\right)^2\right\}. \tag{15.5}$$

15.3.1 Mathematical Formulation through Non-dimensional Form

To design the control equations in non-dimensional form, we employ the subsequent transformations:

$$X=\frac{x}{\text{L}},\ Y=\frac{y}{\text{L}},\ U=\frac{u\text{L}}{\alpha_{\text{f}}},\ V=\frac{v\text{L}}{\alpha_{\text{f}}},\ P=\frac{p\text{L}^2}{\rho_{\text{hnf}}\alpha_{\text{f}}^2},\ \theta=\frac{T-T_{\text{c}}}{T_{\text{h}}-T_{\text{c}}},\ Ra=\frac{g\beta_{\text{f}}\text{L}^3(T_{\text{h}}-T_{\text{c}})}{\nu_{\text{f}}\alpha_{\text{f}}},$$

$$Ha=B\text{L}\sqrt{\frac{\sigma_{\text{f}}}{\rho_{\text{f}}\nu_{\text{f}}}},\ Pr=\frac{\nu_{\text{f}}}{\alpha_{\text{f}}},\ Ec=\frac{\alpha_{\text{f}}^2}{\text{L}^2(C_p)_{\text{f}}(T_{\text{h}}-T_{\text{c}})}.$$

in Equations (15.1)–(15.5)
We get the following equations

$$\frac{\partial U}{\partial X}+\frac{\partial V}{\partial Y}=0, \tag{15.6}$$

$$U\frac{\partial U}{\partial X}+V\frac{\partial U}{\partial Y}=-\frac{\rho_{\text{f}}}{\rho_{\text{hnf}}}\frac{\partial P}{\partial X}+\frac{\mu_{\text{hnf}}}{\rho_{\text{hnf}}\nu_{\text{f}}}Pr\left(\frac{\partial^2 U}{\partial X^2}+\frac{\partial^2 U}{\partial Y^2}\right)$$
$$+\frac{\sigma_{\text{hnf}}}{\sigma_{\text{f}}}\frac{\rho_{\text{f}}}{\rho_{\text{hnf}}}Ha^2 Pr\left(V\sin\gamma\cos\gamma-U\sin^2\gamma\right) \tag{15.7}$$

TABLE 15.2
Thermophysical Properties of Water and Nanoparticles [11]

	ρ (kg·m^{-3})	μ (Pa·s)	c_p (J·kg^{-1}·K^{-1})	k (W·m^{-1}·K^{-1})	σ (Ω·m)$^{-1}$t	α(m^2·s^{-1})
Water	997.1	8.55×10^{-4}	4179	0.613	0.05	1.47×10^{-7}
TiO_2	425	–	686.2	8.9538	2.38×10^{6}	–
$CoFe_2O_4$	4960	–	700	3.7	5.51×10^{9}	–

TABLE 15.3
Theoretical Correlations for the Formulation of Hybrid Nanofluids Properties [11]

Nanofluid Properties	Nanofluids Model
Density	$\rho_{\text{hnf}} = (1-\phi_{\text{hnp}})\rho_{\text{f}} + \phi_{\text{Cu}}\rho_{\text{Cu}} + \phi_{\text{Al}_2\text{O}_3}\rho_{\text{Al}_2\text{O}_3}$
Heat capacitance	$(\rho c_{\text{p}})_{\text{hnf}} = (1-\phi_{\text{hnp}})(\rho c_{\text{p}})_{\text{f}} + \phi_{\text{Cu}}(\rho c_{\text{p}})_{\text{Cu}} + \phi_{\text{Al}_2\text{O}_3}(\rho c_{\text{p}})_{\text{Al}_2\text{O}_3}$
Thermal diffusivity	$(\alpha_{\text{hnf}}) = \dfrac{k_{\text{hnf}}}{(\rho C_{\text{p}})_{\text{hnf}}}$
Thermal expansion coefficient	$(\rho\beta)_{\text{hnf}} = (1-\phi_{\text{hnp}})(\rho\beta)_{\text{f}} + \phi_{\text{hnp}}(\rho\beta)_{\text{hnp}}$
Electrical conductivity	$\dfrac{\sigma_{\text{hnf}}}{\sigma_{\text{f}}} = 1 + \dfrac{3\left(\dfrac{\phi_{\text{Cu}}\sigma_{\text{Cu}} + \phi_{\text{Al}_2\text{O}_3}\sigma_{\text{Al}_2\text{O}_3}}{\sigma_{\text{f}}} - \phi_{\text{hnp}}\right)}{\left(\dfrac{\phi_{\text{Cu}}\sigma_{\text{Cu}} + \phi_{\text{Al}_2\text{O}_3}\sigma_{\text{Al}_2\text{O}_3}}{\sigma_{\text{f}}\phi_{\text{hnp}}} + 2\right) - \left(\dfrac{\phi_{\text{Cu}}\sigma_{\text{Cu}} + \phi_{\text{Al}_2\text{O}_3}\sigma_{\text{Al}_2\text{O}_3}}{\sigma_{\text{f}}} - \phi_{\text{hnp}}\right)}$
Effective thermal conductivity	$\dfrac{k_{\text{hnf}}}{k_{\text{f}}} = \dfrac{\dfrac{\phi_{\text{Cu}}k_{\text{Cu}} + \phi_{\text{Al}_2\text{O}_3}k_{\text{Al}_2\text{O}_3}}{\phi_{\text{hnp}}} + 2k_{\text{f}} + 2(\phi_{\text{Cu}}k_{\text{Cu}} + \phi_{\text{Al}_2\text{O}_3}k_{\text{Al}_2\text{O}_3}) - 2\phi k_{\text{f}}}{\dfrac{\phi_{\text{Cu}}k_{\text{Cu}} + \phi_{\text{Al}_2\text{O}_3}k_{\text{Al}_2\text{O}_3}}{\phi_{\text{hnp}}} + 2k_{\text{f}} - (\phi_{\text{Cu}}k_{\text{Cu}} + \phi_{\text{Al}_2\text{O}_3}k_{\text{Al}_2\text{O}_3}) + \phi k_{\text{f}}}$
Effective viscosity	$\dfrac{\mu_{\text{hnf}}}{\mu_{\text{f}}} = (1-\phi_{\text{hnp}})^{-2.5}$, where $\phi_{\text{hnp}} = \phi_{\text{Cu}} + \phi_{\text{Al}_2\text{O}_3}$

$$\begin{aligned} U\frac{\partial V}{\partial X} + V\frac{\partial V}{\partial Y} = &-\frac{\rho_{\text{f}}}{\rho_{\text{hnf}}}\frac{\partial P}{\partial Y} \\ &+\frac{\mu_{\text{hnf}}}{\rho_{\text{hnf}}\nu_{\text{f}}}Pr\left(\frac{\partial^2 V}{\partial X^2} + \frac{\partial^2 V}{\partial Y^2}\right) \\ &+\frac{\sigma_{\text{hnf}}}{\sigma_{\text{f}}}\frac{\rho_{\text{f}}}{\rho_{\text{hnf}}}Ha^2 Pr\left(U\sin\gamma\cos\gamma - V\cos^2\gamma\right) \\ &+\frac{(\rho\beta)_{\text{hnf}}}{\rho_{\text{hnf}}\beta_{\text{f}}}Ra\,Pr\,\theta, \end{aligned} \tag{15.8}$$

Energy equation

$$U\frac{\partial\theta}{\partial X}+V\frac{\partial\theta}{\partial Y}=\frac{\alpha_{\mathrm{hnf}}}{\alpha_{\mathrm{f}}}\left(\frac{\partial^2\theta}{\partial X^2}+\frac{\partial^2\theta}{\partial Y^2}\right)+\frac{\mu_{\mathrm{hnf}}}{\left(\rho C_p\right)_{\mathrm{hnf}}}\frac{\left(C_p\right)_{\mathrm{f}}}{\nu_{\mathrm{f}}}Ec\,Pr\left[2\left\{\left(\frac{\partial U}{\partial X}\right)^2+\left(\frac{\partial V}{\partial Y}\right)^2\right\}+\left(\frac{\partial V}{\partial X}+\frac{\partial U}{\partial Y}\right)^2\right]. \tag{15.9}$$

15.3.2 Streamlines – Vorticity

Using $U=\frac{\partial\psi}{\partial Y}$, $V=-\frac{\partial\psi}{\partial X}$ and $\zeta=\frac{\partial V}{\partial X}-\frac{\partial U}{\partial Y}$, from Equations (15.6)–(15.9), we get

$$\frac{\partial^2\psi}{\partial X^2}+\frac{\partial^2\psi}{\partial Y^2}=-\zeta. \tag{15.10}$$

$$U\frac{\partial\zeta}{\partial X}+V\frac{\partial\zeta}{\partial Y}=\frac{\mu_{\mathrm{hnf}}}{\rho_{\mathrm{hnf}}\nu_{\mathrm{f}}}Pr\left(\frac{\partial^2\zeta}{\partial X^2}+\frac{\partial^2\zeta}{\partial Y^2}\right)+\frac{\left(\rho\beta\right)_{\mathrm{hnf}}}{\rho_{\mathrm{hnf}}\beta_{\mathrm{f}}}Ra\,Pr\frac{\partial\theta}{\partial X}+\frac{\sigma_{\mathrm{hnf}}}{\sigma_{\mathrm{f}}}\frac{\rho_{\mathrm{f}}}{\rho_{\mathrm{hnf}}}Ha^2Pr\left[\left(\frac{\partial U}{\partial X}\sin\gamma\cos\gamma-\frac{\partial V}{\partial X}\cos^2\gamma\right)-\left(\frac{\partial V}{\partial Y}\sin\gamma\cos\gamma-\frac{\partial U}{\partial Y}\sin^2\gamma\right)\right]. \tag{15.11}$$

$$U\frac{\partial\theta}{\partial X}+V\frac{\partial\theta}{\partial Y}=\frac{\alpha_{\mathrm{hnf}}}{\alpha_{\mathrm{f}}}\left(\frac{\partial^2\theta}{\partial X^2}+\frac{\partial^2\theta}{\partial Y^2}\right)+\frac{\mu_{\mathrm{hnf}}}{\left(\rho C_p\right)_{\mathrm{hnf}}}\frac{\left(C_p\right)_{\mathrm{f}}}{\nu_{\mathrm{f}}}Ec\,Pr\left[2\left\{\left(\frac{\partial U}{\partial X}\right)^2+\left(\frac{\partial V}{\partial Y}\right)^2\right\}+\left(\frac{\partial V}{\partial X}+\frac{\partial U}{\partial Y}\right)^2\right]. \tag{15.12}$$

15.3.3 Dimensional Boundary Specifications

The dimensional boundary conditions manifest as follows:

$$\begin{aligned}
&\text{Left boundary:} && u = v = 0, T = T_c, && \\
&\text{Right boundary:} && u = v = 0, T = T_c, && \\
&\text{Bottom boundary:} && u = v = 0, T = T_h && \text{(for the heating segment),} \\
& && T = T_c && \text{(for the non-heating segment),} \\
&\text{Top boundary:} && u = v = 0, \frac{\partial T}{\partial y} = 0 && \text{(in absence of U-shape).}
\end{aligned} \quad . (15.13)$$

On the U-shaped section:

$$\begin{aligned}
\text{Left boundary: } & u = v = 0, T = T_c, \\
\text{Right boundary: } & u = v = 0, T = T_c, \\
\text{Bottom boundary: } & u = v = 0, T = T_c,
\end{aligned} \tag{15.14}$$

On the baffle(s):

$$u = v = 0, T = T_c$$

15.3.3.1 Non-dimensional Boundary Conditions

With these boundary conditions implemented on the domain boundaries without dimension are:

$$\begin{aligned}
\text{Left boundary: } & U = V = \theta = \psi = 0,\ \zeta = -\frac{\partial^2 \psi}{\partial X^2}, \\
\text{Right boundary: } & U = V = \theta = \psi = 0,\ \zeta = -\frac{\partial^2 \psi}{\partial X^2}, \\
\text{Bottom boundary: } & U = V = \psi = 0,\ \zeta = -\frac{\partial^2 \psi}{\partial Y^2},\ \theta = 1 \text{ (for the heating segment),} \\
& \theta = 0 \text{ (for the non-heating segment),} \\
\text{Top boundary: } & U = V = \psi = 0,\ \frac{\partial \theta}{\partial Y} = 0,\ \zeta = -\frac{\partial^2 \psi}{\partial Y^2} \text{ (Without U-shaped).}
\end{aligned} \tag{15.15}$$

On the U-shaped

$$\begin{aligned}
\text{Left boundary: } & U = V = \psi = 0, \theta = 0,\ \zeta = -\frac{\partial^2 \psi}{\partial X^2} \\
\text{Right boundary: } & U = V = \psi = 0, \theta = 0,\ \zeta = -\frac{\partial^2 \psi}{\partial X^2}, \\
\text{Bottom boundary: } & U = V = \psi = 0, \theta = 0,\ \zeta = -\frac{\partial^2 \psi}{\partial Y^2}.
\end{aligned} \tag{15.16}$$

On the baffle(s)

$$U = V = \psi = \theta = 0, \ \zeta = -\frac{\partial^2 \psi}{\partial X^2} - \frac{\partial^2 \psi}{\partial Y^2}.$$

15.3.4 Heat Transfer Activity

The qualitative thermal efficiency for the considered problem is quantified along the bottom boundary by the local Nusselt number (Nu) [11, 28, 29, 30] defined by Fourier's heat flux as:

$$\mathrm{Nu} = -\left(\frac{k_{\mathrm{hnf}}}{k_{\mathrm{f}}}\right)\left(\frac{\partial \theta}{\partial Y}\right)_{Y=0}, \tag{15.17}$$

and the mean (Nu_{av}) [11, 28] at the thermo-active bottom boundary is specified as

$$\mathrm{Nu}_{\mathrm{av}} = \int_{\text{Heat source}} \mathrm{Nu}\, dX. \tag{15.18}$$

15.4 NUMERICAL PROCEDURE

We have employed a fourth order accurate compact finite difference scheme [31, 32, 33, 34] to handle the governing equations by considering a general second-order semilinear equation as follows:

$$a\frac{\partial^2 \phi}{\partial x^2} + g\frac{\partial^2 \phi}{\partial x \partial y} + b\frac{\partial^2 \phi}{\partial y^2} + c\frac{\partial \phi}{\partial x} + d\frac{\partial \phi}{\partial y} + p\phi = f\left(x, y, \frac{\partial \theta}{\partial x}\right). \tag{15.19}$$

The discretized form of Equation (15.19) as follows:

$$\begin{aligned}
&\textstyle\sum_{k_1=-1}^{1} \sum_{k_2=-1}^{1} w_{i+k_1, j+k_2} \phi_{i+k_1, j+k_2} \\
&\quad = 24F_{i,j} + g_{i,j}\Bigg[\frac{1}{k}\Big\{2(\phi_x)_{i,j-1} - 2(\phi_x)_{i,j+1} - (\phi_x)_{i-1,j-1} \\
&\quad\quad - (\phi_x)_{i+1,j-1} + (\phi_x)_{i+1,j+1} + (\phi_x)_{i-1,j+1}\Big\} \\
&\quad\quad + \frac{1}{h}\Big\{ 2(\phi_y)_{i-1,j} - 2(\phi_y)_{i+1,j} - (\phi_y)_{i-1,j-1} + (\phi_y)_{i+1,j-1} \\
&\quad\quad + (\phi_y)_{i+1,j+1} - (\phi_y)_{i-1,j+1}\Big\}\Bigg]
\end{aligned} \tag{15.20}$$

The system of equations (15.20) can be written in matrix form as

$$M\phi = l \tag{15.21}$$

where the coefficient matrix M is an asymmetric sparse matrix. For a grid of size m× n, M is of size mn×mn and ϕ and l are mn component vectors.

We have used the BiCGStab [35] algorithm for the computation.

The tolerance criterion is

$$\max\left| \chi_{i,j}^{(\epsilon+1)} - \chi_{i,j}^{(\epsilon)} \right| < 0.5\times10^{-5}, \tag{15.22}$$

where, $\chi_{i,j}^{(\epsilon+1)}$ and $\chi_{i,j}^{(\epsilon)}$ quantitatively signify $\psi_{i,j}$, $\zeta_{i,j}$ or $\theta_{i,j}$ at consecutive steps.

15.4.1 Authentication and Grid Independence

Table 15.4 presents the mesh sensitivity analysis for Nu_{av} and $|\psi|$max, considering possible grid structures (31 × 31, 61 × 61, and 121 × 121) for fixed parameters ϕhnp $= 0.04$, $Ha = 40$, $Ra = 10^4$, and $\gamma = 45^0$. In this analysis the deviations Δ of the obtained results in various grids with respect to grid size 61 × 61 is measured. The results indicate that the grid size of 61 × 61 provides optimal computational efficiency.

A comparison with results from Ma et al. [11] for various scenarios, including different heat source positions and Rayleigh numbers, is depicted in Figure 15.2. The obtained results demonstrate excellent agreement, further affirming the accuracy of the computational model.

15.5 RESULTS AND DISCUSSION

Having meticulously validated the computational code and ensured grid independence, we proceed to delve into the thermodynamic phenomena within a domain filled with a water-based TiO_2-$CoFe_2O_4$ hybrid-nanofluid. Throughout our analysis, we systematically vary parameters such as Ra, Ha, γ, Ec, and ϕ, spanning ranges of $10^3 \le Ra \le 10^5$, $0 \le Ha \le 60$, $0^0 \le \gamma \le 45^0$, $0 \le Ec \le 0.03$, and $0 \le \phi \le 0.04$, based on literatures [1, 2] to comprehensively understand hybrid-nanofluid behavior. The flow patterns and thermal behavior are rigorously analyzed through visualization of streamlines (ψ), isotherms (θ), and average Nusselt numbers (Nu_{av}).

In Figure 15.3, we embark on a nuanced exploration of the interplay between thin corner baffle(s) and the Hartmann number (Ha), unraveling their collective influence on the steady-state behavior of streamline patterns within the U-shaped device. While maintaining $Ra = 10^5$, $\gamma = 45^0$, $Ec = 0.001$, and $\phi_{hnf} = 0.04$ constant, we meticulously analyze the intricate shifts in flow dynamics induced by the presence of the baffle(s). Our investigation spans across four specific configurations (Case-I to Case-IV), each representing varying arrangements of baffle placement within the

TABLE 15.4

Errors Associated with the Average Nusselt Number (Nu_{av}) and $|\psi|_{max}$ on the Grids at $Ra = 10^4$, $Ha = 40$, $\gamma = 45^0$, $b = 0.5$, $\phi = 0.04$ and Two Corner Cold Baffles

Ec	Grids	Nu_{av}	$\Delta = \frac{\left\|Nu_{av_{i\times j}} - Nu_{av_{61\times61}}\right\|}{Nu_{av_{61\times61}}} \times 100\%$	$\|\psi\|_{max}$	$\Delta = \frac{\left\| \|\psi\|_{max_{i\times j}} - \|\psi\|_{max_{61\times61}} \right\|}{\|\psi\|_{max_{61\times61}}} \times 100\%$
	31 × 31	4.345	0.7%	0.000806	2.7%
$Ec = 0$	61 × 61	4.315	0%	0.000785	0%
	121 × 121	4.338	0.5%	0.000782	0.4%
	31 × 31	4.339	0.7%	0.000807	2.8%
$Ec = 0.001$	61 × 61	4.309	0%	0.000785	0%
	121 × 121	4.331	0.5%	0.000782	0.4%
	31 × 31	4.326	0.7%	0.000811	3.2%
$Ec = 0.003$	61 × 61	4.296	0%	0.000786	0%
	121 × 121	4.318	0.5%	0.000783	0.4%
	31 × 31	4.313	0.7%	0.000814	3.6%
$Ec = 0.005$	61 × 61	4.283	0%	0.000786	0%
	121 × 121	4.304	0.5%	0.000784	0.3%

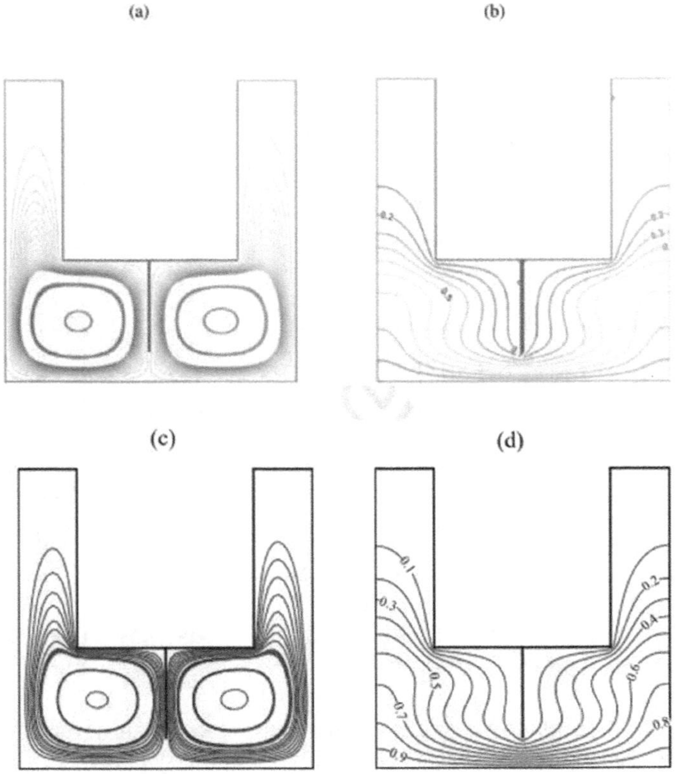

FIGURE 15.2 Comparison of streamlines (a, c) and isotherms (b, d) between Ma et al. [11] (a, b) and the present work (c, d) for $Ra = 10^5$, $Ha = 20$, $AR = 0.6$, and $\phi = 0.05$..

device. In Case-I, where no baffle is present, the flow field demonstrates two distinct counter-rotating circulations for all values of the Hartmann number (Ha), consistent with classical convection patterns under the influence of a magnetic field. These circulations are primarily driven by the buoyancy forces resulting from temperature and concentration gradients within the hybrid nanofluid.

As Ha increases, the Lorentz force induces resistive effects that counteract the buoyancy-driven flow. This not only diminishes the primary circulation strength but also reduces the buoyancy pressure gradient responsible for driving the flow. Consequently, the velocity of the fluid decreases, and the strength of the circulations weakens, facilitating the formation of secondary vortices near the baffles. These secondary vortices signify localized regions of diminished flow activity, which directly correlate with reduced thermal transport efficiency within the enclosure. Consequently, a decrease in $\psi_{\max}$ values is observed with increasing Ha, highlighting the diminishing intensity of flow dynamics under stronger magnetic influences. Specifically, an increase in Ha from 0 to 60 leads to a reduction in $\psi_{\max}$ values by up to 7.56% for Case-I, 4.91% for Case-II, 4.42% for Case-III, and 2.36% for Case-IV.

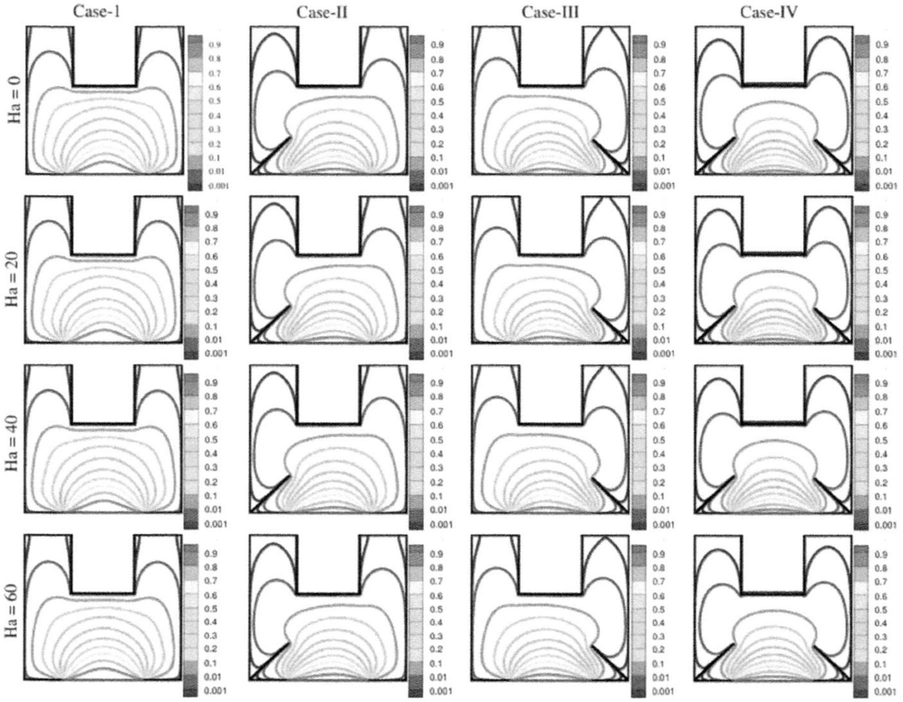

FIGURE 15.3 Streamlines illustrating varying values of the Hartmann number (Ha) across the four cases under study, while maintaining fixed parameters of $Ra = 10^5$, $\gamma = 45°$, $b = 0.5$, $\phi = 0.04$, and $Ec = 0.001$.

These findings reinforce the entangled interplay among baffle placement, magnetic strength, and flow mechanics, elucidating their collective consequences on convective thermal propagation within the U-shaped domain.

Figure 15.4 presents the isotherm profiles corresponding to the four cases under consideration, offering a comprehensive insight into the thermal behavior of the U-type chamber. As the Hartmann number (Ha) varies, the isolines representing thermal layers dynamically shift, reflecting the evolving thermal distribution within the enclosure. Notably, with increasing Ha, the density of the isotherm contours close to the boundaries subjected to fixed temperatures decreases, indicative of a reduction in temperature gradients across the domain. Interestingly, at higher Ha values, the dominance of the conduction heat transfer mechanism becomes more pronounced, overshadowing thermal activity within the device. Consequently, the collective thermal performance of the enclosure diminishes, resulting in a downturn decline in convective heat transfer efficiency. Moreover, the introduction of thin corner baffle(s) from Case-II to Case-IV acts as an impediment to energy transfer within the device, exacerbating the reduction in convective heat transfer. Furthermore, the isotherm distribution highlights distinct thermal characteristics within the device.

In Figure 15.5, we embark on a nuanced exploration of the interplay between thin corner baffle(s) and the heater length (b), unraveling their collective influence on

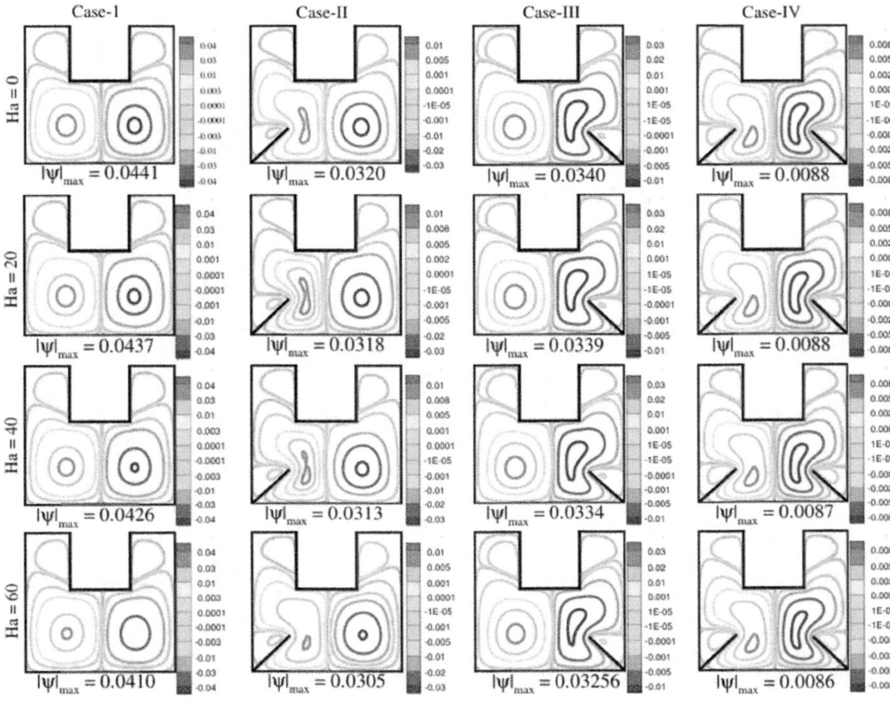

FIGURE 15.4 Isotherms illustrating different Hartmann numbers (*Ha*) across the four cases under investigation, while keeping parameters fixed at $Ra = 10^5$, $\gamma = 45°$, b = 0.5, ϕ = 0.04, and $Ec = 0.001$.

the steady-state behavior of streamline patterns within the U-shaped device. While maintaining $Ra = 10^5$, $\gamma = 45^0$, $Ec = 0.001$, $Ha = 30$ and $\phi_{\text{hnf}} = 0.04$ constant, we meticulously analyze the intricate shifts in flow dynamics induced by the presence of the baffle(s).

In Figure 15.5, we undertake a nuanced examination of the intricate interplay between thin corner baffle(s) and the length of the heating element (b), specifically exploring values of $b = 0.25, 0.5, 0.75$, and 1.0. Through this analysis, we endeavor to unravel their collective influence on the steady-state behavior of streamline patterns within the U-shaped device. It is worth noting that amidst this investigation, we rigorously maintain the constancy of several critical parameters: $Ra = 10^5$, $\gamma = 45^\circ$, $Ec = 0.001$, $Ha = 30$, and $\phi_{\text{hnf}} = 0.04$.

The profound impact of the thin corner baffle(s) on temperature distributions is vividly elucidated in Figure 15.6. This visual depiction provides a compelling narrative of how the presence of the corner baffles intricately shapes the steady-state isotherms across varying heater lengths, as previously expounded in the context of fluid flow patterns. Functioning as disruptive elements within the U-shaped enclosure, the corner baffle(s) wield a profound influence on the thermal energy transfer dynamics from the heated bottom wall to the adjacent cold side walls. Consequently,

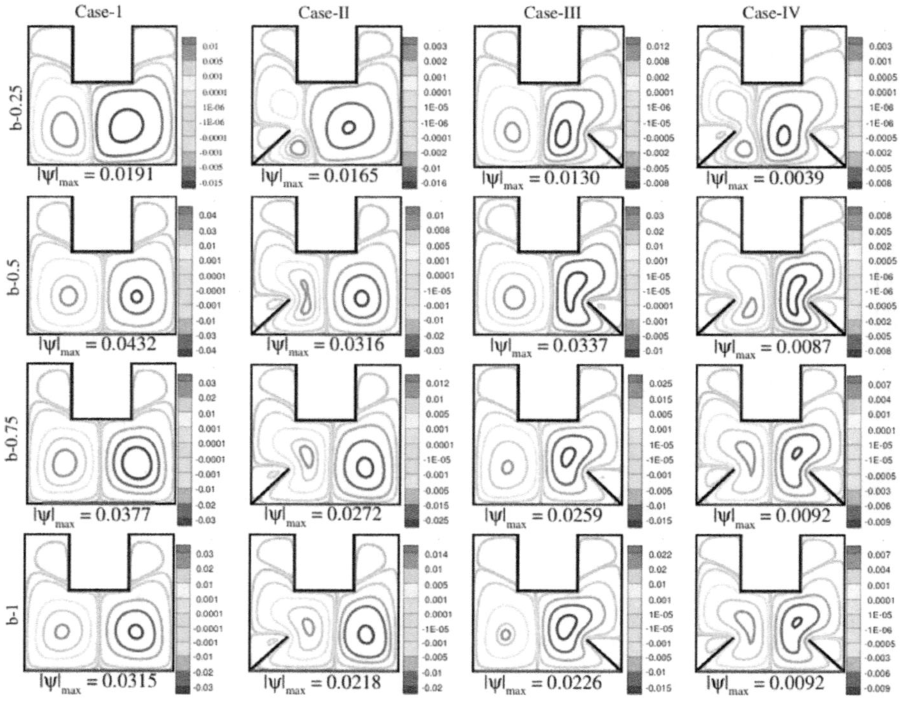

FIGURE 15.5 Streamlines depicting various heater lengths (b) for the four cases under investigation, with fixed parameters of $Ha = 30$, $\gamma = 45°$, $Ra = 10^5$, $\phi = 0.04$, and $Ec = 0.001$.

distinct variations in the isotherm distribution emerge within the upper region of the U-shaped device, independent of heater length. The pronounced clustering of isotherms near the heated bottom wall indicates a region of concentrated heat flux. This phenomenon arises from the introduction of thermal energy at the bottom surface, where the imposed temperature gradient drives buoyancy-induced convection. The localized thermal input causes a strong upward heat transfer, leading to the formation of closely spaced isotherms as heat propagates through the fluid. The dense isotherm clustering near the heated wall reflects the dominance of conductive heat transfer in this region, while the upward movement of heat due to buoyancy contributes to a stratified thermal field in the upper domain, illustrating the combined effects of conduction and convection in shaping the thermal profile within the enclosure.

15.5.1 Thermal Performance: Average Nusselt Number

- **Effect of Rayleigh Number and Hartmann number on Thermal Performance across Four Cases:**

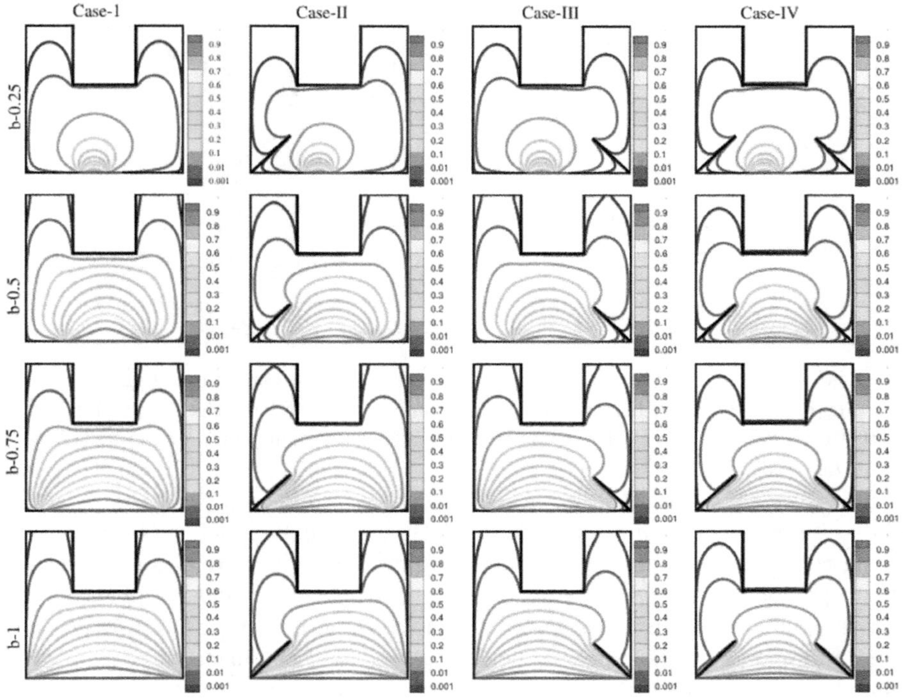

FIGURE 15.6 Isotherms illustrating different heater lengths (b) for the four cases under investigation, with fixed parameters of $Ha = 30$, $\gamma = 45°$, $Ra = 10^5$, $\phi = 0.04$, and $Ec = 0.001$.

In this segment, we delve into the effect of thin corner baffles on convective performance within the U-shaped enclosure. Our analysis focuses on understanding how variations in the buoyancy force and magnetic force impact the mean Nusselt number across four distinct scenarios, as illustrated in Figure 15.7. To elucidate the interrelation between the Rayleigh number (Ra) and the Nusselt number, we depict the ratio of the mean Nusselt number at $Ra = 10^3$ to that at $Ra = 10^3, 10^5$ for each of the four cases in Figure 15.7(a). Remarkably, with the escalation of the Rayleigh number from 10^4 to 10^5, a conspicuous downward trajectory becomes evident in the Nusselt number ratio, manifesting as a decrease of 170.45% in Case-I, 57.30% in Case-II, 59.86% in Case-III, and 18.09% in Case-IV, across all four scenarios in the presence of viscous dissipation. This outcome resonates harmoniously with the conclusions drawn in [22]. Out of the four considerations, Case-IV, characterized by the presence of a double baffle configuration, exhibits the highest heat transfer rate. Transitioning to the impact of the applied magnetic field on the Nusselt number, we consider the ratio between the mean Nusselt number with ($Ha \neq 0$) and mean Nusselt number without the effect of magnetic field ($Ha = 0$), Nusselt number ratio $Nu_{av}\left(\dfrac{Ha \neq 0}{Ha = 0}\right)$ depicted in Figure 15.7(b) for the four distinct cases. In both Case-II and Case-III, a conspicuous upward trajectory unfolds in the Nusselt number ratio, culminating in

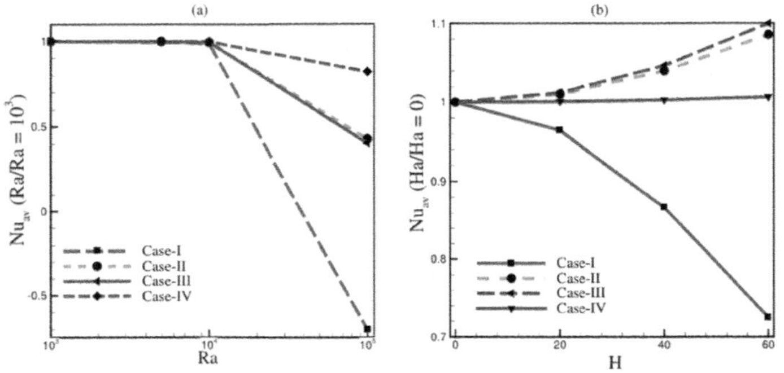

FIGURE 15.7 Average Nusselt number ratio for the four cases under investigation, with (a) varying Rayleigh number (Ra) at fixed parameters $Ha = 30$, $\phi = 0.04$, b = 0.5, $\gamma = 45°$, and $Ec = 0.001$, and (b) varying Hartmann number (Ha) at fixed parameters $Ra = 10^5$, $\gamma = 45°$, $\phi = 0.04$, $Ec = 0.001$, and $b = 0.5$.

a discernible increase of 8.53% in Case-II and 9.97% in Case-III, concomitant with the ascent of the Hartmann (Ha) number from 0 to 60. This trend can be ascribed to the pronounced effects of the applied magnetic field, which induce enhanced fluid motion and convective heat transfer within the enclosure. As the Hartmann number (Ha) increases, the magnetic field exerts a stronger influence on the flow patterns, primarily through the Lorentz force, which acts to suppress chaotic convective motions and stabilize the flow.

- **Effect of magnetic field orientation and nanoparticle's concentration on Thermal Performance across Four Cases:**

In this segment, our focus centers on a nuanced exploration of the interplay between two key variables: the magnetic field orientation (γ) and the solid–fluid volume ratio (ϕ), and their individual influence on the mean Nusselt number. The intricacies of this relationship are expounded across four distinct scenarios, meticulously delineated in Figure 15.8. To explicate the intricate relationship between the magnetic field orientation (γ) and the Nusselt number, we delve into the ratio between the mean Nusselt number with horizontal orientation ($\gamma \neq 0$) and non-horizontal orientation ($\gamma = 0$) of the applied magnetic field, Nusselt number ratio $Nu_{av}\left(\frac{\gamma \neq 0}{\gamma = 0}\right)$ depicted in Figure 15.8(a) for the four distinct cases. A meticulous examination unveils a conspicuous pattern as γ progresses from 0° to 90°. Spanning across Case-II, Case-III, and Case-IV, a notable descent unfolds in the Nusselt number ratio, exhibiting a decline of up to 0.89% in Case-II, 0.97% in Case-III, and 0.08% in Case-IV. This intriguing trend highlights a decreasing relationship between the magnetic field inclination angle (γ) and the Nusselt number, reflecting the complex role that magnetic field orientation plays in shaping convective heat transfer dynamics within the system.

Transitioning to the examination of the influence exerted by nanoparticle concentration (ϕ) on the Nusselt number, it becomes essential to analyze the ratio of the

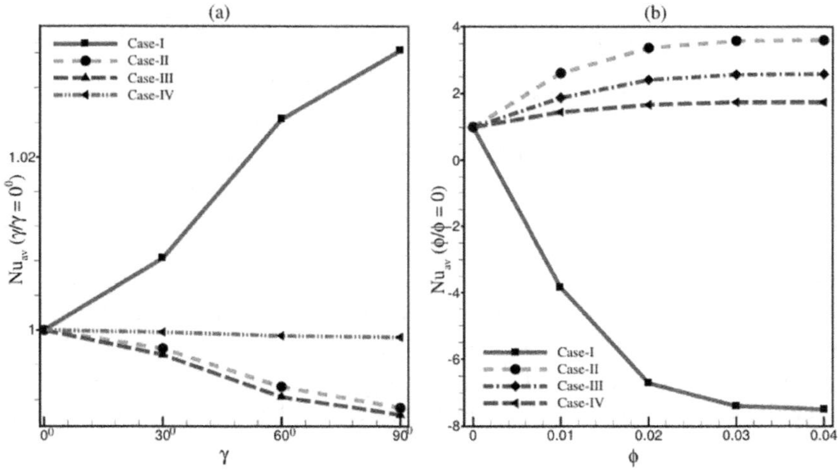

FIGURE 15.8 Average Nusselt number ratio for the four cases under investigation, with (a) varying γ at fixed parameters $Ra = 10^5$, $Ha = 30$, $\phi = 0.04$, $b = 0.5$, and $Ec = 0.001$, and (b) varying ϕ at fixed parameters $Ra = 10^4$, $Ec = 0.001$, $b = 0.5$, and $\gamma = 45^\circ$.

Nu_{av} for a non-zero nanoparticle volume fraction ($\phi \neq 0$) in relation to the Nu_{av} of the base fluid ($\phi = 0$), which is referred to as the Nusselt number ratio $Nu_{av}\left(\dfrac{\phi \neq 0}{\phi = 0}\right)$. This evaluation, illustrated in Figure 15.8(b) for the four unique scenarios, provides significant insights into the system's behavior. The introduction of thin baffle(s) results in considerable modifications, as demonstrated in Figure 15.8(b). Across all three scenarios specifically, Case-II, Case-III, and Case-IV noticeable ascending trend in the Nu_{av} ratio is recorded as the nanoparticle volume fraction ϕ increases from 0 to 0.04. This positive trend can be linked to the improved convective heat transfer enabled by the nanoparticles dispersed within the fluid medium. As the nanoparticle volume fraction rises, the thermal conductivity of the fluid-nanoparticle amalgamation enhances, culminating in more effective heat transfer.

- **Impact of heater size (b) and viscus dissipation (Ec) on thermal execution:**

In this segment, we concentrate on an extensive analysis of the interaction between two pivotal variables: the length of the heater (b) and the viscous dissipation parameter (Ec), alongside their respective effects on the mean Nusselt number. We investigate the nuances of this correlation across four distinct scenarios, comprehensively depicted in Figure 15.9 under the conditions of $Ha = 40$, $\gamma = 45^0$, $Ra = 10^5$, and $\phi_{hnf} = 0.04$. The pronounced impact of the introduced baffles is clearly demonstrated in Figure 15.9. To clarify the intricate relationship between the heater length (b) and the Nusselt number, we examine the ratio between the mean Nusselt number with non-zero heater length ($b \neq 0$) and the mean Nusselt number with a constant length

($b = 0.25$), referred to as the Nusselt number ratio $Nu_{av}\left(\dfrac{b \neq 0.25}{b = 0.25}\right)$, as illustrated in Figure 15.9(a) for the four distinct conditions. Through careful examination, a notable increasing trend is discerned as the heater length b progresses from 0 to 1. In all three scenarios Case-II, Case-III and Case-IV a significant upward trend is evident in the Nusselt number ratio. This behavior can be attributed to the augmented surface area available for heat transfer as the heater length is lengthened. With an extended heater length, there is an increased contact area between the heated surface and the surrounding fluid, thereby facilitating more effective heat transfer. Additionally, the extended heater length may modify the flow patterns within the system, potentially improving convective heat transfer processes.

Now, considering the role of viscous dissipation on the Nusselt number, it's vital to examine the ratio between the average Nusselt number for $Ec \neq 0$ and the average Nusselt number for $Ec = 0$, termed as the Nusselt number ratio $Nu_{\mathrm{av}}\left(\dfrac{Ec \neq 0}{Ec = 0}\right)$. This result, discussed in Figure 15.9(b) for the four cases, provides significant insights into the transport phenomena. The introduction of thin baffle(s) induces significant alterations, as illustrated in Figure 15.9(b). Spanning across all four cases, namely, Case-I, Case-II, Case-III, and Case-IV, a conspicuous downward trend is observed in the Nusselt number ratio. This trend entails a decrease of 36.34% in Case-I, 17.83% in Case-II, 15.70% in Case-III, and 5.23% in Case-IV, across all four scenarios under the influence of viscous dissipation, as the Eckert (Ec) number ascends from 0 to 0.03. This trend suggests that an increase in the viscous dissipation parameter leads to a reduction in the mean Nusselt number. The presence of viscous dissipation introduces additional energy losses within the fluid, thereby hindering convective

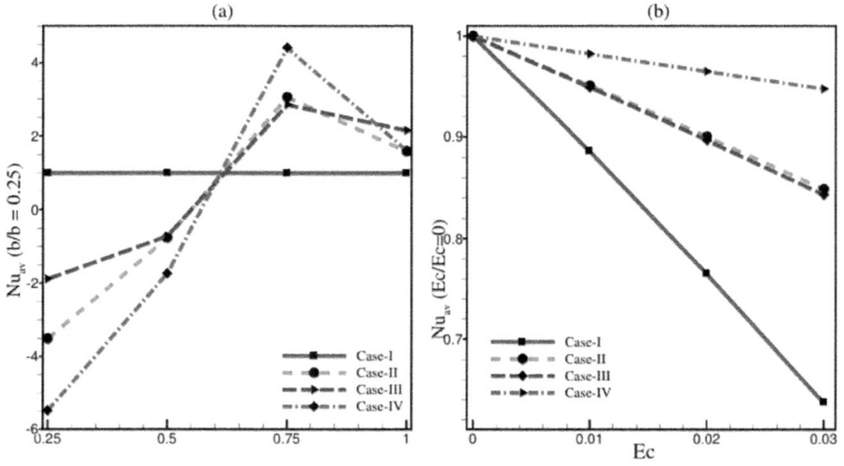

FIGURE 15.9 Average Nusselt number ratio for the four cases under investigation, with (a) varying heater length b at fixed parameters $Ra = 10^5$, $Ha = 30$, $\phi = 0.04$, and $Ec = 0.001$, and (b) varying Eckert number (Ec) at fixed parameters $Ra = 10^4$, $Ha = 30$, $\phi = 0.04$, $b = 0.5$, and $\gamma = 45°$.

heat transfer. As the *Ec* number rises, the viscous forces become increasingly dominant, causing a decrease in the overall heat transfer efficiency. It's noteworthy that Case-IV, characterized by the presence of two corner baffles, yields the maximum Nusselt number ratio, followed by Case-III, Case-II, and Case-I.

15.6 CONCLUSIONS

In the course of this research, we have conducted an extensive investigation into the complex interactions of thermogravitational convection and thermal transport within a U-shaped enclosure that incorporates deliberately positioned corner baffle(s). The chamber was filled with a hybrid nanocomposite of water-TiO_2-$CoFe_2O_4$, and four unique geometrical configurations (Case-I, Case-II, Case-III, and Case-IV) were analyzed, each featuring varying arrangements of baffle structures at the lower corners. Our thorough examination probed the interrelations of several critical parameters, including the Rayleigh number, Hartmann number, Eckert number, the angle of magnetic field inclination, and the solid–fluid concentration of the hybrid nanofluid. From our computational simulations, several significant results were obtained:

(a) The use of a U-type system with corner baffles shows a promising approach for optimizing thermal processes. This configuration offers a versatile framework for enhancing heat transport in applications such as heat exchangers and thermal management systems in electronic devices.
(b) The phenomenon of viscous dissipation was identified as a factor that obstructs the thermal diffusion mechanism, resulting in a reduction of the average Nusselt number for all examined scenarios as the Eckert number escalated.
(c) The inclusion of corner baffles was determined to modulate the interplay between the imposed magnetic field and the fluid dynamics, underscoring their crucial function in affecting thermal transfer.
(d) Across Case-II, Case-III, and Case-IV, an upward trend in the Nusselt number ratio was observed as the nanoparticle volume fraction ϕ increased from 0 to 0.04.
(e) The combination of larger heater sizes (*b*) with corner baffles significantly improves heat transfer efficiency. Larger heaters increase the surface area available for heat exchange, while corner baffles enhance fluid mixing and prevent dead zones, optimizing thermal performance.
(f) Higher nanoparticle concentration values contributed to increased thermal transfer efficiency across all cases.
(g) The inclination angle of the magnetic field was identified as a key factor in strengthening fluid flow circulation, thus influencing heat transfer dynamics within the system.

In summary, our comprehensive research presents notable understanding of the complex dynamics of thermogravitational convection and thermal transmission in

U-shaped devices with corner baffles. These findings not only advance our fundamental understanding of heat transfer processes but also offer practical implications for the framework and refinement of thermal management systems in large-scale engineering applications.

REFERENCES

1. A. Chattopadhyay, K. D. Goswami, S. K. Pandit, and H. F. Öztop, "Exploring the thermal behavior and hydrothermal characteristics of ferrofluid in a complex enclosure under magnetic field and porous media: Influence of surface roughness," *Numeric. Heat Transfer Part A: Appl.* p. 129, 2024. DOI: 10.1080/10407782.2024.2353346.
2. K. D. Goswami, A. Chattopadhya and S. K. Pandit, "Magneto-thermogravitational convection for hybrid nanofluid in a novel shaped enclosure," *Int. J. Mech. Sci.* vol. 234, p. 107674, 2022. DOI: 10.1016/j. ijmecsci.2022.107674.
3. S. K. Pandit, K. D. Goswami, A. Chattopadhyay and H. F. Öztop, "On the analysis of magnetohydrodynamics and magnetic field-dependent viscosity effects on thermo-gravitational convection of hybrid nanofluid in an enclosure with curved walls," *Phys. Fluids.* vol. 33, p. 102010, 2021.
4. S. Suresh, K. Venkitaraj, P. Selvakumar and M. Chandrasekar, "Synthesis of Al_2O_3â€"Cu/water hybrid nanofluids using two step method and its thermo physical properties," *Colloid. Surface* vol. 388, pp. 41–48, 2011. DOI: 10.1016/j.colsurfa.2011.08.005.
5. H. Sayyou, J. Belabid, H. F. Oztop and K. Allali, "Effects of vibration on natural convection in a square inclined porous enclosure filled with Cu-water nanofluid," *Int. J. Numeric. Meth. Heat Fluid Flow* vol. 34, no. 6, pp. 2399–2428, 2024. DOI: 10.1108/HFF-01-2024-0074.
6. E. Bilgen, "Natural convection in cavities with a thin fin on the hot wall," *Int. J. Heat Mass Transf.* vol-48, no. 17, pp. 3493-–3505, 2005. DOI: 10.1016/j.ijheatmasstransfer.2005.03.016.
7. A. Abedini, T. Armaghani and A. J. Chamkha, "MHD free convection heat transfer of a water-Fe_3O_4 nanofluid in a baffled C-shaped enclosure," *J. Therm. Anal. Calor.* vol. 135, pp. 685–695, 2019. DOI: 10.1007/s10973-018-7225-8.
8. A. Rahimi, M. Sepehr, M. J. Lariche, M. Mesbah, A. Kasaeipoor and E. H. Malekshah, "Analysis of natural convection in nanofluid-filled H-shaped cavity by entropy generation and heatline visualization using lattice Boltzmann method," *Physica E: Low-dimen. Systems and Nanostructures* vol. 97, pp. 347–362, 2018. DOI: 10.1016/j.physe.2017.12.003.
9. T. Armaghani, A. Kasaeipoor, N. Alavi and M. M. Rashidi, "Numerical investigation of water-alumina nanofluid natural convection heat transfer and entropy generation in a baffled L-shaped cavity," *J. Mol. Liq.* vol. 223, pp. 243–251, 2016. DOI: 10.1016/j.molliq.2016.07.103.
10. S. K. Pandit, A. Chattopadhyay, R. Malo and K.D. Goswami, "Thermogravitational convection in a multiple baffled enclosure filled with Magneto-hybrid nanofluid subjected to magnetic field dependent viscosity," *J. Nanofluids* vol. 12, no. 7, pp. 1698–1719, 2023. DOI: 10.1166/jon.2023.2051.
11. Y. Ma, R. Mohebbi, M. M. Rashidi, Z. Yang and M. A. Sheremet, "Numerical study of MHD nanofluid natural convection in a baffled U-shaped enclosure," *Int. J. Heat and Mass Transfer* vol. 130, pp. 123–134, 2019. DOI: 10.1016/j.ijheatmasstransfer.2018.10.072.

12. K. Al-Farhany, A. Abdulkadhim, H. K. Hamzah, F. H. Ali and A. Chamkha, "MHD effects on natural convection in a U-shaped enclosure filled with nanofluid-saturated porous media with two baffles," *Progs. Nuclear Energy* vol. 145, pp. 104136, 2022. DOI: 10.1016/j.pnucene.2022.104136.
13. F. Keramat, P. Dehghan, M. Mofarahi and C. H. Lee, "Numerical analysis of natural convection of alumina-water nanofluid in H-shaped enclosure with a V-shaped baffle," *J. Taiwan Instit. Chemic. Eng.* vol. 111, pp. 63–72, 2020. DOI: 10.1016/j.jtice.2020.04.006.
14. F. Zemani, O. Ladjedel and A. Sabeur, "Simulation of CuO-water nanofluid natural convection in a U-shaped enclosure with a T-shaped baffle," *J. Eng. Appl. Sci.* vol. 70, no. 99, p. 99, 2023. DOI: 0.1186/s44147-023-00257-x.
15. W. Shao, M. K. Nayak, S. El-Sapa, A. J. Chamkha, N. Shah and A. M. Galal, "Entropy optimization of non-Newtonian nanofluid natural convection in an inclined U-shaped domain with a hot tree-like baffle inside and considering exothermic reaction," *J. Taiwan Inst. Chemic. Eng.* vol. 148, p. 104990, 2023. DOI: 10.1016/j.jtice.2023.104990.
16. B. Iftikhar, M. A. Siddiqui and T. Javed, "Computational analysis of heat transfer via heatlines for MHD natural convection ferrofluid flow inside the U-shaped cavity," *Europ. Phys. J. Plus* vol. 138, no. 164, p. 164, 2023. DOI: 10.1140/epjp/s13360-023-03769-w.
17. A. A. Pasha, M. K. Nayak, K. Irshad, M. M. Alam, A. S. Dogonchi, A. J. Chamkha and A. M. Galal, "Entropy and hydrothermal analyses of nano-encapsulated phase change materials within a U-shaped enclosure: Impact of diverse structures of baffles and corrugated wall," *J. Energy Stor.* vol. 59, p. 106532, 2023. DOI: 10.1016/j.est.2022.106532.
18. H. A. Nabwey, A. M. Rashad, W. A. Khan and S. I. Alshber, "Effectiveness of magnetize flow on nanofluid via unsteady natural convection inside an inclined U-shaped cavity with discrete heating," *Alex. Eng. J.* vol. 61, no. 11, pp. 8653–8666, 2022. DOI: 10.1016/j.aej.2022.02.010.
19. O. D. Makinde, W. A. Khan and J. R. Culham, "MHD variable viscosity reacting flow over a convectively heated plate in a porous medium with thermophoresis and radiative heat transfer," *Int. J. Heat Mass Transfer* vol. 93, pp. 595–604, 2016. DOI: 10.1016/j.ijheatmasstransfer.2015.10.050.
20. S. K. Pandit and R. Malo, "Thermogravitational magnetonanofluid flow in a square cavity with isotropic heat flow field and viscous dissipation: A compact approach," *Comput. Math. Appl.* vol. 143, pp. 417–434, 2023. DOI: 10.1016/j.camwa.2023.05.019.
21. M. Ghalambaz, M. Sabour and I. Pop, "Free convection in a square cavity filled by a porous medium saturated by a nanofluid: Viscous dissipation and radiation effects," *Eng. Sciens Tech. Int. J.* vol. 19, no. 3, pp. 1244–1253, 2016. DOI: 10.1016/j.jestch.2016.02.006.
22. B. C. Shekar, C. Haritha and N. Kishan, "Magnetohydrodynamic convection in a porous square cavity filled by a nanofluid with viscous dissipation effects. Proceedings of the Institution of Mechanical Engineers," *Part E: J.f Process Mech. Eng.* vol. 233, no. 3, pp. 474–488, 2019. DOI: 10.1177/0954408918765.
23. A. I. Alsabery, H. Saleh and I. Hashim, "Effects of viscous dissipation and radiation on MHD natural convection in oblique porous cavity with constant heat flux," *Advans. Appl. Maths. Mechs.* vol. 9, no. 2, pp. 463–484, 2017. DOI: 10.4208/aamm.2014.m765 .
24. R. Miri, B. Mliki, B. Mohamad, M. A. Abbassi, M. Oreijah, K. Guedri and S. Abderafi, "Entropy generation and heat transfer rate for MHD forced convection of nanoliquid in presence of viscous dissipation term," *CFD Letters* vol. 15, no. 12, pp. 77–106, 2023. DOI: 10.37934/cfdl.15.12.77106.
25. A. Tanveer, S. Ul-Haq and M. B. Ashraf, "Heat transfer neural analysis of bingham-papanastasiou fluid in Lid-driven rectangular cavity with viscous dissipation," (2024). DOI: 10.21203/rs.3.rs-3884960/v1.

26. A. Divya, S. R. Sheri and A. K. Suram, "Finite element analysis of MHD naturally convective flow past an exponentially accelerated plate with viscous dissipation," *Numer. Heat Transfer, Part B: Fundamentals*, vol. 85, no. 9, pp. 1115–1129, 2024. DOI: 10.1080/10407790.2023.2274454.
27. H. Waqas, U. Farooq, R. Naseem, S. Hussain and M. Alghamdi, "Impact of MHD radiative flow of hybrid nanofluid over a rotating disk," *Case Studies in Therm. Eng.* vol. 26, p. 101015, 2021. DOI: 10.1016/j.csite.2021.101015.
28. A. Chattopadhyay, K. D. Goswami and S. K. Pandit, "Thermal performance in transient MHD thermogravitational convection of nanofluid with various heating effects," *J. Therm. Anal. Calor.* vol. 146, pp. 1255–1281, 2021. DOI: 10.1007/s10973-020-10077-3.
29. S. Sivasankaran, A. Malleswaran, J. Lee and P. Sundar, "Hydro-magnetic combined convection in a lid-driven cavity with sinusoidal boundary conditions on both sidewalls," *Int. J. Heat Mass Transfer* vol. 54, pp. 512–525, 2011. DOI: 10.1016/j.ijheatmasstransfer.2010.09.018.
30. P. X. Yu, J. X. Qiu, Q. Qin and Z. F. Tian, "Numerical investigation of natural convection in a rectangular cavity under different directions of uniform magnetic field," *Int. J. Heat Mass Transfer* vol. 67, pp. 1131–1144, 2013. DOI: 10.1016/j.ijheatmasstransfer.2013.08.087.
31. S. K. Pandit and A. Chattopadhay, "A robust higher order compact scheme for solving general second order partial differential equations withy derivative source terms on nonuniform curvilinear meshes," *Comput. Maths. Appls.* vol. 74, no. 6, pp. 1414–1434, 2017. DOI: 10.1016/j.camwa.2017.06.031.
32. S. K. Pandit, J. C. Kalita and D. C. Dalal, "A fourth-order accurate compact scheme for the solution of steady Navier-Stokes equations on non-uniform grids," *Comput. Fluids* vol. 37, no. 2, pp. 121–134, 2008. DOI: 10.1016/j.compfluid.2007.04.002.
33. S. K. Pandit, A. Chattopadhay and H. F. Oztop, "A fourth-order compact scheme for heat transfer problem in porous media," *Comput. Maths. Appls.* vol. 71, no. 3, pp. 805–832, 2016. DOI: 10.1016/j.camwa.2015.12.037.
34. S. K. Lele, "Compact finite difference schemes with spectral like resolution," *J. Comput. Phys.* vol. 103, no. 1, pp. 16–42, 1992. DOI: 10.1016/0021-9991(92)90324-R.
35. H. V. D. Vorst, "BiCGSTAB: A fast and smoothly converging variant of BiCG for the solution of nonsymmetric linear systems," *SIAM J. Sci. Comput.* vol. 13, no. 2, pp. 631–644, 1992. DOI: 10.1137/0913035.

16 AI-Enhanced Simulation of Nanofluid Natural Convection

Sahin Ahmed and Bikash Das

16.1 INTRODUCTION

The study of nanofluids has gained a lot of consideration over the past years due to its variety of implications in the domains of biomedical, optical, and engineering fields. A nanofluid is a novel type of engineered fluid that is prepared by dispersing microscopic particles of size less than 100 nm. Choi (1995) and Khanafer et al. (2003) proposed the notion of nanofluid and were the initial researchers to study how effectively nanofluids transmit thermal energy/heat inside an enclosure. Subsequently, a number of studies have been reported (Sabour et al., 2017; Tahmasebi et al., 2018; Ghalambaz et al., 2019), which examines how nanoparticles influence the fluid's physical properties on which fine particles are incorporated. The books of Nield and Bejan (2013) provide an extensive scope of nanofluid for potential future applications. An innovative kind of engineered nanofluids known as hybrid nanofluids (HNFs) contains two types of nanoparticles that are frequently utilized in thermal management applications due to their exceptional thermal conductivity (Shenoy et al., 2016). Some well-known examples of nanoparticles due to their extensive use are $SWCNT, MgO, MWCNT, Fe_2O_3, Al_2O_3, Ag, Au, Cu, TiO_2$. In recent years, the combination of these nanoparticles has been used in numerous thermal management applications such as cooling systems, thermal exchanger devices, thermal pipes, etc. The improved thermal conductivity of HNF has been extensively investigated by Akilu et al. (2016), Sarkarn et al. (2015), Sidik et al. (2016), and Jana et al. (2007). The study underscores the enhanced heat transport property of CNT, Cu, and gold-based nanofluids; in addition to this, they examined the thermal properties of nanofluids prepared by the combinations of these nanoparticles. The outcomes displayed that the heat transmission property of HNFs including CNT-Au and CNT-Cu was relatively less than the Au and Cu-based nanofluids. The primary reason behind this trend was the poor interaction between CNTs and Cu nanoparticles, which contributed to greater agglomeration when CNTs were added to a mixture of Cu nanoparticles. When CNTs are mixed with a Cu-based nanofluid, the natural convection currents are inhibited, which degrades the heat carrying capability of the suspension since the bulk motion of CNT-based nanofluid is less capable due to

DOI: 10.1201/9781003664345-16

their stability. A study by Jana et al. (2007) examined and documented the improved thermal performance of nano-engineered fluids prepared by mingling two distinct types of nanoparticles. In their work, the thermal efficiency of gold, copper, and CNT nanoparticles suspended in base fluid was examined, along with CNT-Cu/water HNF and CNT-Au/water HNF. The findings demonstrated that the thermal performance of CNT-Au and CNT-Cu HNFs is lower than that of Au and Cu-based mono nanofluids. The primary cause was the weak synergy of Cu nanoparticles and carbon nanotubes, which led to reduced distribution of both types of nanomaterials and higher accumulation when CNTs were added to the Cu nanoparticle solution. Because adding carbon nanotubes (CNTs) to a solution of Cu nanoparticles reduces the suspension's thermal conductivity because the CNTs' stability makes them less susceptible to convection.

Research on using mono base fluid is very common, but a more fascinating scenario arises when base fluids are prepared by mixing two different types of ordinary fluids, for example, water–petrol, water–ethylene glycol, etc. In severe cold climatic conditions, many heat exchanger devices need a lower aqueous freezing point of fluid medium; thus, a mixture of EG and water is widely used in different proportions. Many researchers use 40% ethylene glycol and 60% water by volume to prepare nanofluid (Parker et al., 2000) to enhance heat transfer capability of nanofluids. Various pharmaceutical industries, medical, and chemical processes involving living organisms include the mingling of two different types of ordinary fluids to enhance the mixing of samples and reagents (Ansari et al., 2010; Jeon et al., 2005; Lin et al., 2007; Lin et al., 2011; Wang et al., 2002; Chung et al., 2008), the objective of the investigations was to determine the best suitable geometrical setup that can enhance the mixing of nanoparticles and base fluid mixture. However, the suspension of ethylene glycol (EG) and water reduces the freezing point of water. Mebarek-Oudina et al. (2021) testified better thermal distribution in the annulus region between two homocentric cylinders by utilizing the blend of EG / H_2O. Hamida et al. (2017) analyzed the viscosity of EG / H_2O mixture-based nanofluid and reported that the viscosity of the suspension depends significantly on the volume fraction and the temperature of EG in the blend.

In this analysis, the thermal efficiency of HNF prepared by mixing $SWCNT$ and TiO_2 in EG/water base fluid at different proportions in the annular region of concentric cylinder is explored. The novelty of the investigation is to combine the artificial neural network to forecast the motion and distribution of thermal energy due to heat transmission in the annular region under the influence of Lorentz force. The dimensional governing equations have been transmuted to dimensionless form by adopting appropriate unitless quantities, and the transformed equations have been discretized and solved to obtain the velocity and temperature profile by using the Galerkin finite element method via commercial software COMSOL Multiphysics. The results obtained by using FEM are then used by ANN for predicting the impact of various non-dimensional parameters without simulations and with less computation time. The findings of this investigation will encourage future researchers to find out the suitable propositions of the nanoparticle as well as the blend of base fluid in various thermal management systems.

16.2 MATHEMATICAL FORMULATION

Consider a cylindrical annular cavity with the two cases – with fins attached and without fins as shown in Figure 16.1. The motion of the hybrid nanofluid inside the annular region is due to the disparity in temperature of the exterior cylinder and interior parts. The inner cylinder is equipped with an anchor-shaped fin. The inner cylinder and the fins are supposed to be hot with temperature $(T = T_h)$, while the exterior cylinder is presumed to be cold with temperature $(T = T_c)$. The gravitation g acts vertically downwards, while an unvarying magnetic field of intensity B_0 has been applied which acts in the radial direction. The impact of the induced magnetic field has been overlooked due to the assumption of sufficiently small Reynolds number. The impact of Joule heating, Hall current, chemical reaction, agglomerations, and all slips has been neglected during the simulations process in order to reduce the cost of computations. Assuming the density variations, the Boussinesq approximations are clutched to form the prime governing equations (Kiwan and Zeitoun 2008).

Equation of Continuity:

$$\frac{\partial u}{\partial x} + \frac{\partial v}{\partial y} = 0. \tag{16.1}$$

Momentum Conservation:

$$u\frac{\partial u}{\partial x} + v\frac{\partial u}{\partial y} = -\frac{\partial p}{\partial x} + \nu_{hnf}\left[\frac{\partial^2 u}{\partial x^2} + \frac{\partial^2 u}{\partial^2 y}\right], \tag{16.2}$$

$$u\frac{\partial v}{\partial x} + v\frac{\partial v}{\partial y} = -\frac{\partial p}{\partial y} + \nu_{hnf}\left[\frac{\partial^2 v}{\partial x^2} + \frac{\partial^2 v}{\partial^2 y}\right] + \frac{(\rho\beta)_{hnf}\, g\left(T - T_c\right)}{\rho_{hnf}} - \frac{\sigma_{hnf} B_0^2}{\rho_{hnf}} v. \tag{16.3}$$

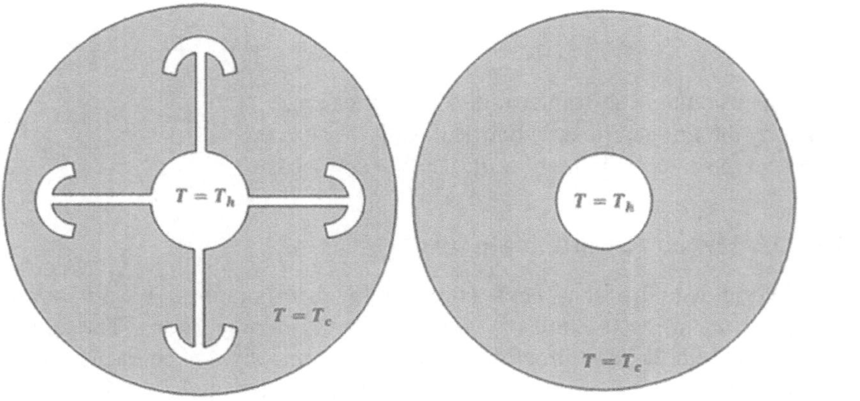

FIGURE 16.1(A-B) Geometrical configuration of the model.

Energy Conservation:

$$\left(\rho C_p\right)_{hnf}\left(u\frac{\partial T}{\partial x}+v\frac{\partial T}{\partial y}\right)=\kappa_{hnf}\left[\frac{\partial^2 T}{\partial x^2}+\frac{\partial^2 T}{\partial^2 y}\right]. \tag{16.4}$$

The requisite constraints at the boundaries are:

Along the fins and inner cylindrical wall: $u = v = 0, T = T_h$.

Along the outer cylindrical wall: $u = v = 0, T = T_c$.

The dimensional equations $(16.1)-(16.4)$ are transmuted to dimensionless equations by means of the subsequent transformations:

$$\left\{\begin{array}{c} X=\frac{x}{L}, Y=\frac{y}{L}, U=\frac{uL}{\alpha_f}, V=\frac{vL}{\alpha_f}, \theta=\frac{T-T_c}{T_h-T_c}, \\ P=\frac{pL^2}{\rho_f\alpha_f^2}, Pr=\frac{\mu_f C_p}{\kappa_f}, Ha=B_0 L\sqrt{\frac{\sigma_f}{\mu_f}}, Ra=\frac{g\beta_f L^3}{\nu_f\alpha_f}\left(T_h-T_c\right) \end{array}\right\} \tag{16.5}$$

By using (16.5), Equations $(16.1)-(16.4)$ reduce to dimensionless form:

$$\frac{\partial U}{\partial X}+\frac{\partial V}{\partial Y}=0, \tag{16.6}$$

$$\Gamma_1\left(U\frac{\partial U}{\partial X}+V\frac{\partial U}{\partial Y}\right)=-\frac{\partial P}{\partial X}+\Gamma_2 Pr\left[\frac{\partial^2 U}{\partial X^2}+\frac{\partial^2 U}{\partial^2 Y}\right] \tag{16.7}$$

$$\Gamma_1\left(U\frac{\partial V}{\partial X}+V\frac{\partial V}{\partial Y}\right)=-\frac{\partial P}{\partial Y}+\Gamma_2 \Pr\left[\frac{\partial^2 V}{\partial X^2}+\frac{\partial^2 V}{\partial^2 Y}\right]+\Gamma_3 Ra \Pr\theta-\Gamma_4 Ha^2 \Pr V, \tag{16.8}$$

$$U\frac{\partial \theta}{\partial X}+V\frac{\partial \theta}{\partial Y}=\frac{\Gamma_5}{\Gamma_6}\left[\frac{\partial^2\theta}{\partial X^2}+\frac{\partial^2\theta}{\partial^2 Y}\right]. \tag{16.9}$$

The dimensionless constraints at the boundaries are:

Along the fins and inner cylindrical wall: $U = V = 0, \theta = 1$.

Along the outer cylindrical wall: $U = V = 0, \theta = 0$.

16.2.1 Hybrid Nanofluid Mixture

The hybrid nanofluid has been prepared by incorporating $SWCNT$ and Al_2O_3 nanoparticles in the mixture of EG / H_2O in proper proportions. The addition of nanoparticles in the mixture of fluid alters the thermophysical characteristics of the engineered fluid; thus, an appropriate model has been adopted to evaluate the effective physical properties of the HNF. Table 16.1 illustrates the adopted models to compute the properties as in Raza et al. (2016) and White (2006).

TABLE 16.1
Models for Computing Thermophysical Properties of HNF

$\Gamma_1 = \rho_{hnf} / \rho_f$	$(1-\phi_2)\left[(1-\phi_1)+\phi_1\left(\frac{\rho_1}{\rho_f}\right)\right]+\phi_2\rho_2\rho_f$
$\Gamma_2 = \frac{\mu_{hnf}}{\mu_f}$	$\frac{1}{(1-\phi_1-\phi_2)}$
$\Gamma_6 = \frac{(\rho C_p)_{hnf}}{(\rho C_p)_f}$	$(1-\phi_2)\left[(1-\phi_1)+\phi_1\frac{(\rho C_p)_1}{(\rho C_p)_2}\right]+\phi_2\frac{(\rho C_p)_2}{(\rho C_p)_f}$
$\Gamma_5 = \frac{\kappa_{hnf}}{\kappa_f}$	$\frac{\frac{\phi_1\kappa_1+\phi_2\kappa_2}{\phi_1+\phi_2}+2\kappa_f+2(\phi_1\kappa_1+\phi_2\kappa_2)-2\kappa_f(\phi_1+\phi_2)}{\frac{\phi_1\kappa_1+\phi_2\kappa_2}{\phi_1+\phi_2}+2\kappa_f-2(\phi_1\kappa_1+\phi_2\kappa_2)+2\kappa_f(\phi_1+\phi_2)}$
$\Gamma_4 = \frac{\sigma_{hnf}}{\sigma_f}$	$1+\frac{3\sigma_1\phi_1+\sigma_2\phi_2-3\sigma_f(\phi_1+\phi_2)}{\sigma_1(1-\phi_1)+\sigma_2(1-\phi_2)+\sigma_f(2+\phi_1+\phi_2)}$
$\Gamma_3 = \frac{(\rho\beta)_{hnf}}{(\rho\beta)_f}$	$(1-\phi_2)\left(1-\phi_1+\phi_1\frac{(\rho\beta)_1}{(\rho\beta)_f}\right)+\phi_2\frac{(\rho\beta)_2}{(\rho\beta)_f}$

where the subscript f stands for base fluid, and the above thermophysical properties are valid for mono-base fluid (EG and *water*, respectively). In this investigation, the base fluid is a blend of EG (60%) and water (40%) by weight. The respective physical characteristics of both the base fluid EG and water at 300 K are provided in Table 16.2. Since the base fluid is a blend of two different types of fluid, the law of mixture has been utilized to compute the effective thermophysical properties of hybrid nanofluid.

$$(T_p)_{mixture} = \phi_{water}(T_p)_{EG} + (1-\phi_{water})(T_p)_{water}, \tag{16.10}$$

where $(T_p)_{mixture}$ is the thermophysical property of mixture, $\phi_{water}\left(=\frac{\frac{W_{water}}{\rho_{water}}}{\frac{W_{water}}{\rho_{water}}+\frac{W_{EG}}{\rho_{EG}}}\right)$ is the volume fraction of water, $(T_p)_{EG}$ is the effective thermophysical property of EG based nanofluid, and $(T_p)_{water}$ is the effective thermophysical property of water based nanofluid.

TABLE 16.2
Physical Properties of *SWCNT*, *TiO*$_2$, *water* and *EG*.

Physical Properties	SWCNT	TiO_2	Water	EG
$C_p(\mathbf{J/kgK})$	425	686.2	4179	2415
$\rho(\mathbf{kg/m^3})$	2600	4250	997.1	1114
$\kappa(\mathbf{W/mK})$	6600	8.9538	0.613	0.252
$\beta\times10^{-5}(1/\mathbf{K})$	27	0.9	21	57
σ	10^6-10^7	10^{-12}	0.05	
$\mu(\mathbf{kgm^{-1}s^{-1}})$			0.001003	

16.3 METHODOLOGY

COMSOL Multiphysics is a commercial software that can be used for the simulation of various physics ranging from fluid mechanics to solid mechanics. In this analysis, the very famous Galerkin Finite Element Method has been adopted for the discretization of non-linear governing equations with the aid of COMSOL Multiphysics. This commercial software also allows the user to couple various physics with boundary conditions, allowing for the visualization and efficient simulations of the physics of flow. In traditional numerical schemes such as the finite difference method, which requires manual meshing and discretization of governing equations, this software doesn't require manual coding, which reduces human labor. One of the key reasons for the adoption of COMSOL Multiphysics in this analysis is the complex geometry considered, which requires good quality user-defined mesh for an accurate and convergent solution. The current investigation is semi-experimental as the flow model considered in this investigation has been developed by considering various assumptions which are mentioned in the latter section. The transformation of equations into non-dimensional form makes it semi-experimental as the impact of thermophysical properties of the fluid is not dealt with directly; rather, it is considered with the help of non-dimensional parameters. The solution of the problem has been plotted for the impact of the non-dimensional parameters with the help of contours and streamlines.

16.3.1 Code Validation

Mesh independency test is one of the most crucial parts of the analysis; this test certifies the stability and convergence of the numerical schemes with higher accuracy. In this section, four types of mesh sizes have been considered, and the average Nusselt number has been computed. Table 16.3 portrays the numerical values of the average rate of heat transfer (Nusselt number) at the outer cylindrical surface for different mesh sizes. It is evident that the fine mesh size is the best mesh size, so the fine mesh size has been adopted for the simulation because increasing the density of the

TABLE 16.3
Result of Grid Independency Test on Average Nusselt Number at $Ra = 10^5$ and $Ha = 50$

Mesh Size	Number of Elements	Average Nusselt Number
	Case 1: Geometry without Fins	
Coarse	1056 Elements 864 Triangles 194 Quadrilaterals	1.1059
Fine	2162 Elements 1882 Triangles 280 Quadrilaterals	1.0814
Extra fine	16692 Elements 15508 Triangles 1184 Quadrilaterals	1.0551
	Case 2: Geometry with Fins	
Coarse	3156 Elements 2580 Triangles 576 Quadrilaterals	3.6191
Fine	6678 Elements 5790 Triangles 888 Quadrilaterals	3.6001
Extra fine	39530 Elements 36482 Triangles 3048 Quadrilaterals	3.5762

mesh enhances the computation cost, as well as the accuracy of the numeric results varying significantly. Figure 16.2(a-b) demonstrates the mesh distribution within the computational domain.

Figure 16.3(a) demonstrates the streamlines generated by COMSOL Multiphysics and, in contrast, Figure 16.3(b) illustrates the streamlines predicted using artificial neural networks by using the data generated by the numerical method via COMSOL Multiphysics. A comparison between the two results highlights the ability of the ANN to capture and analyze the already existing data to reproduce the characteristic streamline patterns with reasonable accuracy. This illustrates the potential of ANN as a reliable substitute for conventional high-cost computational methods for predicting flow structures.

16.4 RESULTS AND DISCUSSION

In this section, the impact of numerous non-dimensional parameters over the streamline patterns, isothermal contours, and Nusselt number has been illustrated and discussed in detail. The effective thermal conductivity of nanofluid of single base fluid and blend of two base fluids has been depicted in Table 16.4. It is reported

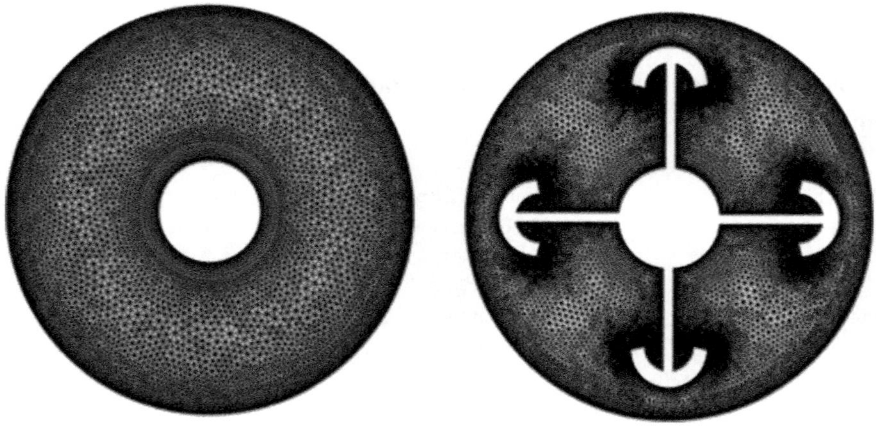

FIGURE 16.2 Mesh distribution (a) geometry without fins; (b) geometry with fins.

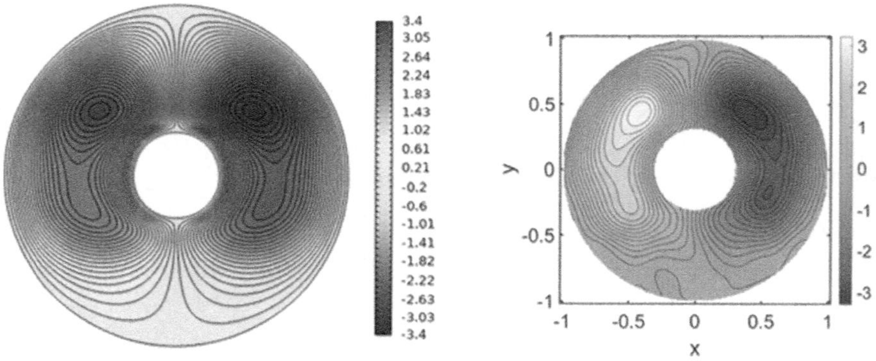

FIGURE 16.3 Streamlines generated by (a) COMSOL Multiphysics; (b) ANN.

that $\kappa_{SWCNT\,(EG)}$ is higher than $\kappa_{TiO_2(EG)}$ for all the cases of ϕ, while $\kappa_{SWCNT\,(water)}$ is greater than $\kappa_{SWCNT\,(EG)}$. The thermal conductivity of nanofluid of two base fluids lies between the thermal conductivities of single base fluid.

Figure 16.4 shows the impact of $Ha = 0, 50, 100, 200$ over the streamlines for the two cases of geometry. The presence of magnetic force introduces a retarding force in an electrically conducting fluid due to which the momentum of the fluid molecules degrades; as a result, the magnitude of the stream function degrades. Figure 16.4(a) compares the streamlines for the two cases of geometry in the absence of Lorentz force. For $Ha = 0$ (without magnetic field), two vortices of same intensity are formed within the annular region; the red color zone indicates the vortex in which fluid molecules experience clockwise direction with positive values of stream function, while the blue zone indicates the anticlockwise motion of hybrid nanofluid with negative values of stream function. It is worth citing that the intensity of vortices reduces for

TABLE 16.4
Effective Thermal Conductivity of $SWCNT-EG, SWCNT-water, SWCNT-(EG, water), TiO_2-EG$, $TiO_2-water, TiO_2-(EG, water)$ Nanofluids at Different Volume Fractions

$$\frac{\kappa_{nf}}{\kappa_f}=\frac{\left(\kappa_1+2\kappa_f-2\phi_1\left(\kappa_f-\kappa_1\right)\right)}{\left(\kappa_1+2\kappa_f+2\phi_1\left(\kappa_f-\kappa_1\right)\right)}$$

ϕ	$\kappa_{SWCNT(EG)}$	$\kappa_{TiO_2(EG)}$	$\kappa_{SWCNT\,(water)}$	$\kappa_{TiO_2(water)}$	$\kappa_{SWCNT\,(EG,water)}$	$\kappa_{TiO_2(EG,\,water)}$
0.01	0.25963	0.25902	0.63156	0.62819	0.41844	0.41692
0.025	0.27138	0.26979	0.66013	0.65145	0.43722	0.4273
0.05	0.29178	0.28845	0.70975	0.69155	0.47025	0.4620

the case of geometry with fins in the absence of magnetic field, while in the presence of magnetic field, the strength of vortices is greater in the geometry having fins attached to the inner cylinder. The application of fins in the geometry introduces more constraints to the fluid to flow freely; as a result, four vortices have been observed in which the top two vortices show clockwise motion while the lower two vortices show anticlockwise motion. The increase in the number of vortices can be beneficial in preventing agglomeration of nanoparticles within the fluid domain; as a result, it may enhance the convective mode of heat transfer, while reduced intensity of vortices can lead to stability in the flow structure. The strength of vortices can be reduced by 25.73% by introducing fins in the inner hot cylinder in the absence of Lorentz force. However, in the presence of magnetic field, the action of Lorentz force is more in the geometry without fins. The increase in the intensity of vortices for the case of $Ha = 50, 100, 200$ is 24.41%, 125%, and 334%, respectively, in the geometry without fins.

The consequence of the applied magnetic field over isothermal contours has been plotted in Figure 16.5(a-d) and compared for the two cases of geometry. The application of the magnetic field reduces the motion of nanofluid molecules, due to which the convection mode of heat transfer declines and an increase in the conductive mode of heat transfer dominates. In the absence of magnetic field $(Ha = 0)$, the isothermal lines originate from the hot inner cylinder and spread towards the upper part, as shown in the geometry without fins attached. However, the isothermal contours show interesting behavior in the geometry with fins attached; the thermal energy from the hot inner cylinder and fins spreads throughout the fluid domain. An increase in the Hartmann number reduces this transmission; as a result, the isothermal contours shrink towards the hot cylinder.

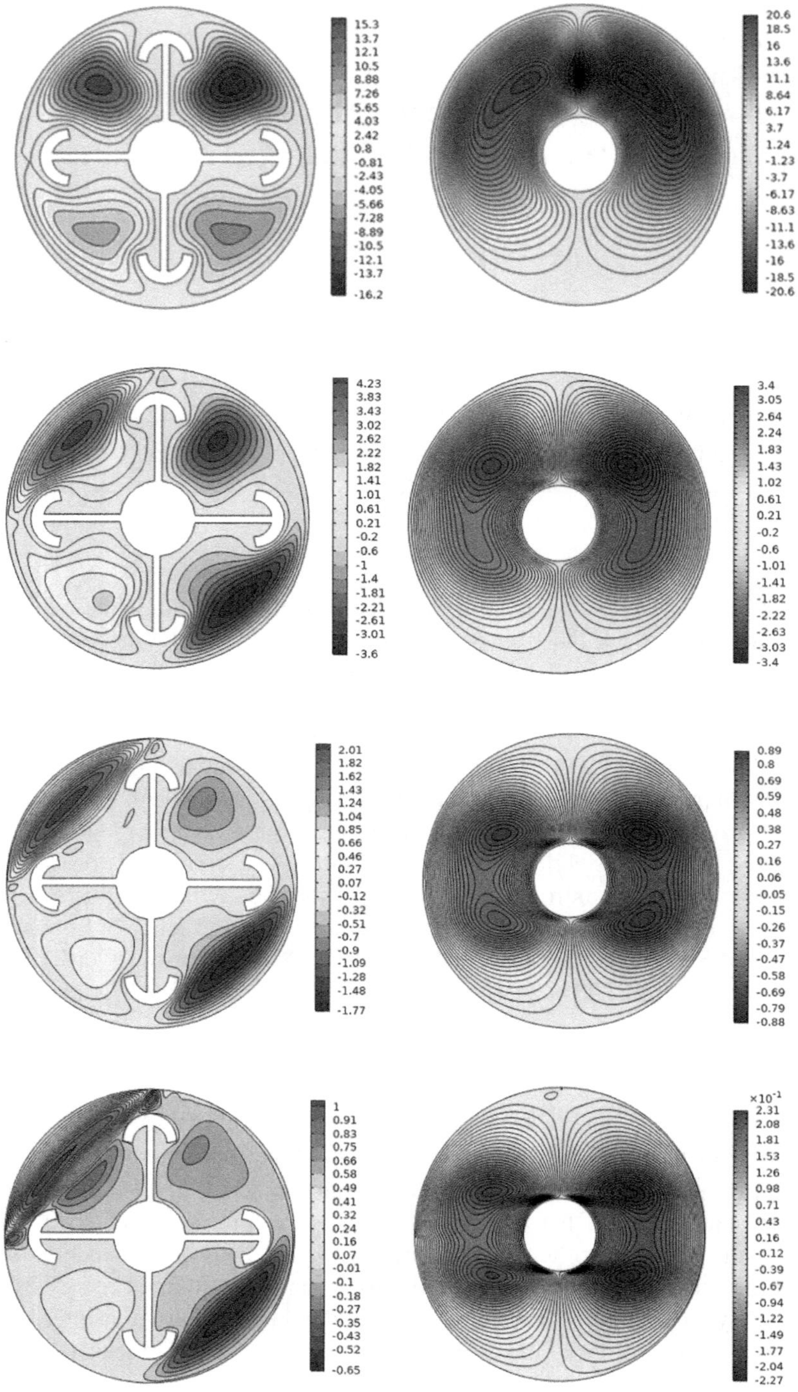

FIGURE 16.4 Impact of Ha over streamlines for the two cases of geometry.

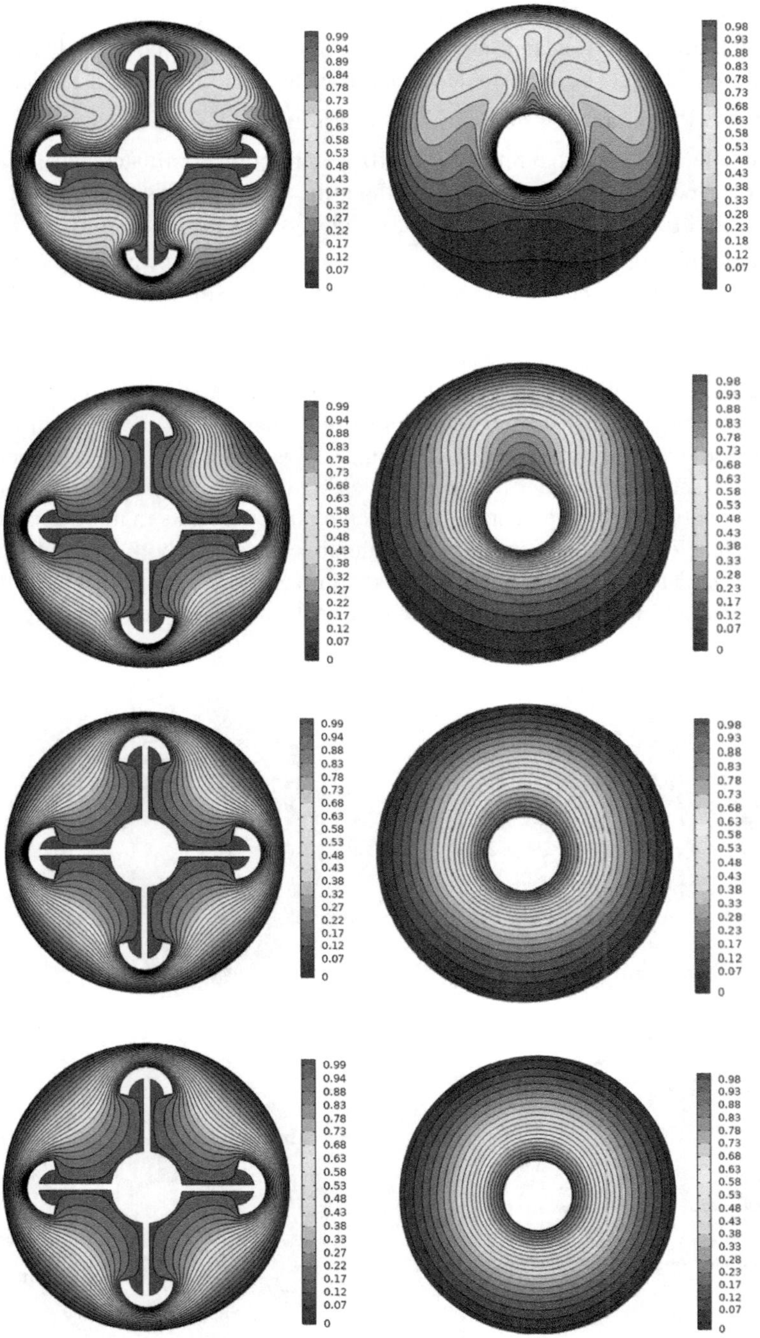

FIGURE 16.5(A-D) Impact of Ha over isothermal contours for the two cases.

The impact of Rayleigh number over the isothermal contours has been plotted in Figure 16.6(a, b). Rayleigh number is a non-dimensional number that expresses the nature of the convection mode of heat transfer. Below a critical Rayleigh number $\left(Ra < 10^3\right)$conduction mode of heat transmission is dominated, while above a critical value of Rayleigh number $\left(Ra > 10^3\right)$ convection dominates. For $Ra = 10$ the isothermal contours in the geometry without fins, the contours are compact and are not spread throughout the fluid domain, which signifies the conduction mode of heat transfer, while for $Ra = 10^5$, the thermal energy distributes towards the outer cylinder, which marks the convection mode of heat transfer. Similarly, with fins attached, more thermal energy is distributed in the fluid medium for higher Ra.

The impression of various non-dimensional parameters over the rate of heat transfer has been illustrated in Table 16.5. For the case of geometry without fins, the numerical values of the Nusselt number less than one signify the dominance of conduction over convection, while the numerical values of Nu in the geometry having fins attached are greater than one, which signifies the dominance of convection over conduction. Enhancement in the Rayleigh number results in more convection of fluid molecules, due to which the Nusselt number boosts for both cases of geometry. The increment in the ϕ of nanoparticles signifies more nanoparticles in the base fluid, which offers a greater surface area for heat transfer as well as greater thermal

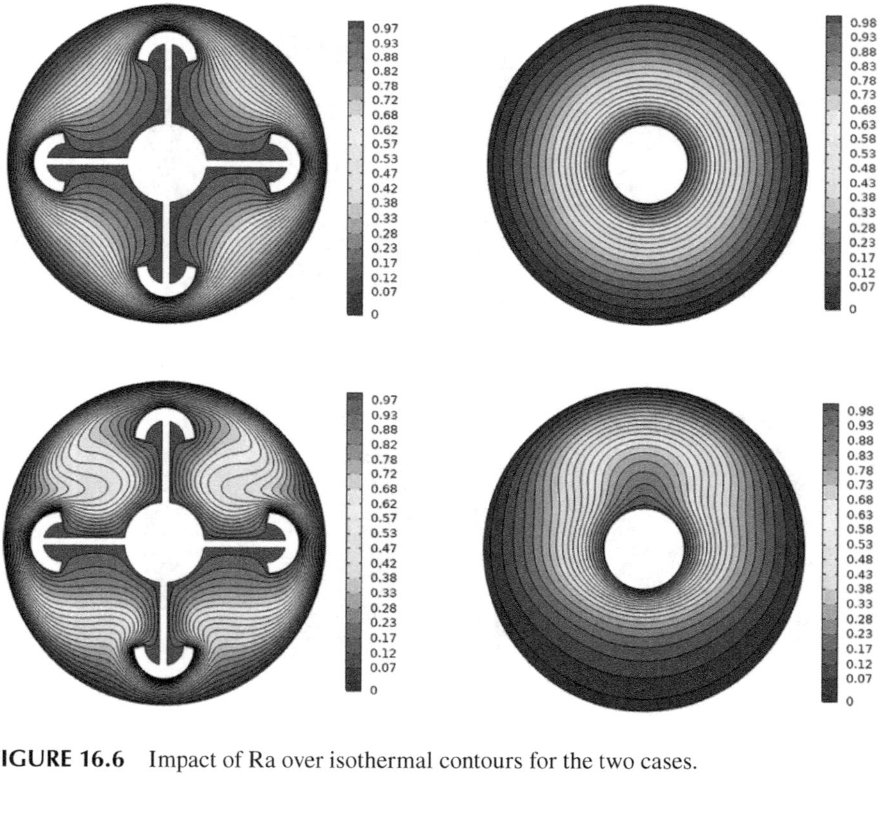

FIGURE 16.6 Impact of Ra over isothermal contours for the two cases.

TABLE 16.5
Impact of Various Parameters over Rate of Heat Transfer at Outer Cylinder

$(40\%\, W, 60\%\, EG)$

Ra	ϕ_1	ϕ_2	Ha	$Nu(Without\ Fins)$	$Nu(With\ Fins)$	Increased %
10	0.025	0.025	50	0.13168	3.4084	2488.396
10^2				0.13168	3.4084	2488.396
10^5				0.16793	3.5762	2029.578
	0			0.12238	3.1676	2488.331
	0.025			0.13168	3.4084	2488.396
	0.05			0.14146	3.6615	2488.364
		0		0.12386	3.2058	2488.245
		0.025		0.13168	3.4084	2488.396
		0.05		0.13985	3.6197	2488.273
			$0(Ra=10)$	0.13168	3.4084	2488.396
			$50(Ra=10)$	0.13168	3.4084	2488.396
			$100(Ra=10)$	0.13168	3.4084	2488.396
			$0(Ra=10^5)$	0.35839	4.7181	1233.211
			$50(Ra=10^5)$	0.16793	3.5762	2029.578
			$100(Ra=10^5)$	0.13499	3.4239	2436.41

conductivity of the hybrid nanofluid, due to which enhancement in the Nusselt number has been reported in Table 16.5. It is worth noting that the impact of the Hartmann number at lower Rayleigh number is very insignificant, while at higher Rayleigh number, an increase in the Hartmann number resulted in the decaying of the Nusselt number due to the resistive force offered by the magnetic field, which degrades the convection of fluid molecules.

16.5 CONCLUSIONS AND FUTURE SCOPE

The article investigates free convection in the annular region of concentric cylinders under the influence of Lorentz force filled with a hybrid nanofluid of EG/water mixture-based fluid. The non-linear dimensionless equations have been discretized and resolved using FEM via COMSOL Multiphysics and ANN. Following are the key findings of the current investigation –

1. Fins attached geometry provides better and higher number of vortices within the domain in the absence of magnetic field, while geometry without fins generates two main vortices of less strength in the presence of magnetic field.
2. Hartmann number enhances the conductive mode of heat transmission by reducing the convective motion of fluid molecules.
3. Higher Rayleigh number boosts convection, thus enhancing the cooling rate.
4. Augmented values of volume fraction of nanoparticles augment the rate of heat transmission for both cases of geometry.
5. At lower Rayleigh number $(Ra = 10)$, augmented values of Ha does not have any significant change in the rate of heat transfer, but at higher Rayleigh number $Ra = 10^5$, improved values of Ha reduces the rate of heat transfer due to less convection.

Current investigation deals with the motive to decide a better geometry which provides a better rate of heat transmission. A mixture of base fluid has been used for the preparation of hybrid nanofluid, while neglecting the chemical reactions between the nanoparticles and the base fluid. In this analysis, the impact of the magnetic field has been considered while ignoring the other physical effects such as Brownian motion, mass diffusion, viscous heat dissipation, etc. Future researchers can adopt other geometries and base fluids at different propositions to explore the advanced heat transfer capabilities of this type of hybrid nanofluid.

REFERENCES

Akilu, S., Sharma, K.V., Baheta, A.T., and Mamat, R. (2016). A review of thermophysical properties of water based composite nanofluids. *Renewable and Sustainable Energy Reviews*, 66, 654–678.

Ansari, M.A., Kim, K.Y., and Kim, S.M. (2010). Numerical study of the effect on mixing of the position of fluid stream interfaces in a rectangular microchannel. *Microsystem Technologies*, 16, 1757–1763.

Chiam, H.W., Azmi, W.H., Usri, N.A., Mamat, R., and Adam, N.M. (2017). Thermal conductivity and viscosity of A12O3 nanofluids for different based ratio of water and ethylene glycol mixture. *Experimental Thermal and Fluid Science*, 81, 420–429.

Choi, S.U.S. (1995). Enhancing thermal conductivity of fluids with nanoparticles. Proceedings of the 1995 ASME International Mechanical Engineering Congress and Exposition, FED 231/MD 66, pp. 99–105.

Chung, C.K., and Shih, T.R. (2008). Effect of geometry on fluid mixing of the rhombic micromixers. *Microfluidics and Nanofluidics*, 4, 419–425.

Ghalambaz, M., Sheremet, M.A., Mehryan, S.A.M., Kashkooli, F.M., and Pop, I. (2019). Local thermal non–equilibrium analysis of conjugate free convection within a porous enclosure occupied with Ag–MgO hybrid nanofluid. *Journal of Thermal Analysis and Calorimetry*, 135(2), 1381–1139.

Jana, S., Saheli–Khojin, A., and Zhong, W.H. (2007). Enhancement of fluid thermal conductivity by the addition of single and hybrid nano–additives. *Thermochimica Acta*, 462, 45–55.

Jeon, M.K., Kim, J.H., Noh, J., Kim, S.H., Park, H.G., and Woo, S.I. (2005). Design and characterization of a passive recycle micromixer. *Journal of Micromechanics and Microengineering*, 15, 346–350.

Khanafer, K., Vafai, K., and Lightstone, M. (2003). Buoyancy–driven heat transfer enhancement in a two–dimensional enclosure utilizing nanofluids. *International Journal of Heat and Mass Transfer*, 46(19), 3639–3653.

Kiwan, S., and Zeitoun, O. (2008). Natural convection in a horizontal cylindrical annulus using porous fins. *International Journal of Numerical Methods for Heat & Fluid Flow*, 18(5), 618–634.

Lin, Y., Yu, X., Wang, Z., Tu, S.T., and Wang, Z. (2011). Design and evaluation of an easily fabricated micromixer with three-dimensional periodic perturbation. *Chemical Engineering Journal*, 171, 291–300.

Lin, Y.C., Chung, Y.C., and Wu, C.Y. (2007). Mixing enhancement of the passive microfluidic mixer with J-shaped baffles in the tee channel. *Biomedical Microdevices*, 9, 215–221.

McQuiston, F.C., Parker, J.D., and Spitler, J.D. (2000). *Heating Ventilating and Air-Conditioning*. John Wiley & Sonc Inc, New York.

Mebarek-Oudina, F., Hussein, A.K., Younis, O., Rostami, S., and Nikbakhti, R. (2021). Natural convection enhancement in the annuli between two homocentric cylinders by using ethylene glycol/water based titania nanofluid. *Journal of Advanced Research in Fluid Mechanics and Thermal Sciences*, 80(2), 56–73.

Nield, D.A., and Bejan, A. (2013). *Convection in Porous Media*, 4th ed. Springer, New York, NY.

Raza, J., Rohni, A.M., and Omar, Z. (2016). Numerical investigation of copper-water (Cu-water) nanofluid with different shapes of nanoparticles in a channel with stretching wall: Slip effects. *Mathematical and Computational Applications*, 21(4), 43.

Sabour, M., Ghalambaz, M., and Chamkha, A. (2017). Natural convection of nanofluids in a cavity: Criteria for enhancement of nanofluids. *International Journal of Numerical Methods for Heat and Fluid Flow*, 27, 1504–1534.

Sarkarn, J., Ghosh, P., and Adil, A. (2015). A review on hybrid nanofluids: Recent research, development and applications. *Renewable and Sustainable Energy Reviews*, 43, 164–177.

Shenoy, A., Sheremet, M., and Pop, I. (2016). *Convective Flow and Heat Transfer from Wavy Surfaces: Viscous Fluids, Porous Media and Nanofluids*. CRC Press, Taylor and Francis Group, New York, NY.

Sidik, N.A.C., Adamu, I.M., Jamil, M.M., Kefayati, G.H.R., Mamat, R., and Najafi, G. (2016). Recent progress on hybrid nanofluids in heat transfer applications: A comprehensive review. *International Communications in Heat and Mass Transfer*, 78, 68–79.

Tahmasebi, A., Mahdavi, M., and Ghalambaz, M. (2018). Local thermal nonequilibrium conjugate natural convection heat transfer of nanofluids in a cavity partially filled with porous media using Buongiorno's model. *Numerical Heat Transfer, Part A: Applications*, 73(4), 254–276.

Wang, H., Iovenitti, P., Harvey, E., and Masood, S. (2002). Optimizing layout of obstacles for enhanced mixing in microchannels. *Smart Materials and Structures*, 11, 662–667.

White, F.M., and Majdalani, J. (2006). *Viscous Fluid Flow*. McGraw-Hill, New York, Vol. 3, pp. 433–434.

Index

For Product Safety Concerns and Information please contact our EU representative GPSR@taylorandfrancis.com Taylor & Francis Verlag GmbH, Kaufingerstraße 24, 80331 München, Germany

Batch number: 10831019

Printed by Printforce, the Netherlands